VOLUME ONE HUNDRED AND FIFTY THREE

Progress in MOLECULAR BIOLOGY AND TRANSLATIONAL SCIENCE

WNT Signaling in Health and Disease

VOLUME ONE HUNDRED AND FIFTY THREE

PROGRESS IN MOLECULAR BIOLOGY AND TRANSLATIONAL SCIENCE

WNT Signaling in Health and Disease

Edited by

JUAN LARRAÍN
Department of Cellular and Molecular Biology
Pontificia, Universidad Catolica de Chile, Santiago, Chile

GONZALO OLIVARES
Institute of Biomedical Sciences, Faculty of Medicine,
Universidad de Chile, Santiago, Chile

ELSEVIER

ACADEMIC PRESS
An imprint of Elsevier

Academic Press is an imprint of Elsevier
125 London Wall, London EC2Y 5AS, United Kingdom
525 B Street, Suite 1800, San Diego, CA 92101-4495, United States
50 Hampshire Street, 5th Floor, Cambridge, MA 02139, United States
The Boulevard, Langford Lane, Kidlington, Oxford OX5 1GB, United Kingdom

First edition 2018

ISBN: 978-0-12-811429-2
ISSN: 1877-1173

Publisher: Zoe Kruze
Acquisition Editor: Ashlie M. Jackman
Editorial Project Manager: Joanna Collett
Production Project Manager: James Selvam
Cover Designer: Greg Harris

Typeset by Thomson Digital

CONTENTS

CONTRIBUTORS

Caroline M. Alexander
University of Wisconsin-Madison, Madison, WI, United States

Nick Barker
A*STAR Institute of Medical Biology, Singapore;
Kanazawa University, Kanazawa, Japan;
Centre for Regenerative Medicine, University of Edinburgh, Edinburgh, United Kingdom

Konrad Basler
Institute of Molecular Life Sciences, University of Zurich, Zurich, Switzerland

Bahar Degirmenci
Institute of Molecular Life Sciences, University of Zurich, Zurich, Switzerland

Dustin J. Flanagan
Doherty Institute of Infection and Immunity, University of Melbourne, Melbourne, VIC, Australia

Francesco Girardi
Sorbonne Universitésy, UPMC Université Paris 06, INSERM UMRS974, Centre of Research in Myology, Paris, France

Stephanie Grainger
University of California, La Jolla, San Diego, CA, United States;
Sanford Consortium for Regenerative Medicine, La Jolla, San Diego, CA, United States

George Hausmann
Institute of Molecular Life Sciences, University of Zurich, Zurich, Switzerland

Nibaldo C. Inestrosa
Center for Aging and Regeneration (CARE-UC), Pontifical Catholic University of Chile, Santiago, Chile;
Center for Healthy Brain Ageing, University of New South Wales, Sydney, NSW, Australia;
Center of Excellence in Biomedicine of Magallanes (CEBIMA), University of Magallanes, Punta Arenas, Chile

Michael Kahn
Beckman Research Institute of the City of Hope, Duarte, CA, United States

Michael Kühl
Institute of Biochemistry and Molecular Biology, Ulm University, Ulm, Germany

Fabien Le Grand
Sorbonne Universitésy, UPMC Université Paris 06, INSERM UMRS974, Centre of Research in Myology, Paris, France

Youhua Liu
National Clinical Research Center of Kidney Disease, Nanfang Hospital, Southern Medical University, Guangzhou, People's Republic of China;
University of Pittsburgh School of Medicine, Pittsburgh, PA, United States

Lawrence Lum
University of Texas Southwestern Medical Center, Dallas, TX, United States

Carla Montecinos-Oliva
Center for Aging and Regeneration (CARE-UC), Pontifical Catholic University of Chile, Santiago, Chile;
Interdisciplinary Institute for Neuroscience (IINS), University of Bordeaux, Bordeaux, France

Jean M. Moselen
Doherty Institute of Infection and Immunity, University of Melbourne, Melbourne, VIC, Australia

Carolina A. Oliva
Center for Aging and Regeneration (CARE-UC), Pontifical Catholic University of Chile, Santiago, Chile

Astrid S. Pfister
Institute of Biochemistry and Molecular Biology, Ulm University, Ulm, Germany

Toby J. Phesse
Doherty Institute of Infection and Immunity, University of Melbourne, Melbourne, VIC, Australia;
European Cancer Stem Cell Research Institute, Cardiff University, Cardiff, United Kingdom

Renate H.M. Schwab
Doherty Institute of Infection and Immunity, University of Melbourne, Melbourne, VIC, Australia

Si Hui Tan
A*STAR Institute of Medical Biology, Singapore

Bang M. Tran
Doherty Institute of Infection and Immunity, University of Melbourne, Melbourne, VIC, Australia

David Traver
University of California, La Jolla, San Diego, CA, United States

Tomas Valenta
Institute of Molecular Life Sciences, University of Zurich, Zurich, Switzerland

Elizabeth Vincan
Doherty Institute of Infection and Immunity, University of Melbourne, Melbourne, VIC, Australia;
Curtin University, Perth, WA, Australia

Yongping Wang
National Clinical Research Center of Kidney Disease, Nanfang Hospital, Southern Medical University, Guangzhou, People's Republic of China

Karl Willert
University of California, La Jolla, San Diego, CA, United States;
Sanford Consortium for Regenerative Medicine, La Jolla, San Diego, CA, United States

Li-shu Zhang
University of Texas Southwestern Medical Center, Dallas, TX, United States

Chengji J. Zhou
University of California Davis, Sacramento, CA, United States

PREFACE

Since the discovery of the first Wnt gene 35 years ago (Nusse and Varmus, 1982), Wnt signaling has been the focus of extensive research because of its key role in many aspects of developmental biology, tissue homeostasis, stem cells differentiation, and disease (especially cancer). The aim of this book is to summarize the current knowledge on Wnt signaling in tissue development and stem cells (e.g., muscle, skin, intestine, etc.). Additionally, due to its impact on human health, each chapter describes the pathologies associated to dysregulation of Wnt signaling and the therapeutic strategies that have been developed. Therefore, this book is going to be of interest to students and researchers on the biological sciences, physicians, pharmacologists, and any one interested to understand how an evolutionary conserved signaling pathway regulates tissue homeostasis and how its dysregulation gives rise to disease and the current strategies to target pharmacologically this signaling pathway.

The book consists of 11 chapters. Chapter 1 explains the role of Wnt signaling on intestine and liver homeostasis and disease, focusing on stem cells niche and cancer. Chapter 2 reviews the role of Wnt signaling on epithelial tissue, specifically on the regulation of stem cells and pathology (i.e., cancer). Chapter 3 reviews the current understanding of the role of Wnt signaling on central nervous system development and homeostasis, considering also the pathologies associated to the pathway malfunctioning. Chapter 4 defines the novel function of Wnt signaling in regulating ribosome biogenesis. Chapter 5 discusses Wnt signaling function on skeletal muscle development and disease. Wnt signaling role on kidney development and malfunctioning is the subject of Chapter 6. Wnt signaling function on stem cells and cancer stem cells at the molecular level is the focus of Chapter 7. In Chapter 8, authors review the current therapeutic strategies devoted to treat cancer associated to dysregulation of Wnt signaling. Chapter 9 reviews the current knowledge on Wnt signaling and mammary stem cells and breast cancer. Chapter 10 reviews the role of Wnt signaling in organoid culture and its importance on personalized medicine. Chapter 11 defines the role of Wnt signaling on hematological development and disease.

We are tremendously grateful of all contributors for their interest to review the advances and impact of Wnt signaling on different systems and for the time dedicated to collaborate with this volume. We would like to acknowledge to the Elsevier serials team who help and guide us on this amazing endeavor: Helene Kabes (former Senior Editorial Project Manager), Joanna Collett (Editorial Project Manager), and Ashlie Jackman-Juler (Acquisition Editor, Serials and Handbooks).

We would like to devote this book to the memory of Dr. Michael Conn who passed away on December 2016. Dr. Conn was instrumental on the conception of this book's topic as he visualized the relevance to summarize the role of Wnt signaling of health and disease for the scientific community.

CHAPTER ONE

Wnt Ligands as a Part of the Stem Cell Niche in the Intestine and the Liver

Bahar Degirmenci, George Hausmann, Tomas Valenta, Konrad Basler[1]
Institute of Molecular Life Sciences, University of Zurich Zurich, Switzerland
[1] Corresponding author. E-mail address: kb@imls.uzh.ch

Contents

Abstract

The term "Wnt signaling" does not refer to one uniform signal transduction cascade. Instead, it describes the multiple discrete signals elicited by Wnt ligands following their interaction with distinct receptor complexes. The interaction of stem cells with niche cells is coordinated by the involvement of different signaling pathways, including Wnt signaling. The stem cell populations are highly sensitive to modulation of Wnt pathway activity. Wnt signaling is of paramount importance for stem cell self-renewal, survival, proliferation, differentiation, movement, and cell polarity. Aberrant activation of Wnt/β-catenin signaling is associated with the pathology of many types of cancer, such as colorectal cancer and hepatocellular carcinoma. Importantly, although often initiated by mutation(s) downstream of the Wnt-receptor complex, the progression of colorectal cancer still seems to be augmented by Wnt ligand-mediated signaling. This

Progress in Molecular Biology and Translational Science, Volume 153
ISSN 1877-1173
http://dx.doi.org/10.1016/bs.pmbts.2017.11.011

chapter focuses on the role of Wnt ligands in the intestine and the liver during homeostasis and cancer.

1. INTRODUCTION

Almost 40 years ago Raymond Schofield proposed the concept of a stem cell niche for mammalian haemocytes.[1] Schofield suggested that hematopoietic stem cell populations are maintained in conducive microenvironments within the bone marrow. If the progeny of a stem cell is in the niche it will become a stem cell; outside of the niche it differentiates. Recently, the location and function of stem cell niches in many other tissues including intestine, liver, and pancreas have been characterized.[2–5] The interaction of stem cells with the niche cells has to ensure the balance between the demands of stem cell self-renewal and tissue renewal. The system has to be flexible to adapt to the changes in demand seen during times of normal tissue homeostasis and in response to tissue injury. This is achieved by the coordinated action of different signaling pathways, such as Notch signaling,[6] Hedgehog signaling,[7] and Wnt signaling.[2,8] In addition to the developmental and homeostatic contexts, the stem cell niche concept is also seen as critical for tumor growth and progression. Cancer stem cells (CSCs) also occupy niches, which help them to keep their unique plasticity and enhance their metastatic potential.[9] In this chapter, we discuss the role of Wnt ligands in stem cell niches. We focus on the niches found in the intestine and liver, both during homeostasis and in cancer.

2. AN OVERVIEW OF WNT SIGNALING

Wnt signaling is of paramount importance for stem cell self-renewal, survival, proliferation, differentiation, movement, and cell polarity. The term "Wnt signaling" does not refer to one uniform signal transduction cascade. Instead, it describes the multiple discrete signals elicited by Wnt ligands following their interaction with distinct receptor complexes. Wnts are secreted proteins. In mice and humans, there are 19 Wnt proteins that activate different branches of Wnt signaling.

Wnt protein synthesis is started by their translation in the endoplasmic reticulum and is followed by a number of maturation and modifications steps.

Two crucial modifications of the Wnt proteins are acylation and glycosylation. Although glycosylation may promote the efficiency of secretion of some Wnts; it seems to be dispensable for the function of mature Wnt proteins: glycosylation deficient Wnts can still trigger the signaling.[10–13] Acylation on the other hand is absolutely required for Wnt function; this is demonstrated by the fact that mutations in Porcupine hinder Wnt signaling pathway. Porcupine is the endoplasmic reticulum membrane bound O-acyl-transferase that catalyzes the addition of lipid to Wnt proteins.[14] Palmitoylated Wnt proteins are bound by the multispan transmembrane protein Wntless (Wls) that binds and escorts Wnts to the plasma membrane via the Golgi.[15–17] Upon Wnt release from the cell, Wls is recycled via the retromer complex to the Golgi. Ablation of either Wls expression or its recycling via the retromer results in a Wnt secretion defect.[18–20] Loss of Wls leads to embryonic lethality in mice.[21]

The signaling cascades activated by Wnts are classified as β-catenin–-dependent and β-catenin–independent pathways. The pathway(s) activated by Wnts are dictated by the cellular context, in particular: the receptor cohort that is present. However, some Wnts, such as Wnt1, Wnt2, Wnt2b, Wnt3, Wnt6, and Wnt9b, are thought to activate only the β-catenin–dependent pathway. Wnt5a and Wnt11 trigger the β-catenin–-independent pathway.

To initiate the β-catenin–dependent pathway, in mammals, Wnt ligands bind to the seven-pass transmembrane receptors of the Frizzled (FZD) family (10 members), which interact with single-pass low-density lipoprotein receptor-related protein 5 or 6 (LRP5/6). Zinc/RING finger protein 3 (ZNRF3) and RING finger protein 43 (RNF43) are ubiquitin ligases that continually downregulate FZDs via ubiquitination. Binding of R-spondins to ZNRF3 and RNF43 and the Lgr4/5/6 receptor prevents ZNRF3 and RNF43 activity, and hence stabilizes FZDs. The recruitment of Dishevelled (Dvl) and Axin by the Wnt-FZD-LRP5/6 complex triggers a series of events that results in inhibition of β-catenin phosphorylation and degradation, as a consequence β-catenin accumulates.[22] In the nucleus, the accumulating β-catenin binds to the transcription factors T-cell factor/lymphoid enhancing factor (TCF/LEF) to activate Wnt responsive gene expression.

In the absence of Wnt ligands, β-catenin is phosphorylated by the so-called "destruction complex" which is composed of adenomatous polyposis coli (APC), Axin, and the kinases the glycogen synthase kinase (GSK3) and casein kinase 1 alpha (CKIα). Phosphorylation marks β-catenin for subsequent ubiquitination and degradation. The activity of the β-catenin destruction complex is tightly regulated and efficiently keeps the level of nuclear β-catenin low in the cell in the absence of Wnts.

The β-catenin–independent pathways comprise two major branches: planar cell polarity (PCP) pathway and the Wnt/calcium pathway. PCP is an important biological phenomenon defining the polarity of epithelial cells within the plane of cells perpendicular to the apical–basal axis. Wnt5a and Wnt11 are key regulators of PCP signaling in mammals. The GTPases RhoA and Rac, Rho kinase (ROCK) and PKC are the cytoplasmic proteins that are regulated by PCP signaling.[23] Notably, the Wnt ligand Wnt3a, which was believed to specifically regulate the β-catenin–dependent pathway, is also able to stimulate RhoA and Rac1.[24,25] This demonstrates the context dependence, as well as potential flexibility, of the pathways that Wnt ligands can trigger. The distinctive feature of the Wnt/calcium pathway is a Wnt-mediated increase of the intracellular calcium ion concentration. Sensing of Wnt3a, Wnt5a, and Wnt11 can lead to the intracellular calcium release.[26,27] The Wnt/calcium pathway is pivotal for various developmental processes including regulation of cell migration during gastrulation[28,29] and migration of neural crest cells.[30] It may also have role during tumor development and progression, especially in metastasis.[31,32]

Wnt/TOR signaling is another β-catenin–independent branch and it regulates the cell growth. GSK3β is the negative regulator of mTOR signaling. GSK3β phosphorylates TSC2, causing mTOR inhibition via Rheb. Hence, Wnt-induced inhibition of GSK3β leads to activation of mTOR and downstream components. The Wnt-triggered activation of mTOR enhances cell growth.[33]

Although the β-catenin–dependent pathway is comparatively well characterized, the more diverse β-catenin–independent pathways are less understood. Earlier we described the general components of both signaling pathways and refer the interested reader these excellent reviews.[34,35]

3. THE INTESTINE

The intestinal tract consists of the small intestine and the colon. The small intestine is mostly responsible for nutrient absorption, whereas the colon resorbs water. The structural and functional differences are due to the cellular composition of their epithelium (Fig. 1). The overall structure of the mammalian intestine is composed of four distinct layers. The outermost layer is the serosa, which is a single layer of mesothelial cells. Next is the muscularis

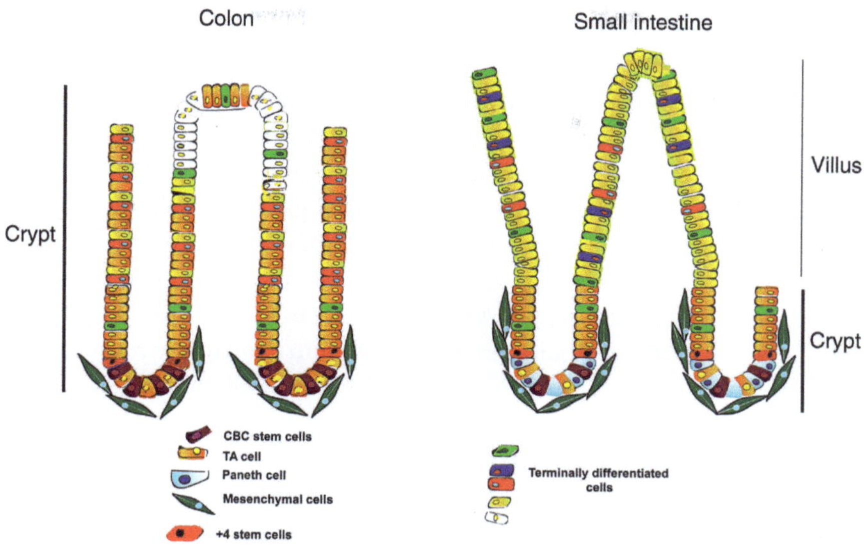

Fig. 1 Schematic representation of the colon and the small intestine. The colonic epithelium consists of only crypts, whereas the small intestinal epithelium is composed of crypt and villi. The epithelial Lgr5^{+} crypt base columnar cells (CBC) are stem cells positioned at the bottom of the crypt. They give rise to various other specialized types of epithelial cells. Paneth cells reside at the bottom of the small intestinal crypt, and constitute a niche for the stem cells. In addition to the CBC cells, the +4 positioned stem cells, contribute to the regenerative response upon tissue injury. Underlying the epithelial layer are mesenchymal cells.

propria, which comprises the smooth muscle cells that drive the rhythmic peristaltic movements of the intestine. The submucosa is a fibrous connective layer between muscularis mucosae and muscularis propria. It encompasses fibroblasts, nerve fibers, blood and lymph vessels, and various immune cells. The innermost layer is the mucosa composed of epithelium and lamina propria. Lamina propria underlies the epithelium and contains connective tissue, lymphatics, and capillaries. The single cell layer intestinal epithelium is composed of various cell types: stem cells, Paneth cells, goblet cells, enteroendocrine cells, enterocytes, and tuft cells (Fig. 1). The absorptive surface area of the small intestine is increased by villi, protrusions that point toward inner lumen, and tubular invaginations into the submucosa called as crypt. The colonic epithelium consists of only crypt without Paneth cells, whereas the small intestinal epithelium is composed of crypt and villi (Fig. 1). A mouse intestinal epithelium is a highly dynamic tissue and is renewed every 4–5 days.

3.1 The Wnt Ligands are Part of Essential Niche for the Intestinal Stem Cells

The driving force behind the active renewal of the intestinal epithelium is the intestinal epithelial stem cells that are positioned at the base of the crypt. One pool of stem cells are the crypt base columnar cells (CBC) expressing Lgr5 (Fig. 1).[36] Gene expression analysis revealed that $Lgr5^+$ stem cells also express other Wnt target genes, such as Olfm4 and Ascl2.[37] In addition to the fast cycling $Lgr5^+$ CBC cells, an alternative pool of stem cells are the +4 positioned cells, which are marked by Bmi1, mTert, Hopx, and Lrig1 (Fig. 1).[38–41] These more quiescent stem cells contribute to the regenerative response upon tissue injury.[42,43] The $Lgr5^+$ stem cells differentiate into rapidly dividing transit-amplifying cells that migrate upward and further give rise to various other specialized types of epithelial cells. These cells continue to migrate to the tops of the villi where they finally undergo apoptosis. Paneth cells represent the only terminally differentiated lineage intermingled with $Lgr5^+$ stem cells; they reside at the bottom of the small intestinal crypt, and constitute a niche for the stem cells (Fig. 1).[2,44] They provide Wnts (Wnt3, Wnt6, and Wnt9b), EGF, and Notch ligands (Dll1 and Dll4).[2] They also secrete antimicrobial peptides and lysozyme. Lineage tracing experiments with $Lgr5^+$ stem cells revealed that these stem cells divide symmetrically and their stemness is determined by their physical contact to Paneth cells.[45] The precursor of the Paneth cell lineage and mature Paneth cells are capable of stem cell label retaining, extensive proliferation, and can give rise to main epithelial cell types upon intestinal injury.[43–46] Renewal of the intestinal epithelial stem cells relies on the Wnt/β-catenin signaling. The stem cell populations are highly sensitive to modulation of Wnt pathway activity. The systemic block of Wnt secretion in the adult mouse leads to lethality due to the loss of epithelial stem cells and perturbations in the intestinal epithelial renewal.[8] During normal intestinal homeostasis, Paneth cells secrete the Wnt ligands and thus serve as a niche for intestinal epithelial stem cells. The Wnt signals control stem cell renewal and regulate cell fate determination.[47] Wnt signaling is also essential for the development of early secretory cell progenitors.[48] Notably, in vivo, depletion of Paneth cells did not abrogate the intestinal epithelial cell renewal, suggesting that extra-epithelial Wnts can substitute the niche for stem cells.[49,50] Extra-epithelial mesenchymal cells secrete diverse Wnt ligands.[8,51–53] Wnt2b, produced by an unknown subepithelial mesenchymal population, could mediate the self-renewal of intestinal crypt in the absence of Paneth cells.[8]

3.2 Deregulation of Wnt Signaling in Colorectal Cancer

Colorectal cancer is the second most leading cause of cancer mortality and third most common cancer.[54] Colorectal cancer progresses to malignancy by metastasizing to the lymph nodes and liver in patients with an advanced disease stage. Wnt pathway–activating mutations are the main drivers of the tumorigenesis in colon. In 1991, mutations in the APC gene were discovered to cause familial adenomatous polyposis (FAP).[55,56] These patients develop many polyps in their colon, some of which progress to adenocarcinoma if they are not treated. APC mutations were also found to cause most cases of sporadic colorectal cancers.[57] Loss of APC function results in the accumulation of β-catenin, which translocates to the nucleus where it forms a complex with TCF/LEF transcription factors and triggers the activation of Wnt target genes.[58,59] The genome sequencing studies have revealed that 51% of hypermutated tumors and 81% of nonhypermutated tumors have APC mutations and about 5% harbor β-catenin–activating mutations.[60] The receptor FZD10 is also overexpressed in 17% of colorectal tumors.[60] On the other hand, the significant upregulation of Wnt2 and Wnt5a in some colorectal tumors suggests that Wnt ligands may also play a role in malignant colon cancer progression.[61–63] Thus, there appears to be a subset of colorectal tumors driven by enhanced Wnt ligand activity, which highlights the therapeutic potential of pharmacological interventions in Wnt secretion or activity.

3.3 Wnt Ligands are Also Part of the Stem Cell Niche in the Colon

In contrast to the small intestinal epithelium, the colonic epithelium does not express any Wnt ligands.[51,53] Several Wnt ligands are normally secreted from the subepithelial layer of colon. However, many colon cancer studies reported that epithelial colon cancer cells also start to secrete Wnt ligands.[61,63] In 1996 Vider et al.[63] reported that low amounts of Wnt2 are expressed in healthy colonic mesenchyme but its expression dramatically increases in colonic tumors. Wnt5a is expressed both in healthy colon stroma and tumor tissue. Later, Holcombe et al.[61] confirmed these findings and extended the list of Wnts that are expressed in both healthy and tumor tissue with addition of Wnt1, Wnt4, Wnt5b, Wnt6, Wnt7b, and Wnt10b. More recently, Voloshanenko et al.[64] reported that colon cancer cells with mutations in APC or β-catenin still depend on Wnt ligands to maintain maximal level of Wnt activity. Various colon cancer cell lines with Wnt pathway–activating mutations express canonical Wnt ligands; the most notable ones are Wnt3 and

Wnt10b. Elimination of Wls in these cell lines causes the abrogation of canonical Wnt pathway activity.[64] CSCs show high Wnt activity to preserve their stemness. They could be dependent on Wnts to keep Wnt signaling level high even in the presence of an APC mutation. It was shown that Wnt2 secreted by subepithelial fibroblasts enhances colon cancer progression.[65] Furthermore, Wnt2b with FZD7 potentiate the mesenchymal to epithelial transition in colon cancer cells.[62] The subepithelial cells in both the small intestine and the colon express Wnt2b closely to stem cells.[8] In colon cancer it is tempting to speculate that cancer cells start to secret Wnt ligands and together with subepithelial cells they establish their own Wnt-dependent niche. Since Wnt ligands produced by niche cells are crucial during intestinal homeostasis, it is conceivable that CSCs are also dependent on Wnt ligands during cancer initiation and progression. Malignancy of colon cancer is due to the progression of the disease to metastatic stage. Stromal Wnt ligands may play crucial roles during epithelial to mesenchymal transition (EMT) which is generally considered to be a critical step in metastasis formation. If stromal Wnt ligands are essential during EMT is still an open question. There are many questions that need to be answered to better understand the colon cancer initiation and progression to malignancy.

3.4 RNF43/ZNRF3 Mutations and RSPO2/3 Translocations are Markers for Wnt Dependent Tumors

The E3 ubiquitin ligases RNF43 and ZNRF3 promote ubiquitination and degradation of the Wnt receptors FZD and LRP6. Ubiquitination-dependent turnover of Wnt receptors is an important mechanism for regulating Wnt pathway activity since it orchestrates the cell response to Wnt ligands both in Wnt/β-catenin and Wnt/PCP signaling.[66]

R-spondins (RSPO 1-4) are secreted proteins that enhance Wnt signaling. R-spondins simultaneously bind to the extracellular domains of RNF43/ZNRF3 and Lgr4/5/6 and lead to auto-ubiquitination, internalization, and degradation of RNF43/ZNRF3. This results in increased levels of FZD receptors on the cell surface. R-spondins bind to Lgr4/5/6 through the Furin2 domain and interact with RNF43/ZNRF3 through the Furin 1 domain. To be functional, R-spondins must bind both Lgr4/5/6 and RNF43/ZNRF3; mutations abolishing the interactions abrogate the Wnt stimulatory activity of R-spondins.[67,68] RNF43 and ZNRF3 negatively regulate Wnt signaling and mutations in these proteins sensitize cell to Wnt ligands. Recent whole exome sequencing of 185 colorectal tumors shows a large number of nonsilent somatic mutations in RNF43.[69] Additionally, reanalysis of The Cancer Genome Atlas (TCGA) data showed the existence of nonsilent RNF43 mutations in 17.6% of 222 colorectal tumors.[69] Most RNF43 mutations result in truncation of protein, which is consistent with the current view of RNF43 as a tumor suppressor gene. Inactivating RNF43 mutations result in increased Wnt signaling due to an increased level of the Wnt FZD receptors. Mutations in APC are mutually exclusive with RNF43 mutations in colon tumors highlighting their shared consequences in activating Wnt/β-catenin signaling pathway.[69]

In an analysis of 68 colorectal cancer cases, RSPO2 and RSPO3 translocations were found in 3 and 8% of the cases.[70] The RSPO2 and RSPO3 form fusions with EIF3E and PTPRK, respectively. The fusion transcripts give rise to the functional RSPO2/3 proteins under the control of the stronger EIF3E and PTPRK promoters.[71] Tumors with such fusions have higher expression levels of RSPO2/3 proteins compared to tumors without fusions.[70,71] Similar to RNF43/ZNRF3 mutations; RSPO2/3 translocations are mutually exclusive with APC and β-catenin mutations. Thus, to obtain significant and continuous levels of Wnt/β-catenin signaling, cancer cells can bypass the negative regulatory power of RNF43/ZNRF3 with mutations of RNF43/ZNRF3 or translocations of R-spondin proteins. Importantly, such tumors are addicted to Wnt ligands.[72] Along with the classical APC loss of function and β-catenin gain of function mutations, RNF43/ZNRF3 mutations and R-spondin translocations are markers for activated Wnt signaling in colorectal cancers. In apparent contrast to the classical mutants, these subsets of colorectal cancers are driven by enhanced cellular sensitivity to Wnt ligands. The identity of the Wnts involved is under investigation.

Wnt production and the interaction of Wnt ligands with their receptors represent exciting therapeutic targets for treating colorectal tumors dependent on continued Wnt ligand activity.[73] A potent Porcupine inhibitor LGK974 has been shown to be effective in mice to target Wnt ligand dependent colorectal cancers.[73] The small molecule has reached to the clinical trials.[73]

Another focus is on inhibition of the interaction between Wnts and their receptors. Anti-FZD antibody OMP-18R5 (Vantictumab) binds to 5 of the 10 FZD receptors.[74] A similar approach is to use soluble Frizzled related proteins (sFRP) that hinder the interaction of Wnt ligands with their receptors. OMP-54F28 (Ipafricept) is such a recombinant protein and is under clinical trials.[75] Finally, neutralizing antibodies against RSPOs have been also investigated to target tumors which express RSPO fusion proteins.[76]

4. THE LIVER

4.1 Adult Liver Homeostasis

The liver is an important organ for systemic homeostasis and accomplishes a variety of functions including metabolism, glycogen storage, drug detoxification, and bile secretion. One of the most impressive and interesting capabilities of liver is the capacity to indefinitely regenerate and regulate its own size. Most of the functions are performed by hepatocytes that form approximately 60% of the total liver cells and comprise 80% of the volume of the organ. Hepatocytes are terminally differentiated epithelial cells (Fig. 2). Bile secreted from mature hepatocytes is exported through bile canaliculi to the bile ducts. The bile duct is composed of a distinct type of epithelial cell called a cholangiocyte (Fig. 2). The situation in terms of stem/progenitor cells in the adult liver is complicated and remains puzzling. Even the presence of stem cells in the liver and their role in liver regeneration remains hotly

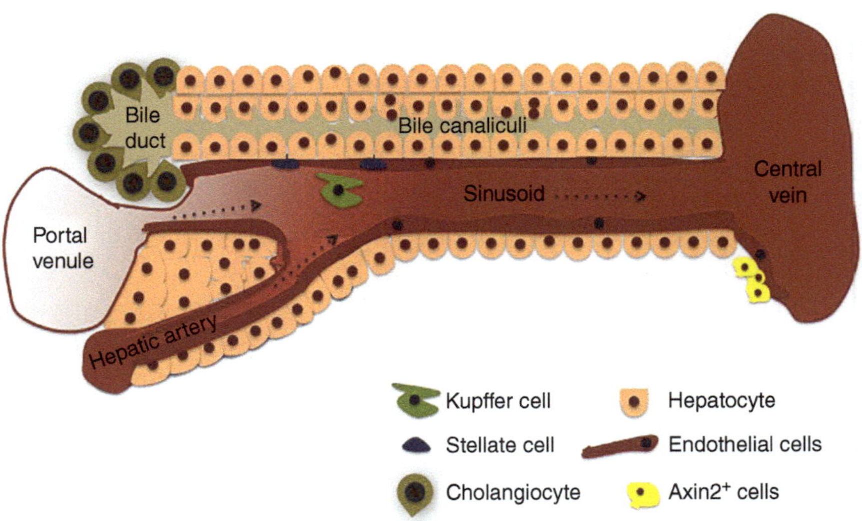

Fig. 2 Structure of a liver lobule. Hepatocytes are terminally differentiated epithelial cells. Bile secreted from mature hepatocytes is exported through bile canaliculi to the bile ducts. The bile duct is composed of a distinct type of epithelial cell called a cholangiocyte. Sinusoids carry blood from portal triad that consists of a portal venule, hepatic artery and bile duct. Hepatocyte cords are divided by sinusoids. Kupffer cells are, macrophages of the liver, located in sinusoids. Axin2^{+} cells are located close to the central vein and are surrounded by endothelial cells secreting Wnt2 and Wnt9b.

debated.[77–80] In adult tissues stem cells have two functions: homeostatic tissue maintenance/renewal under normal physiological conditions and tissue repair/regeneration upon injuries. In contrast to the intestine, in which mature differentiated cells have short lifetime and their continued renewal is needed to maintain tissue integrity, the turnover of mature hepatocytes takes a much longer time.[81,82] Until recently, the most widely accepted view in the field was that maintenance of hepatic tissue under normal physiological conditions is achieved by the proliferation of mature hepatocytes. However, a population of Axin2$^+$ proliferating and self-renewing cells residing close to the central vein has been identified (Fig. 2).[3] These cells express the early progenitor marker Tbx3 and they are diploid whereas majority of the mature hepatocytes are polyploid. By lineage tracing experiments, it was shown that the progeny of these cells differentiate into Tbx3-negative polyploidy hepatocytes and can mark all hepatocytes in the liver during homeostatic maintanence.[3] Remarkably, Axin2$^+$ cells located close to the central vein are surrounded by endothelial cells secreting Wnt2 and Wnt9b. Hence endothelial cells positioned at the central vein serve as a Wnt-producing niche for Axin2$^+$ stem cells (Fig. 2).[3] Similar to other stem cell niches, Wnt ligand mediated stimulation of the stem cells acts in the short-range, thereby keeping the stem cells very close to the niche.[3] Wnt ligands are lipid modified proteins and therefore hydrophobic. As a consequence it is thought that they do not travel freely in the extracellular space and thus have a short range of action.[83]

4.2 Stem Cells and Liver Regeneration

The liver can regenerate after either a partial hepatectomy (PH), in humans the original size of the tissue will be restored within 3–6 months,[84] or when recovering from injury, such as massive cell death induced by toxins or viral infection. In a healthy liver, most cells are quiescent and in the G0 phase of cell cycle. Following a PH or chemical injury, cells enter cell cycle and all types of liver cell are involved in proliferation during blood vessels rebuilding, biliary tree, and extracellular matrix replacement. After PH or chemical injury, hepatocytes are unique in that they undergo cell division without losing their differentiated phenotype and restore liver mass. The majority of liver regeneration involves proliferation of hepatocytes. In situations where hepatocytes are unable to divide, the liver has hepatic progenitor cells, so-called the oval cells, that undergo cell division and expansion followed by differentiation into hepatocytes.[85] They are positioned in the terminal bile ducts known as the Canals of Herring.[86,87] Notably, the oval cells are essential players during regeneration in response to hepatic cell death due to viral infection or chemical toxicity.[88]

Active β-catenin dependent Wnt signaling can be distinguished in mouse hepatocytes at periportal regions after 1–3 days of PH. Abolishing the β-catenin activity in hepatocytes leads to decreased hepatocyte proliferation after PH and increased hepatocyte apoptosis. Thus Wnt/β-catenin signaling is required for hepatocyte proliferation after PH. Overexpressing Wnt8 or stabilized β-catenin in hepatocytes elevated the hepatocyte proliferation.[89–92] Liver sinusoidal endothelial cells (LSECs) occupying the dense capillary region in the liver were reported to be a niche regulating hepatocyte proliferation after PH.[93] They secrete Wnt2b. Toxin-induced liver injury stimulates the oval cells. Similarly, PH combined with the drug 2 acetylaminofluorene (2-AAF) which blocks hepatocyte proliferation in the rat activates the oval cells in liver.[94] Prominently, the expression of Wnt7a, Wnt7b, and Wnt10 is robustly enhanced in mouse livers injured with hepatotoxins; however the cells secreting these Wnt ligands were not identified.[95] Later, it was shown that macrophages surrounding the hepatic progenitor cells in liver injured mice secrete Wnt3a and Wnt7a.[96] Additionally, Wnt6 is upregulated in toxin-induced liver damage.[97] Liver sinusoidal endothelial cells and macrophages form a Wnt-producing niche to maintain hepatocyte and oval cell proliferation. Lgr5, which is a stem cell marker in the intestine, stomach, and hair follicles, is not expressed in uninjured liver tissue; however, $Lgr5^+$ cells are identified around the bile duct after toxin-induced injury. In in vivo lineage tracing experiments, $Lgr5^+$ cells were shown to reconstitute the hepatocytes and biliary duct cells.[4] Moreover, liver organoids obtained from single $Lgr5^+$ cell displayed the features of oval stem cells and active Wnt/β-catenin signaling.[4] Recently, it was shown that healthy human bile duct derived bipotent progenitor cells, but not hepatocytes, could be cultured as organoids and express the stem cell marker Lgr5, as well as ductal and hepatocytes markers. These cells can differentiate into functional hepatocytes in vitro and give rise to hepatocytes upon transplantation.[98] The Wnt ligands involved in this process are still not known.

4.3 Wnt Signaling in Hepatocellular Carcinoma

Hepatocellular carcinoma (HCC) is the most common primary liver malignancy and a major cause of cancer related death. Initiation and progression of HCC is a multistep process; however, the molecular mechanisms that underpin HCC formation are poorly understood. Certain risk factors are associated with the development of HCC: viral hepatitis B and C, alcohol abuse, metabolic liver disorder, and α-toxin intoxication. Over longer periods of time these insults result in cirrhosis; the associated inflammation is a well-documented step in the development of HCC.[99] Similar to colorectal cancer, the Wnt/β-catenin signaling is important contributor to HCC. A mutation analysis of HCC indicated that Wnt pathway components are mutated or shows aberrant copy numbers in 44% HCC.[100] However in contrast to what is seen in colorectal cancers, activating β-catenin mutations are a late event in HCC, suggesting that deregulation of the Wnt/β-catenin pathway is involved in the progression of HCC but not its initiation. Wnt signaling, β-catenin–dependent and independent, may also play a role in early stages of HCC since increased expression of Wnt1, Wnt3, Wnt4, Wnt5a, and Wnt10b has been reported in HCC tumors and in the surrounding nontumorigenic liver tissues.[101–105] It is tempting to speculate that Wnt secreting cells may serve as a niche for activated liver progenitor cells in HCC, analogous to the Wnt-producing niche formed by endothelial cells positioned at the central vein for the $Axin2^+$ cells in healthy liver. Notably, animals expressing a constitutively active form of β-catenin do not form HCC but active β-catenin facilitates the progression of HCC.[106,107] The β-catenin activation may cooperate with other genomic changes, including TERT promoter modification, TP53 inactivation, and activating mutations of the RTK/RAS/PI(3)K pathway, to promote HCC oncogenesis.[108,109]

There is a great interest in developing of new therapeutic approaches to treat HCC. Recombinant sFZD7 peptides blocked Wnt/β-catenin signaling and decreased proliferation in particular HCC cell lines.[110] The effectiveness of the molecule increased in combination with Doxorubicin.[110] PMED1, which blocks the interaction of β-catenin and CBP, suppressed Wnt/β-catenin signaling targets in HCC cells.[111]

5. CONCLUDING REMARKS

Wnt ligands are an integral part of the stem cell niches in intestine and liver homeostasis, during regeneration and in cancer. In recent years, our knowledge on the contribution of Wnt ligands as a stem cell niche component has increased. In addition to the scenarios we have described for colorectal cancer and HCC, Wnt ligands are also involved in the cancer stem cell niches in pancreatic and lung cancers.[112,113] The interactions between the cancer stem cell and niche cell are complex. We have sketched the role of Wnt ligands in this process but additional signals are also involved, such as Notch and Hedgehog. Unraveling the individual contribution of these pathways, as well as their interaction, is the critical next step.

REFERENCES

1. Schofield R. The relationship between the spleen colony-forming cell and the haemopoietic stem cell. *Blood Cells.* 1978;4(1–2):7–25.
2. Sato T, van Es JH, Snippert HJ, et al. Paneth cells constitute the niche for Lgr5 stem cells in intestinal crypts. *Nature.* 2011;469(7330):415–418.
3. Wang B, Zhao L, Fish M, Logan CY, Nusse R. Self-renewing diploid Axin2(+) cells fuel homeostatic renewal of the liver. *Nature.* 2015;524(7564):180–185.
4. Huch M, Dorrell C, Boj SF, et al. In vitro expansion of single Lgr5+ liver stem cells induced by Wnt-driven regeneration. *Nature.* 2013;494(7436):247–250.
5. Huch M, Bonfanti P, Boj SF, et al. Unlimited in vitro expansion of adult bi-potent pancreas progenitors through the Lgr5/R-spondin axis. *EMBO J.* 2013;32(20):2708–2721.
6. Gering M, Patient R. Notch in the niche. *Cell Stem Cell.* 2008;2(4):293–294.
7. Blank U, Karlsson G, Karlsson S. Signaling pathways governing stem-cell fate. *Blood.* 2008;111(2):492–503.
8. Valenta T, Degirmenci B, Moor AE, et al. Wnt ligands secreted by subepithelial mesenchymal cells are essential for the survival of intestinal stem cells and gut homeostasis. *Cell Rep.* 2016;15(5):911–918.
9. Plaks V, Kong N, Werb Z. The cancer stem cell niche: how essential is the niche in regulating stemness of tumor cells? *Cell Stem Cell.* 2015;16(3):225–238.
10. Komekado H, Yamamoto H, Chiba T, Kikuchi A. Glycosylation and palmitoylation of Wnt-3a are coupled to produce an active form of Wnt-3a. *Genes Cells.* 2007;12(4): 521–534.
11. Kurayoshi M, Yamamoto H, Izumi S, Kikuchi A. Post-translational palmitoylation and glycosylation of Wnt-5a are necessary for its signalling. *Biochem J.* 2007;402(3):515–523.
12. Doubravska L, Krausova M, Gradl D, et al. Fatty acid modification of Wnt1 and Wnt3a at serine is prerequisite for lipidation at cysteine and is essential for Wnt signalling. *Cell Signal.* 2011;23(5):837–848.
13. Tang X, Wu Y, Belenkaya TY, et al. Roles of N-glycosylation and lipidation in Wg secretion and signaling. *Dev Biol.* 2012;364(1):32–41.
14. Zhai L, Chaturvedi D, Cumberledge S. Drosophila wnt-1 undergoes a hydrophobic modification and is targeted to lipid rafts, a process that requires porcupine. *J Biol Chem.* 2004;279(32):33220–33227.
15. Bänziger C, Soldini D, Schütt C, Zipperlen P, Hausmann G, Basler K. Wntless, a conserved membrane protein dedicated to the secretion of Wnt proteins from signaling cells. *Cell.* 2006;125(3):509–522.
16. Bartscherer K, Pelte N, Ingelfinger D, Boutros M. Secretion of Wnt ligands requires Evi, a conserved transmembrane protein. *Cell.* 2006;125(3):523–533.
17. Goodman RM, Thombre S, Firtina Z, et al. Sprinter: a novel transmembrane protein required for Wg secretion and signaling. *Development.* 2006;133(24):4901–4911.
18. Belenkaya TY, Wu Y, Tang X, et al. The retromer complex influences Wnt secretion by recycling wntless from endosomes to the trans-Golgi network. *Dev Cell.* 2008;14 (1):120–131.
19. Franch-Marro X, Wendler F, Guidato S, et al. Wingless secretion requires endosome-to Golgi retrieval of Wntless/Evi/Sprinter by the retromer complex. *Nat Cell Biol.* 2008;10(2):170–177.

20. Port F, Kuster M, Herr P, et al. Wingless secretion promotes and requires retromer-dependent cycling of Wntless. *Nat Cell Biol.* 2008;10(2):178–185.
21. Fu J, Jiang M, Mirando AJ, Yu HM, Hsu W. Reciprocal regulation of Wnt and Gpr177/mouse Wntless is required for embryonic axis formation. *Proc Natl Acad Sci USA.* 2009;106(44):18598–18603.
22. Valenta T, Hausmann G, Basler K. The many faces and functions of β-catenin. *EMBO J.* 2012;31(12):2714–2736.
23. Seifert JR, Mlodzik M. Frizzled/PCP signalling: a conserved mechanism regulating cell polarity and directed motility. *Nat Rev Genet.* 2007;8(2):126–138.
24. Endo Y, Wolf V, Muraiso K, et al. Wnt-3a-dependent cell motility involves RhoA activation and is specifically regulated by dishevelled-2. *J Biol Chem.* 2005;280(1):777–786.
25. Kishida S, Yamamoto H, Kikuchi A. Wnt-3a and Dvl induce neurite retraction by activating Rho-associated kinase. *Mol Cell Biol.* 2004;24(10):4487–4501.
26. Kühl M, Sheldahl LC, Malbon CC, Moon RT. Ca(2+)/calmodulin-dependent protein kinase II is stimulated by Wnt and Frizzled homologs and promotes ventral cell fates in Xenopus. *J Biol Chem.* 2000;275(17):12701–12711.
27. Nalesso G, Sherwood J, Bertrand J, et al. WNT-3A modulates articular chondrocyte phenotype by activating both canonical and noncanonical pathways. *J Cell Biol.* 2011;193(3):551–564.
28. Kühl M, Geis K, Sheldahl LC, Pukrop T, Moon RT, Wedlich D. Antagonistic regulation of convergent extension movements in Xenopus by Wnt/beta-catenin and Wnt/Ca2+ signaling. *Mech Dev.* 2001;106(1–2):61–76.
29. Du SJ, Purcell SM, Christian JL, McGrew LL, Moon RT. Identification of distinct classes and functional domains of Wnts through expression of wild-type and chimeric proteins in Xenopus embryos. *Mol Cell Biol.* 1995;15(5):2625–2634.
30. Garriock RJ, Krieg PA. Wnt11-R signaling regulates a calcium sensitive EMT event essential for dorsal fin development of Xenopus. *Dev Biol.* 2007;304(1):127–140.
31. Kremenevskaja N, von Wasielewski R, Rao AS, Schöfl C, Andersson T, Brabant G. Wnt-5a has tumor suppressor activity in thyroid carcinoma. *Oncogene.* 2005;24(13):2144–2154.
32. Weeraratna AT, Jiang Y, Hostetter G, et al. Wnt5a signaling directly affects cell motility and invasion of metastatic melanoma. *Cancer Cell.* 2002;1(3):279–288.
33. Inoki K, Ouyang H, Zhu T, et al. TSC2 integrates Wnt and energy signals via a coordinated phosphorylation by AMPK and GSK3 to regulate cell growth. *Cell.* 2006;126(5):955–968.
34. Anastas JN, Moon RT. WNT signalling pathways as therapeutic targets in cancer. *Nat Rev Cancer.* 2013;13(1):11–26.
35. Nusse R, Clevers H. Wnt/β-catenin signaling, disease, and emerging therapeutic modalities. *Cell.* 2017;169(6):985–999.
36. Barker N, van Es JH, Kuipers J, et al. Identification of stem cells in small intestine and colon by marker gene Lgr5. *Nature.* 2007;449(7165):1003–1007.
37. Muñoz J, Stange DE, Schepers AG, et al. The Lgr5 intestinal stem cell signature: robust expression of proposed quiescent '+4' cell markers. *EMBO J.* 2012;31(14):3079–3091.
38. Sangiorgi E, Capecchi MR. Bmi1 is expressed in vivo in intestinal stem cells. *Nat Genet.* 2008;40(7):915–920.
39. Montgomery RK, Carlone DL, Richmond CA, et al. Mouse telomerase reverse transcriptase (mTert) expression marks slowly cycling intestinal stem cells. *Proc Natl Acad Sci USA.* 2011;108(1):179–184.
40. Takeda N, Jain R, LeBoeuf MR, Wang Q, Lu MM, Epstein JA. Interconversion between intestinal stem cell populations in distinct niches. *Science.* 2011;334(6061):1420–1424.

41. Powell AE, Wang Y, Li Y, et al. The pan-ErbB negative regulator Lrig1 is an intestinal stem cell marker that functions as a tumor suppressor. *Cell.* 2012;149(1):146–158.
42. Tian H, Biehs B, Warming S, et al. A reserve stem cell population in small intestine renders Lgr5-positive cells dispensable. *Nature.* 2011;478(7368):255–259.
43. Buczacki SJ, Zecchini HI, Nicholson AM, et al. Intestinal label-retaining cells are secretory precursors expressing Lgr5. *Nature.* 2013;495(7439):65–69.
44. van Es JH, Jay P, Gregorieff A, et al. Wnt signalling induces maturation of Paneth cells in intestinal crypts. *Nat Cell Biol.* 2005;7(4):381–386.
45. Snippert HJ, van der Flier LG, Sato T, et al. Intestinal crypt homeostasis results from neutral competition between symmetrically dividing Lgr5 stem cells. *Cell.* 2010;143(1):134–144.
46. Roth S, Franken P, Sacchetti A, et al. Paneth cells in intestinal homeostasis and tissue injury. *PLoS One.* 2012;7(6):e38965.
47. Korinek V, Barker N, Moerer P, et al. Depletion of epithelial stem-cell compartments in the small intestine of mice lacking Tcf-4. *Nat Genet.* 1998;19(4):379–383.
48. Pinto D, Gregorieff A, Begthel H, Clevers H. Canonical Wnt signals are essential for homeostasis of the intestinal epithelium. *Genes Dev.* 2003;17(14):1709–1713.
49. Kim TH, Escudero S, Shivdasani RA. Intact function of Lgr5 receptor-expressing intestinal stem cells in the absence of Paneth cells. *Proc Natl Acad Sci USA.* 2012;109(10):3932–3937.
50. Durand A, Donahue B, Peignon G, et al. Functional intestinal stem cells after Paneth cell ablation induced by the loss of transcription factor Math1 (Atoh1). *Proc Natl Acad Sci USA.* 2012;109(23):8965–8970.
51. Gregorieff A, Pinto D, Begthel H, Destrée O, Kielman M, Clevers H. Expression pattern of Wnt signaling components in the adult intestine. *Gastroenterology.* 2005;129(2):626–638.
52. Klostermeier UC, Barann M, Wittig M, et al. A tissue-specific landscape of sense/antisense transcription in the mouse intestine. *BMC Genomics.* 2011;12:305.
53. Farin HF, Van Es JH, Clevers H. Redundant sources of Wnt regulate intestinal stem cells and promote formation of Paneth cells. *Gastroenterology.* 2012;143(6):1518–1529.
54. Cromheecke M, de Jong KP, Hoekstra HJ. Current treatment for colorectal cancer metastatic to the liver. *Eur J Surg Oncol.* 1999;25(5):451–463.
55. Groden J, Thliveris A, Samowitz W, et al. Identification and characterization of the familial adenomatous polyposis coli gene. *Cell.* 1991;66(3):589–600.
56. Nishisho I, Nakamura Y, Miyoshi Y, et al. Mutations of chromosome 5q21 genes in FAP and colorectal cancer patients. *Science.* 1991;253(5020):665–669.
57. Powell SM, Zilz N, Beazer-Barclay Y, et al. APC mutations occur early during colorectal tumorigenesis. *Nature.* 1992;359(6392):235–237.
58. Morin PJ, Sparks AB, Korinek V, et al. Activation of beta-catenin-Tcf signaling in colon cancer by mutations in beta-catenin or APC. *Science.* 1997;275(5307):1787–1790.
59. Korinek V, Barker N, Morin PJ, et al. Constitutive transcriptional activation by a beta-catenin-Tcf complex in APC-/- colon carcinoma. *Science.* 1997;275(5307):1784–1787.
60. Network CGA. Comprehensive molecular characterization of human colon and rectal cancer. *Nature.* 2012;487(7407):330–337.
61. Holcombe RF, Marsh JL, Waterman ML, Lin F, Milovanovic T, Truong T. Expression of Wnt ligands and Frizzled receptors in colonic mucosa and in colon carcinoma. *Mol Pathol.* 2002;55(4):220–226.
62. Schwab RHM, Amin N, Flanagan DJ, Johanson TM, Phesse TJ, Vincan E. Wnt is necessary for mesenchymal to epithelial transition in colorectal cancer cells. *Dev Dyn.* 2017 doi:10.1002/dvdy.24527.
63. Vider BZ, Zimber A, Chastre E, et al. Evidence for the involvement of the Wnt 2 gene in human colorectal cancer. *Oncogene.* 1996;12(1):153–158.

64. Voloshanenko O, Erdmann G, Dubash TD, et al. Wnt secretion is required to maintain high levels of Wnt activity in colon cancer cells. *Nat Commun.* 2013;4:2610.
65. Kramer N, Schmöllerl J, Unger C, et al. Autocrine WNT2 signaling in fibroblasts promotes colorectal cancer progression. *Oncogene.* 2017;36(39):5460–5472.
66. Hao HX, Xie Y, Zhang Y, et al. ZNRF3 promotes Wnt receptor turnover in an R-spondin-sensitive manner. *Nature.* 2012;485(7397):195–200.
67. Kim KA, Wagle M, Tran K, et al. R-Spondin family members regulate the Wnt pathway by a common mechanism. *Mol Biol Cell.* 2008;19(6):2588–2596.
68. Zebisch M, Xu Y, Krastev C, et al. Structural and molecular basis of ZNRF3/RNF43 transmembrane ubiquitin ligase inhibition by the Wnt agonist R-spondin. *Nat Commun.* 2013;4:2787.
69. Giannakis M, Hodis E, Jasmine Mu X, et al. RNF43 is frequently mutated in colorectal and endometrial cancers. *Nat Genet.* 2014;46(12):1264–1266.
70. Shinmura K, Kahyo T, Kato H, et al. RSPO fusion transcripts in colorectal cancer in Japanese population. *Mol Biol Rep.* 2014;41(8):5375–5384.
71. Seshagiri S, Stawiski EW, Durinck S, et al. Recurrent R-spondin fusions in colon cancer. *Nature.* 2012;488(7413):660–664.
72. Madan B, Ke Z, Harmston N, et al. Wnt addiction of genetically defined cancers reversed by PORCN inhibition. *Oncogene.* 2016;35(17):2197–2207.
73. Liu J, Pan S, Hsieh MH, et al. Targeting Wnt-driven cancer through the inhibition of Porcupine by LGK974. *Proc Natl Acad Sci USA.* 2013;110(50):20224–20229.
74. Gurney A, Axelrod F, Bond CJ, et al. Wnt pathway inhibition via the targeting of Frizzled receptors results in decreased growth and tumorigenicity of human tumors. *Proc Natl Acad Sci USA.* 2012;109(29):11717–11722.
75. Jimeno A, Gordon MS, Chugh R, et al. A first-in-human Phase 1 study of the anti-cancer stem cell agent ipafricept (OMP-54F28), a decoy receptor for Wnt ligands, in patients with advanced solid tumors. *Clin Cancer Res.* 2017 doi:10.1158/1078-0432.CCR-17-2157.
76. Storm EE, Durinck S, de Sousa e Melo F, et al. Targeting PTPRK-RSPO3 colon tumours promotes differentiation and loss of stem-cell function. *Nature.* 2016;529(7584):97–100.
77. Duncan AW, Dorrell C, Grompe M. Stem cells and liver regeneration. *Gastroenterology.* 2009;137(2):466–481.
78. Riehle KJ, Dan YY, Campbell JS, Fausto N. New concepts in liver regeneration. *J Gastroenterol Hepatol.* 2011;26(Suppl 1):203–212.
79. Miyajima A, Tanaka M, Itoh T. Stem/progenitor cells in liver development, homeostasis, regeneration, and reprogramming. *Cell Stem Cell.* 2014;14(5):561–574.
80. Itoh T. Stem/progenitor cells in liver regeneration. *Hepatology.* 2016;64(2):663–668.
81. Macdonald RA. "Lifespan" of liver cells. Autoradio-graphic study using tritiated thymidine in normal, cirrhotic, and partially hepatectomized rats. *Arch Intern Med.* 1961;107:335–343.
82. Magami Y, Azuma T, Inokuchi H, et al. Cell proliferation and renewal of normal hepatocytes and bile duct cells in adult mouse liver. *Liver.* 2002;22(5):419–425.
83. Clevers H, Loh KM, Nusse R. Stem cell signaling. An integral program for tissue renewal and regeneration: Wnt signaling and stem cell control. *Science.* 2014;346(6205):1248012.
84. Fausto N, Campbell JS, Riehle KJ. Liver regeneration. *Hepatology.* 2006;43(2 Suppl 1):S45–S53.
85. Wang X, Foster M, Al-Dhalimy M, Lagasse E, Finegold M, Grompe M. The origin and liver repopulating capacity of murine oval cells. *Proc Natl Acad Sci USA.* 2003;100(Suppl 1):11881–11888.
86. Thorgeirsson SS. Hepatic stem cells in liver regeneration. *FASEB J.* 1996;10(11):1249–1256.

87. Tatematsu M, Ho RH, Kaku T, Ekem JK, Farber E. Studies on the proliferation and fate of oval cells in the liver of rats treated with 2-acetylaminofluorene and partial hepatectomy. *Am J Pathol.* 1984;114(3):418–430.
88. Fausto N. Liver regeneration and repair: hepatocytes, progenitor cells, and stem cells. *Hepatology.* 2004;39(6):1477–1487.
89. Nejak-Bowen KN, Monga SP. Beta-catenin signaling, liver regeneration and hepatocellular cancer: sorting the good from the bad. *Semin Cancer Biol.* 2011;21(1): 44–58.
90. Nejak-Bowen KN, Thompson MD, Singh S, et al. Accelerated liver regeneration and hepatocarcinogenesis in mice overexpressing serine-45 mutant beta-catenin. *Hepatology.* 2010;51(5):1603–1613.
91. Nejak-Bowen K, Monga SP. Wnt/beta-catenin signaling in hepatic organogenesis. *Organogenesis.* 2008;4(2):92–99.
92. Goessling W, North TE, Lord AM, et al. APC mutant zebrafish uncover a changing temporal requirement for wnt signaling in liver development. *Dev Biol.* 2008;320(1): 161–174.
93. Ding BS, Nolan DJ, Butler JM, et al. Inductive angiocrine signals from sinusoidal endothelium are required for liver regeneration. *Nature.* 2010;468(7321):310–315.
94. Fausto N, Campbell JS. The role of hepatocytes and oval cells in liver regeneration and repopulation. *Mech Dev.* 2003;120(1):117–130.
95. Itoh T, Kamiya Y, Okabe M, Tanaka M, Miyajima A. Inducible expression of Wnt genes during adult hepatic stem/progenitor cell response. *FEBS Lett.* 2009;583(4): 777–781.
96. Boulter L, Govaere O, Bird TG, et al. Macrophage-derived Wnt opposes Notch signaling to specify hepatic progenitor cell fate in chronic liver disease. *Nat Med.* 2012;18(4):572–579.
97. Huch M, Boj SF, Clevers H. Lgr5(+) liver stem cells, hepatic organoids and regenerative medicine. *Regen Med.* 2013;8(4):385–387.
98. Huch M, Gehart H, van Boxtel R, et al. Long-term culture of genome-stable bipotent stem cells from adult human liver. *Cell.* 2015;160(1–2):299–312.
99. Schuppan D, Afdhal NH. Liver cirrhosis. *Lancet.* 2008;371(9615):838–851.
100. Network CGA. Comprehensive and integrative genomic characterization of hepatocellular carcinoma. *Cell.* 2017;169(7):1327–1341.
101. Kim M, Lee HC, Tsedensodnom O, et al. Functional interaction between Wnt3 and Frizzled-7 leads to activation of the Wnt/beta-catenin signaling pathway in hepatocellular carcinoma cells. *J Hepatol.* 2008;48(5):780–791.
102. Lee HH, Uen YH, Tian YF, et al. Wnt-1 protein as a prognostic biomarker for hepatitis B-related and hepatitis C-related hepatocellular carcinoma after surgery. *Cancer Epidemiol Biomarkers Prev.* 2009;18(5):1562–1569.
103. Yoshikawa H, Matsubara K, Zhou X, et al. WNT10B functional dualism: beta-catenin/Tcf-dependent growth promotion or independent suppression with deregulated expression in cancer. *Mol Biol Cell.* 2007;18(11):4292–4303.
104. Bengochea A, de Souza MM, Lefrançois L, et al. Common dysregulation of Wnt/Frizzled receptor elements in human hepatocellular carcinoma. *Br J Cancer.* 2008;99(1): 143–150.
105. Wei W, Chua MS, Grepper S, So SK. Blockade of Wnt-1 signaling leads to anti-tumor effects in hepatocellular carcinoma cells. *Mol Cancer.* 2009;8:76.
106. Cadoret A, Ovejero C, Saadi-Kheddouci S, et al. Hepatomegaly in transgenic mice expressing an oncogenic form of beta-catenin. *Cancer Res.* 2001;61(8):3245–3249.
107. Harada N, Miyoshi H, Murai N, et al. Lack of tumorigenesis in the mouse liver after adenovirus-mediated expression of a dominant stable mutant of beta-catenin. *Cancer Res.* 2002;62(7):1971–1977.

108. Schulze K, Imbeaud S, Letouzé E, et al. Exome sequencing of hepatocellular carcinomas identifies new mutational signatures and potential therapeutic targets. *Nat Genet.* 2015;47(5):505–511.
109. Totoki Y, Tatsuno K, Covington KR, et al. Trans-ancestry mutational landscape of hepatocellular carcinoma genomes. *Nat Genet.* 2014;46(12):1267–1273.
110. Wei W, Chua MS, Grepper S, So SK. Soluble Frizzled-7 receptor inhibits Wnt signaling and sensitizes hepatocellular carcinoma cells towards doxorubicin. *Mol Cancer.* 2011;10:16.
111. Delgado ER, Yang J, So J, et al. Identification and characterization of a novel small-molecule inhibitor of β-catenin signaling. *Am J Pathol.* 2014;184(7):2111–2122.
112. Steinhart Z, Pavlovic Z, Chandrashekhar M, et al. Genome-wide CRISPR screens reveal a Wnt-FZD5 signaling circuit as a druggable vulnerability of RNF43-mutant pancreatic tumors. *Nat Med.* 2017;23(1):60–68.
113. Tammela T, Sanchez-Rivera FJ, Cetinbas NM, et al. A Wnt-producing niche drives proliferative potential and progression in lung adenocarcinoma. *Nature.* 2017;545(7654):355–359.

CHAPTER TWO

Wnt Signaling in Adult Epithelial Stem Cells and Cancer

Si Hui Tan*, Nick Barker*,†,‡,1
* A*STAR Institute of Medical Biology, Singapore
† Kanazawa University, Kanazawa, Japan
‡ Centre for Regenerative Medicine, University of Edinburgh, Edinburgh, United Kingdom
1 Corresponding author. E-mail address: nicholas.barker@imb.a-star.edu.sg

Contents

Abstract

Wnt/β-catenin signaling is integral to the homeostasis and regeneration of many epithelial tissues due to its critical role in adult stem cell regulation. It is also implicated in many epithelial cancers, with mutations in core pathway components frequently present in patient tumors. In this chapter, we discuss the roles of Wnt/β-catenin signaling and Wnt-regulated stem cells in homeostatic, regenerative and cancer contexts of the intestines, stomach, skin, and liver. We also examine the sources of Wnt ligands that form part of the stem cell niche. Despite the diversity in characteristics of various tissue stem cells, the role(s) of Wnt/β-catenin signaling is generally

Progress in Molecular Biology and Translational Science, Volume 153
ISSN 1877-1173
http://dx.doi.org/10.1016/bs.pmbts.2017.11.017

coherent in maintaining stem cell fate and/or promoting proliferation. It is also likely to play similar roles in cancer stem cells, making the pathway a salient therapeutic target for cancer. While promising progress is being made in the field, deeper understanding of the functions and signaling mechanisms of the pathway in individual epithelial tissues will expedite efforts to modulate Wnt/β-catenin signaling in cancer treatment and tissue regeneration.

1. INTRODUCTION

Epithelia comprise the largest tissue type in our body. They serve as barriers against external agents, absorptive surfaces for water and nutrients, and/or glands secreting important hormones or enzymes. Most epithelia line surfaces of our body that come into contact with the external environment (e.g., skin, cornea), or tissue cavities within our body that could be exposed to external agents (e.g., gastrointestinal tract, lung, urogenital tissues). As a result, it is necessary for the tissues to turnover frequently to replace lost or damaged cells so as to maintain the normal function and integrity of the epithelial tissues. Small populations of multipotent adult epithelial stem cells are critical for effecting this relentless tissue renewal throughout life.

1.1 Adult Stem Cells

Adult stem cells fuel tissue renewal during both homeostasis and regeneration. They constitute specialized classes of cells, each representing a specific lineage of their resident organs that give rise to rapidly proliferating progenitors responsible for producing the various differentiated cells that carry out the requisite physiological tissue functions. Stem cells are the only population that can persist long-term, thereby sustaining the supply of differentiated cells over the organism's lifetime. The two hallmarks of adult stem cells are: (1) infinite self-renewal capability as a population and (2) an ability to generate all differentiated lineages of an organ.[1]

Stem cells can be broadly classified as homeostatic and facultative. Homeostatic stem cells are responsible for the day-to-day maintenance of the tissue; whereas facultative stem cells (a.k.a. quiescent stem cells) are not typically active during homeostasis, but are capable of performing stem cell functions when called to duty following injury. These two types of stem cells are not mutually exclusive and can coexist as distinct populations in the same tissue.[1]

1.2 Cancer

Cancer, a disease of unrelenting cell proliferation and compromised normal tissue function, is a frequent and deadly disease of the modern world. Epithelial cancers, a.k.a. carcinomas, are the most common type of cancer, as highlighted by the fact that the 10 cancers with highest incidence worldwide in 2012 are all carcinomas, regardless of geography.[2]

Cancers arise from a series of molecular changes that endow the tumors with eight distinct survival advantages, now famously referred to as "Hallmarks of Cancer."[3] The cells in which these changes initially accumulate are the cell-of-origin (a.k.a. tumor initiating cell) of the cancer. In many carcinomas, the cell-of-origin is often thought to be a stem or progenitor cell that has been transformed through the acquisition of multiple genetic mutations, equipping it with a survival and proliferative advantage.[4] Once established, the progression of the tumor then becomes fueled by cancer stem cells (CSCs), which may or may not be the same as the cancer cell-of-origin.

1.3 Wnt/β-Catenin Signaling

Wnt/β-catenin signaling is instrumental in directing tissue development, homeostasis, and regeneration, particularly in the stem cell context. Unsurprisingly, it is also commonly subjugated by tumor cells to drive cancer. There are two other known Wnt pathways, commonly referred to as the planar cell polarity pathway and Wnt/Ca^{2+} pathway.[5] Here, we will be focusing on Wnt/β-catenin pathway, which has been established as a principal signaling pathway in stem cell and cancer biology.

The detailed mechanics of Wnt/β-catenin signaling are extensively described elsewhere.[6–8] To summarize (Fig. 1), in the absence of Wnt ligands, a destruction complex, which includes adenomatous polyposis coli (Apc), glycogen synthase kinase-3 (Gsk-3β), and Axin, binds to cytoplasmic β-catenin and marks it for proteosomal degradation, thereby preventing Wnt/β-catenin signaling activity in the nucleus. At the same time, the T-cell factor (Tcf)/lymphoid-enhancing factor (Lef)/Groucho nuclear complex actively represses transcription of Wnt/β-catenin target genes (Fig. 1A). When Wnt ligands bind to Frizzled (Fzd) and LDL-receptor-related protein 5/6 (Lrp5/6) receptors in the Wnt-responding cells, the receptors sequester Axin, causing the destruction complex to disassemble, thereby allowing cytosolic β-catenin to accumulate. β-Catenin subsequently enters the nucleus, and converts the Tcf/Lef complex from a repressor to a transcriptional activator,[9–12] inducing expression of target genes, such as *Axin2*[13] and

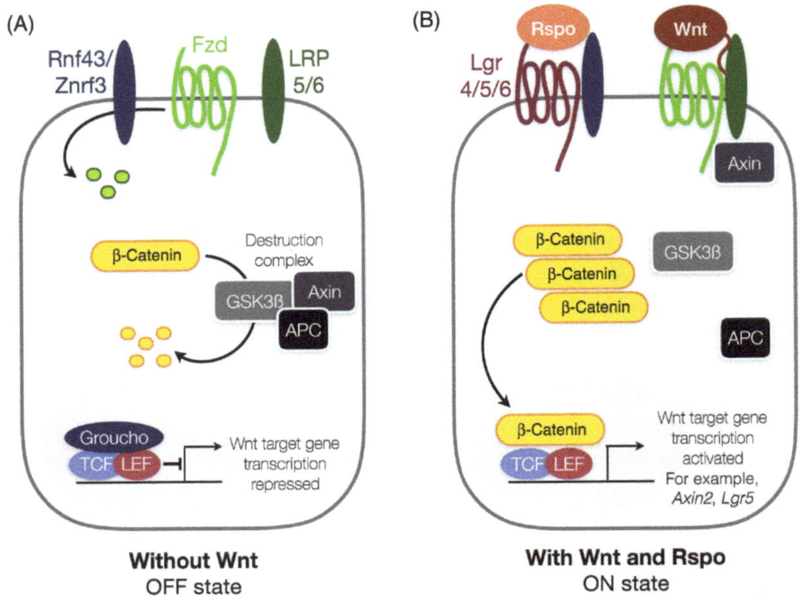

Fig. 1 Wnt/β-catenin signaling. (A) In the absence of Wnt ligand, the Frizzled (Fzd) receptor is targeted for degradation by Rnf43 and Znrf3. In addition, cytosolic β-catenin is also marked for degradation by the destruction complex comprising of Axin, Gsk-3β, and adenomatous polyposis coli (Apc). In the absence of β-catenin, the Tcf/Lef/Groucho complex in the nucleus acts as a transcriptional repressor and the Wnt/β-catenin targets are not expressed. (B) When Wnt ligand binds to the Fzd/Lrp receptor complex, they sequester Axin and cause the destruction complex to disintegrate leaving β-catenin molecules to accumulate in the cytosol and translocate into the nucleus. In the nucleus, β-catenin associates with Tcf/Lef complex at promoters/enhancers of Wnt/β-catenin target genes, such as *Axin2* and *Lgr5* to activate their transcription. When Lgr5 is expressed at the membrane, binding of R-spondin ligand sequesters Rnf43/Znrf3, thereby inhibiting Fzd degradation and potentiating Wnt/β-catenin signaling activity. *LEF*, Lymphoid-enhancing factor; *TCF*, T-cell factor.

leucine-rich repeat containing G protein–coupled receptor 5 (*Lgr5*, a.k.a. *Gpr49*)[14] (Fig. 1B). Nuclear localization of β-catenin is commonly used as an indicator of high Wnt/β-catenin signaling activity. Interestingly, while Axin2 is a weak negative feedback regulator of Wnt/β-catenin signaling, Lgr5 is a potentiator of Wnt signaling. When Lgr5 (or closely related family members Lgr4 and Lgr6) binds to its secreted ligand R-spondin (there are four R-spondins in humans and mice), they form a complex with the membrane-bound E3 ligases zinc and ring-finger 43 (Znrf3) and ring-finger protein 43 (Rnf43), thereby preventing them from ubiquitinating Fzd receptors, which would target them for degradation.[15–17] Hence, when Lgr4/5/6 is expressed in the presence of R-spondins, Fzd receptors accumulate on the cell membrane to enhance Wnt/β-catenin signaling.

In vivo lineage tracing models employing Cre-lox technology have identified *Lgr5* and/or *Axin2* as markers of adult stem cell populations in a wide range of tissues, such as the intestines and hair follicles. In brief, lineage tracing models comprise two essential components: an inducible CreERT2 cassette whose expression is controlled by an endogenous *Lgr5* or *Axin2* promoter, and a conditional reporter allele, such as Rosa26-LacZ,[18] mTmG,[19] Confetti,[20] or Ai6/9/14.[21] Administration of tamoxifen allows the Cre recombinase to translocate to the nucleus,[22] where it catalyzes the recombination of a genomic region flanked by loxP sites, such as the transcriptional stop cassette that precedes the reporter gene. Consequently, cells that express *Axin2* or *Lgr5* when the tamoxifen was administered are labeled with an indelible, heritable genetic mark that is subsequently passed on to all their progeny in a living tissue. Long-term labeling of all differentiated cell lineages in a given tissue identifies the Cre-expressing $Lgr5^+$ or $Axin2^+$ cells as self-renewing, multipotent stem cells.

1.4 Sources of Wnt Ligands are Essential Stem Cell Niche Components

The stem cell niche is the microenvironment that supports the stem cells and regulates their behavior during tissue homeostasis and regeneration. As Wnt/β-catenin signaling is one of the critical pathways for stem cell maintenance, the Wnt ligand-producing cells define a core component of the niche. As Wnt ligands are palmitoylated, they have limited diffusion range in the extracellular matrix and therefore typically act as short-range signals.[23] Wnt ligands can theoretically be produced by any of the cell types present in the niche, such as fibroblasts, hematopoietic cells, endothelial cells, or other epithelial cells, which would consequently form a paracrine signaling axis. Alternatively, autocrine signaling is possible when Wnt-responding stem cells supply the Wnt ligands for their own use.

Regardless of the precise mode of signaling, the production of mature Wnt ligands and their transport to the membrane for secretion requires two proteins—Porcupine (Porcn) and Wntless (Wls, a.k.a. Gpr177, Evenness interrupted/Eva, MIG-14) (Fig. 2). Porcn catalyzes the palmitoylation, which is required for the function of all Wnt ligands,[24–26] as the modification likely facilitates binding to the Fzd receptors.[27,28] Once palmitoylated, the

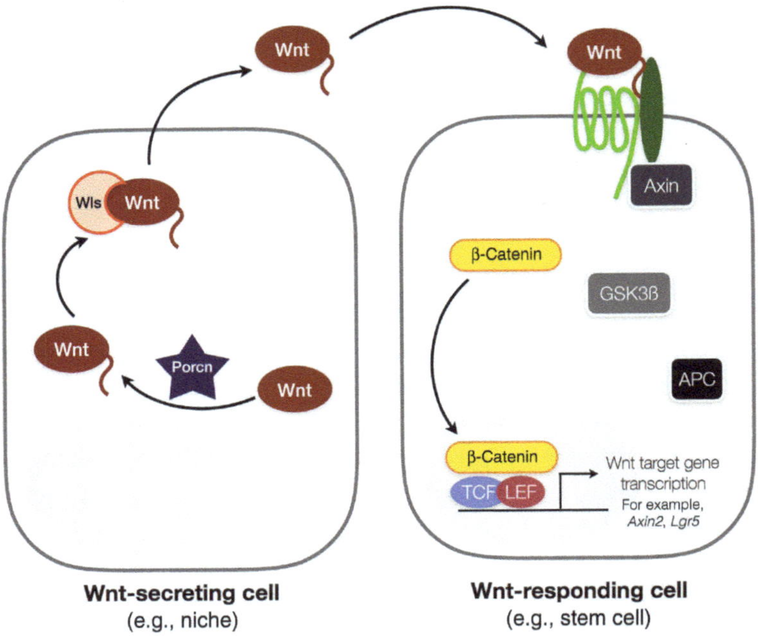

Fig. 2 Posttranslational modification of Wnt ligands. In the Wnt-producing cells (left), nascent Wnt proteins are palmitoylated by the enzyme Porcupine (Porcn). The lipid modification is required for their secretion and function. Thereafter, the lipidated Wnt molecule associates with a multispan transmembrane protein Wntless (Wls) in a secretory vesicle, and is secreted from the cell via exocytosis.

Wnts are insoluble in cytosol and require transport within exocytic vesicles for secretion, where Wls acts as a chaperone for the docking of the Wnt ligands to the vesicular membrane.[29–31] Therefore, inactivating Wls or Porcn in Wnt-producing niche cells using pharmacological inhibitors or via genetic mouse models effectively disrupts the supply of functional Wnts to the stem cells, attenuating Wnt/β-catenin signaling in vivo. This has proven to be a useful approach for functionally examining the role of Wnt-producing niches in stem-cell-driven homeostasis, regeneration, and cancer. Interestingly, small molecules that inhibit Porcn activity have already shown promising results for treating Wnt-driven cancers.[32–37] Some of these are currently in cancer clinical trials, indicating that targeting Wnt-producing cells in cancer is potentially a viable therapeutic strategy.

The study of Wnt/β-catenin signaling in stem cells has proven useful in two major aspects. First, we learn about the underlying mechanism of stem cell and tissue homeostasis in vivo, as well as the more ominous pathway alterations that lead to tumorigenesis and carcinogenesis. Second, the list of Wnt/β-catenin target genes has yielded useful adult stem cell markers by studying the central role of this pathway in effecting stem cell maintenance in vivo. Using these markers, the stem cells can be identified and isolated for a range of applications, including ex vivo expansion, organoid technology, manipulation using DNA editing technologies, and expression profiling. These advances have facilitated major advances for the stem cell field and regenerative medicine.

In this chapter, we will discuss the role of Wnt/β-catenin signaling in several prototypical epithelial stem cell systems, focusing on its roles in adult stem cells and cancer, with a particular emphasis on adult stem cells marked by Wnt target genes, such as *Axin2* and *Lgr5*.

2. GASTROINTESTINAL STEM CELLS AND CANCER

The gastrointestinal tract is made up of a single layer of columnar epithelial cells, and exhibits the highest turnover in the body: the gastric and intestinal epithelia renew every 7–10 and 3–14 days, respectively.[38,39] The gastrointestinal system, which spans the mouth, esophagus, stomach, and intestines, is primarily responsible for the digestion and absorption of nutrients and water from our food intake. In this section, we will examine stem cells of the intestine and stomach.

2.1 Intestine

The intestinal epithelium is undoubtedly where the function and mechanism of Wnt/β-catenin signaling has been characterized most comprehensively, and is a classical prototype for the study of adult stem cells (reviewed in Clevers).[40] Effecting this rapid turnover during homeostasis are intestinal stem cells (ISCs) governed by Wnt/β-catenin signaling, which can also be the causative force in intestinal cancer when dysregulated.

2.1.1 Wnt/β-Catenin Signaling in Intestinal Crypts are Crucial During Homeostasis

The functional unit of the small intestinal mucosa comprises the crypt and villus. Stem and progenitor cells located in the crypts of Lieberkühn generate differentiated progeny: absorptive enterocytes, enteroendocrine cells, goblet cells, tuft cells, or Paneth cells. Except for Paneth cells, which remain at the

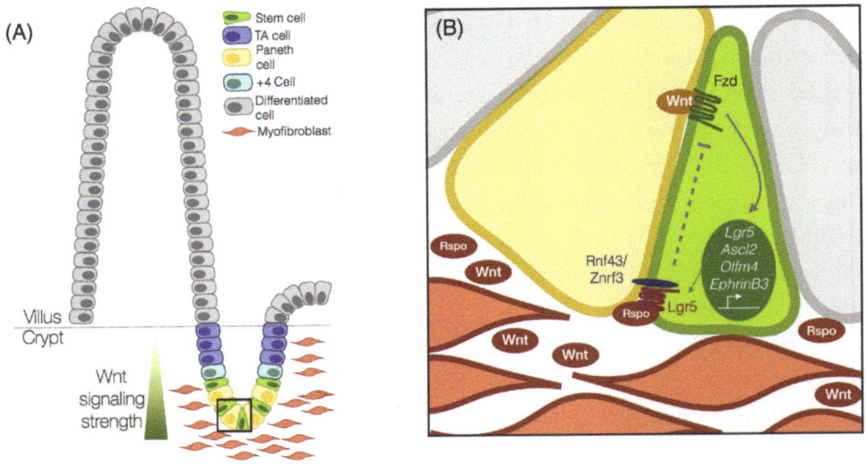

Fig. 3 Small intestinal stem cell. (A) Villi of the small intestine contain differentiated cells, which originate from the stem and transit amplifying (TA) cells in the crypts. Wnt/β-catenin signaling, which is strongest at the base of the crypts, maintain Lgr5-expressing intestinal stem cells (ISCs), which are the crypt base columnar (CBC) cells. The ISCs are interdigitated between Paneth cells, which are their progeny and a component of their niche. (B) Paneth cells secrete Wnts, which are immediately received by Fzd receptor on the adjacent ISCs, leading to expression of target genes, such as *Lgr5*, *Ascl2*, and *EphrinB3*, which are widely used as ISC markers. In particular, Lgr5 expression potentiates Wnt/β-catenin signaling as explained in Fig. 1, reinforcing the robust Wnt/β-catenin signaling status at the crypt base. Stromal myofibroblasts surrounding the crypt epithelial cells are also constituents of the ISC niche as they secrete Wnts to complement those from the Paneth cells.

base of the crypts, all the other differentiated cells migrate upward along the villi toward the lumen (Fig. 3A). While it has long been known that the cellular source of the intestinal epithelium resided in the crypts, the exact identity of the ISCs and their genetic markers were not defined until the late 2000s (reviewed in Barker *et al.*).[41] Nonetheless, the importance of Wnt/β-catenin signaling for maintenance of the intestinal epithelium was revealed through the use of the Villin-CreERT2 driver, which facilitated genetic studies in the entire intestinal epithelium spanning the crypt/villus axis.[42,43] Wnt/β-catenin signaling activity in the crypt exists as a gradient, with robust levels at the crypt base tapering off toward the crypt-villus junction (Fig. 3A). Indeed, nuclear β-catenin, a hallmark of active Wnt signaling, is confined to the crypt base columnar (CBC) cells and Paneth cells populating the crypt base.[44]

Abrogating Wnt/β-catenin signaling in the intestinal epithelium severely perturbs homeostasis and survival. Administering the Wnt antagonist Dkk1 or knocking out critical effectors, such as β-catenin or *Tcf4* (*Tcf7l2*) impairs progenitor/transit amplifying cell proliferation in the crypts and blocks differentiation toward secretory lineages, resulting in severely compromised villi comprising exclusively absorptive enterocytes and increased mortality.[45–49] Similarly, *Tcf4*-null mutant mice lack proliferative cells and enteroendocrine cells in intestinal crypts at birth.[50] In contrast, knocking out *Tcf1* and *Tcf3* in the intestinal epithelium had no phenotype,[49] identifying Tcf4 as the workhorse of the Tcf family during Wnt-driven intestinal homeostasis.

Conversely, hyperactivating Wnt/β-catenin signaling increases transit amplifying (TA) cell proliferation, accompanied by β-catenin stabilization leading to crypt expansion, and impaired differentiation toward the enterocyte lineage.[51–55] Such hyperactivation is typically achieved through forced expression of a β-catenin mutant that is not targeted for degradation or injection of the agonist protein R-spondin1. Interestingly, R-spondin1 administration protected ISCs against various adverse challenges to the intestines, such as chemoradiation[54,56] and graft versus host disease (from allogeneic bone marrow transplantation),[57] implying that amplifying Wnt/β-catenin signaling could increase robustness of the ISCs, and hence the health of the intestinal tissue as a whole.

2.1.2 ISCs Express Wnt/β-Catenin Target Genes

Blocking dysregulated Wnt/β-catenin signaling in colorectal cancer cell lines via inducible expression of dominant negative TCF4 (dnTCF4) revealed a list of potential direct Wnt/β-catenin target genes.[58] Following validation of these target genes in normal murine intestinal tissue and human intestinal tumor samples, the target genes were shortlisted and classified into three categories: (1) genes associated with ISCs (e.g., *Lgr5*, *Ascl2*, *Troy*), (2) genes associated with proliferating stem and transit amplifying progenitors (e.g., *Axin2*, *Myc*, *CD44*, *CyclinD1*), and (3) genes expressed in differentiated Paneth cells (e.g., *Mmp7*, *EhpB3*).[14,59] The ISC marker genes were subsequently confirmed by an independent study that profiled transcriptomes and proteomes of $Lgr5^+$ ISCs.[60] From there, *Lgr5*, *EphB3*, and *Achaete-scute complex homolog 2* (*Ascl2*, discussed later) have emerged as the most prominent and widely used ISC markers.

Rapidly cycling CBC cells were identified as the Lgr5-expressing ($Lgr5^+$) ISCs in the small intestine,[59] in agreement with Cheng and Leblond's proposal in 1974.[61] Similarly, colonic crypt base cells are the $Lgr5^+$ ISCs of the colon.[59] In vivo lineage tracing using the Lgr5-EGFP-ires-CreERT2 knockin mouse showed that these $Lgr5^+$ cells in small intestine and colon can continuously give rise to all differentiated lineages over a mouse's lifespan.[59] Additionally, as $Lgr5^+$ cells in the Lgr5-EGFP-ires-CreERT2 knockin mouse express GFP, $Lgr5^+$ cells can be isolated by flow cytometry for downstream experiments, such as in vitro organoid culture and transcriptome profiling.[62,63] $Lgr5^+$ ISCs were designated as the population with highest GFP expression ($Lgr5^{Hi}$), and could be distinguished from their immediate daughter cells, which contained lower levels of GFP protein that they had inherited from the parent ISC.[62,63] Hence, $Lgr5^+$ cells are the *bona fide* adult intestinal stem cells during homeostasis. The use of Lgr5 as a marker of homeostatic and facultative adult stem cell marker was subsequently extended to a plethora of tissues (Table 1), some of which will be reviewed in later sections.

Ephrin signaling governs cell positioning along the crypt-villus axis by imposing an upward migratory force from the crypt base to maintain discrete compartments within the intestinal architecture. Wnt/β-catenin signaling inversely regulates the expression of the EphB receptors and their ephrinB ligands.[44,58] The B-subfamily of Ephrins comprises the receptors EphB2 and EphB3, and their ligands ephrinB-1 and ephrinB-2. EphB3 expression is enriched in $Lgr5^+$ ISCs; whereas EphB2 is expressed throughout the proliferative compartment of the crypt base, with the exception of terminally differentiated Paneth cells.[44] The ephrinB-1 and ephrinB-2 ligands are highly expressed at the crypt-villus junction and in the lower third of the villus. Deletion of the EphB2 and EphB3 receptors leads to mislocalization of the Paneth cells along the crypt-villus axis, and erroneous intermingling of cells in the crypts, but is not lethal. Furthermore, human colonic stem cells expressing high levels of EPHB2 can be isolated using antibodies against EPHB2, and propagated in vitro long term, while retaining their multilineage potential,[97] presenting opportunities for downstream research, diagnostic and therapeutic applications.

2.1.3 Use of ISC Marker Lgr5 Reveals ISC Dynamics and Properties

Following the identification of Lgr5 as an ISC marker, questions about ISC dynamics and properties have been widely studied. For instance, are these ISCs required during homeostasis and regeneration? How does this population reconcile with the other purported ISC population at the +4 position? What are the population dynamics of the $Lgr5^+$ ISCs? What are the

Table 1 Wnt-Responding Adult Tissue Stem Cells

Epithelial Tissue	Wnt-Responding Adult Stem Cell Type	Homeostatic/ Facultative	Wnt Marker	Capability to Form Organoids or be Transplanted?	Cancer Cell-of-Origin?	Aberrant Wnt/β-Catenin Pathway Involved in Cancer? (Selected References)
Small intestine and colon	Crypt columnar base cells	Homeostatic	Lgr5[59]	Organoid[62]	Yes[64]	Yes[65]
Gastric corpus	Chief cells	Facultative	Lgr5[66]	Organoid[66]	Yes[66]	Yes[67]
	Chief cells (and parietal cells)	Facultative	Troy[68]	Organoid[68]	NT	—
Gastric pylorus	Basal cells	Homeostatic	Lgr5[39]	Organoid[39]	Yes[39,69]	Yes[67]
Interfollicular epidermis	Basal cells	Homeostatic	Axin2[70]	NT	NT	Yes[71,72]
Hair follicle	Outer bulge cells	Homeostatic	Axin2[73,74], Lgr5[75]	Transplantation[75]	NT	Yes[71,72]
Liver	Pericentral hepatocytes	Homeostatic	Axin2[76]	NT	NT	Yes[77]
	Parenchymal hepatocytes	Homeostatic	Lgr4[78]	NT	NT	
	Biliary duct cells	Facultative	Lgr5[79]	Organoid[79]	NT	
Mammary gland	Basal cells	Homeostatic	Axin2,[80] Lgr5,[81,82] Procr[83]	Transplantation[82–84]	NT	Yes[85]
Ovary	Hilum ovarian surface epithelial cell	Homeostatic	Lgr5[86]	NT	NT	—

(Continued)

Table 1 Wnt-Responding Adult Tissue Stem Cells—cont'd.

Epithelial Tissue	Wnt-Responding Adult Stem Cell Type	Homeostatic/ Facultative	Wnt Marker	Capability to Form Organoids or be Transplanted?	Cancer Cell-of-Origin?	Aberrant Wnt/ β-Catenin Pathway Involved in Cancer? (Selected References)
Inner ear cochlea	Tympanic border cells	Homeostatic	Axin2[87]	NT	NT	—
Cornea	Limbal epithelial stem cells	Homeostatic	Lgr5, suggested but not proven[88]	NT	NT	—
Tongue (posterior)	Base of circumvallate papillae	Homeostatic	Lgr5[75,89–91]	Organoid[90]	NT	—
Nail	Nail matrix cells	Homeostatic	Lgr6[92]	NT	NT	—
Kidney	Nephron segment-specific stem cell	Homeostatic	Axin2,[93] Lgr5[94]	NT	NT	—
Pancreas	Pancreatic duct cells	Facultative	Lgr5[95]	Organoid[95]	NT	—
Prostate	Luminal cells	Homeostatic	Lgr5[96]	Transplantation[96]	NT	—

NT, Not tested.

constituents of the ISC niche? The resulting collection of studies stemming from the discovery of $Lgr5^+$ ISCs demonstrates the utility of a robust marker in understanding stem cell biology.

In vivo ablation of endogenous $Lgr5^+$ stem cells using Lgr5-DTR-EGFP mice has been used to study the role of $Lg5^+$ stem cells during homeostasis and regeneration in adult tissues. In this mouse model, $Lgr5^+$ ISCs selectively express diphtheria toxin receptor (DTR), which renders them susceptible to diphtheria toxin.[98] Surprisingly, intestinal $Lgr5^+$ ISCs were found to be dispensable during homeostasis.[98] Also, damage induced by a near lethal dose of irradiation (12 Gy) in wild-type intestines permitted regeneration despite transient loss of $Lgr5^+$ ISCs.[55] These suggest two possibilities—the presence of an alternative stem cell population, and/or plasticity within the crypt cells to rapidly replace the lost $Lgr5^+$ ISCs.

The alternative stem cell population hypothesis was supported by the ability of a purported Lgr5-independent, $Bmi1^+$ population to regenerate all intestinal epithelial cells, including the $Lgr5^+$ ISCs.[98] Bmi1,[99] along with Hopx,[100] Lrig1,[101] and Tert,[102] are proposed markers of a slower-cycling ISC population at the +4 position (the fourth cell from the crypt base). Potten and coworkers had originally proposed crypt cells at the +4 position as ISCs due to their radiosensitivity and label retention property.[103,104] Contrary to more recent studies, Potten and coworkers showed that +4 cells are actively cycling, but demonstrate label retention because of asymmetric DNA strand segregation during division.[105]

It is ambiguous whether the +4 markers indeed label Lgr5-independent ISC population. Expression of the +4 markers are not confined to the +4 position, but found throughout the crypts.[60,106] In fact, their expression patterns overlap with Lgr5's by transcriptome analysis and multiplex in situ hybridization.[60,106] Additionally, single cell qRT-PCR of label-retaining cells (LRCs, occupying positions +3 to +6) ascertained their overlap with $Lgr5^+$ ISCs.[107] Thus, it is possible that tracing using these +4 markers was initiated from $Lgr5^+$ ISCs coexpressing these markers. In fact, the relatively short timeline of the emergence of crypt-to-villus tracing units for Bmi1[99] and Lrig1[101] is similar to that of $Lgr5^+$ ISCs and contrasts with the behavior expected from a quiescent nature (short-term tracing dynamics were not shown for Hopx[100] and Tert).[102] Therefore, it is unclear whether these +4 markers truly label a stem cell population distinct from the $Lgr5^+$ ISCs.

The rapid replacement of $Lgr5^+$ ISCs upon ablation could also be explained by inherent plasticity of cell populations residing within the crypts.

Tetteh *et al.* found that differentiated enterocytes can be induced to dedifferentiate upon loss of $Lgr5^+$ cells to restore the ISC pool in vivo.[108] Furthermore, $Hopx^+$ cells can give rise to $Lgr5^+$ CBCs and vice versa.[100] Using an inducible H2B-YFP mouse, Buczacki *et al.*[107] labeled a population of LRCs that overlaps with the $Dll1^+$ cells identified by van Es *et al.*[109] During homeostasis, both populations contain secretory precursors and do not generate multilineage clones. However, epithelial injury induced by a sublethal dose of irradiation (6 Gy) activates the stem potential of LRCs and $Dll1^+$ precursors to repopulate the small intestine with all known differentiated lineages. Since $Dll1^+$ cells originate from $Lgr5^+$ ISCs, plasticity built within the crypts serves to maintain intestinal survival during times of assault.[109] Additionally, recent studies purport that $Bmi1^+$ cells are enteroendocrine precursors distinct but closely related to $Lgr5^+$ ISCs, and are able to dedifferentiate into $Lgr5^+$ ISCs through epigenetic remodelling induced by injury.[109a,109b] These suggest that ISC identity is determined by its location in the crypt, that is, the ISC niche. Strikingly, although wildtype intestines can recover from a near lethal dose of irradiation (10–14 Gy), intestinal regeneration is severely impeded if $Lgr5^+$ cells were ablated prior to irradiation.[110] This suggests that following sublethal damage events, endogenous $Lgr5^+$ ISCs are still necessary for complete and efficient regeneration of the intestine.

The identification of $Lgr5^+$ ISCs also facilitated a detailed investigation of their endogenous population dynamics using the reporter Confetti, which is a stochastic multicolor reporter for clonal output analysis.[20] $Lgr5^+$ ISCs initially formed multicolored crypt bases, which gave way to single colored crypts, indicating that crypts drift toward clonality over time.[20] Mathematical modeling of the clonal dynamics indicate that the ISCs divide symmetrically, and engage in neutral competition in a stochastic manner, a behavior subsequently ascertained in human colon samples.[20,111] Such clonal dynamics indicate that $Lgr5^+$ ISCs are functionally equivalent, in agreement with results from one single cell transcriptome study of $Lgr5^+$ cells.[112] In vivo live imaging of $Lgr5^+$ cells and their progeny reveal a more complex picture; while the $Lgr5^+$ cells may be intrinsically equivalent, their fates are determined by their positions in the crypt. Specifically, $Lgr5^+$ ISCs positioned at the bottom of the crypt (positions 0 to +2) are more likely to generate progeny that will remain as stem cells at the base; whereas those along the side of the crypts (positions +3 and +4) are biased toward leaving the ISC niche and differentiating,[113] a behavior that concords with two other recent single cell transcriptome studies showing two distinct subpopulations of

Lgr5+ ISCs.[109b,113a] However, the positions the cells occupy are not fixed due to constant transfer between these two groups. Therefore, Lgr5+ cells are functionally equivalent to one another. This underscores that stemness and longevity of the ISC are determined by the position it occupies, rather than its intrinsic properties.

It has been postulated that niche space is a major defining factor for ISC population size as ISCs compete for niche space to maintain their stemness. The *Lgr5*+ ISCs are nested between Paneth cells, descendants of the ISCs, forming a regular geometric pattern such that every ISC contacts a Paneth cell[20,59,114] (Fig. 3). Paneth cells secrete Wnt3, 6, and 9b, acting as a source of Wnt ligands for the adjacent ISCs.[114–116] ISCs directly sequester Wnt ligands from adjacent Paneth cells via direct binding to Fzd receptors rather than relying primarily on intercellular diffusion of Wnt ligands.[117] As Fzd protein levels are negatively regulated by Rnf43/Znrf3,[16] this suggests that availability of Fzd receptor would be the rate limiting factor for Wnt/β-catenin signaling activity. The main Fzd receptor in the ISCs is Fzd7, as it is the most highly expressed Fzd receptor and knocking it out in ISCs causes ISC loss and impairs regeneration.[118] In the colon where Paneth cells are lacking, Reg4+ deep crypt secretory cells serve as the stem cell niche and are required for colon stem cell survival.[119] Interestingly, while these deep crypt secretory cells do not secrete Wnt ligands like the Paneth cells, active Wnt/β-catenin signaling is required for their specification.

However, ablating Paneth cells[120,121] or removing epithelial sources of Wnt via Villin-Cre-mediated deletion of Porcn[122] did not impair stem cell-driven epithelial homeostasis or injury in the intestine, indicating a nonepithelial contribution to the ISC niche in vivo. While *Porcn*-deficient intestinal crypts could not establish themselves in culture, coculturing them with stromal myofibroblasts supported organoid growth in the absence of exogenous R-spondin1.[122] These myofibroblasts are marked by Gli1 expression, and highly express Wnt2b, R-spondin2, and R-spondin3, enabling them to support the growth of *Porcn* or *Wls*-knockout organoids in vitro.[122,123] However, concomitantly deleting *Porcn* from epithelial cells and myofibroblasts did not yield a phenotype.[124] In contrast, systemic deletion of *Wls* using Rosa26-CreERT2 caused depletion of crypts and ISCs,[123] and systemic inhibition of Porcn using a pharmacological small molecule C59 during homeostasis and postinjury caused crypt loss or atrophy together with villi blunting.[122] Recently, CD34+gp38+ nonmyofibroblastic mesenchymal cells has been identified as producer of Wnt2b and R-spondin1,[124a] constituting a third component of the ISC niche. Other possible candidates include endothelial cells, lymphocytes, and nerve cells. Furthermore, it would be interesting to determine if the constituents of the ISC niche are functionally equivalent, or if various niche constituents function in specific contexts (e.g., homeostasis or injury).

2.1.4 Wnt/β-Catenin Signaling Regulates Multiple Aspects of ISC Behavior

During homeostasis, Wnt/β-catenin plays multiple roles to ensure a healthy population of ISCs for efficient tissue renewal throughout the organism's lifetime. The major functions of Wnt/β-catenin signaling can be summarized as: (1) maintaining stemness, (2) inducing proliferation of stem cells, and (3) preserving genomic integrity.

First, high levels of Wnt/β-catenin signaling reinforce the stem cell state through induction of its target gene *Ascl2*,[14,125] which has been proposed as the master regulator of ISC identity. This sets up a positive feedback loop as Ascl2, Tcf4, and β-catenin act synergistically to effect transcription of the Wnt program.[125,126] Forced expression of Ascl2 resulted in increased proliferation, increased c-Myc expression, and expansion of crypt structures and the ISC compartment as marked by $Lgr5^+$ and $Olfm4^+$ (a commonly used ISC marker).[63] However, Ascl2 is not an oncogene as its overexpression in a mouse model with predisposition to intestinal neoplasia (Apc^{Min}; multiple intestinal neoplasia (Min); described in detail later) does not accelerate tumorigenesis.[127] Conversely, deletion of *Ascl2* reduced ISC population size, although there was no notable phenotype due to "escaper" wildtype crypts.[63] Forced expression of Ascl2 rescues *Tcf4* knockout organoids from lethality, underscoring Ascl2's role as an effector to induce expression of vital stem cell-associated genes to promote stemness. The positive feedback loop between Wnt/β-catenin and Ascl2 expression ensures that Ascl2-mediated regulation of ISC identity is confined to the crypt base, where Wnt/β-catenin signaling is strongest.

The Ascl2/Tcf/Lef complex competes with the Smad/Hdac complex activated by bone morphogenetic protein (BMP) signaling for occupancy of promoters of ISC marker genes, such as *Lgr5* and *Ascl2*.[128] BMP signaling represses Wnt/β-catenin signaling in the crypts and consequently blocks ISC proliferation and renewal, as evidenced by the fact that *BMP receptor 1a* (*Bmpra1*) null mutant crypts exhibit expansion of the $Lgr5^+/Olfm4^+$ stem compartment and epithelial hyperproliferation that manifested as polyps.[128,129] Therefore, BMP signaling counterbalances Wnt/β-catenin–-derived forces and limits the size of ISC population in the crypt. However, it is unclear whether BMP signaling also restricts ISC population size by actively driving their differentiation.[128,130]

As mentioned above, high levels of Wnt/β-catenin signaling at the crypt base are critical for ISC maintenance. Wnt/β-catenin signaling renders ISCs responsive to R-spondin/Lgr5 signaling by providing Lgr5 receptors on the cell surface, enabling them to respond to available R-spondin ligands, which in turn further augments Wnt/β-catenin signaling.[131] Furthermore, concomitant deletion of *Lgr4* and *Lgr5* caused loss of crypts and ISCs, eventually leading to mortality.[17] Hence, potentiation of Wnt/β-catenin signaling by R-spondin/Lgr signaling axis is indispensable for ISC self-renewal and ensures that Wnt/β-catenin signaling is the highest at the crypt base.

Second, Wnt/β-catenin signaling induces proliferation of ISCs and the TA compartment through its mitogenic effects exerted via downstream target genes *CyclinD1* and *Myc*. c-Myc is a transcriptional repressor of cell cycle inhibitors p21 and p27,[58,132,133] while CyclinD1 promotes G1 entry and progression by binding to Cyclin-dependent kinases 3/4/6[134]. Dysregulation of upstream regulators of CyclinD1 and c-Myc, such as Apc results in their overexpression, leading eventually to cancer formation[58]. In contrast, *Myc*-deficient crypts are smaller and proliferate more slowly, indicating that Wnt/β-catenin signaling is involved in controlling ISC proliferation.[135] CyclinD1, Cyclin E1, and c-Myc are G1/S regulators, which can be phosphorylated by GSK-3β for degradation.[136–138] As Wnt/β-catenin signaling inhibits GSK-3β activity, activating the pathway also blocks proteosomal degradation of these G1/S regulators, resulting in their accumulation. Collectively, Wnt pathway activity delivers higher expression and increased stability of the G1/S regulators to promote cell cycle progression in the crypt. While Wnt/β-catenin signaling has also been shown to play a role in mitosis, it is yet to be demonstrated in the intestine.[139]

Third, Wnt/β-catenin signaling is involved in telomere maintenance to preserve integrity of the genetic information, which is especially important given that ISCs cycle every 24 h[59,140]. In the intestine, β-catenin and Tert, the catalytic subunit of telomerase, regulate each other's expression reciprocally. β-Catenin binds to the *Tert* promoter in Lgr5$^+$ ISCs from normal mouse small intestine.[141] In turn, Tert interacts with Tcf-binding elements of promoters of Wnt/β-catenin target genes, including *Axin2*, *CyclinD1*, and *Myc* in whole intestinal lysates; thus serving as a transcriptional effector of the Wnt/β-catenin signaling pathway and establishing a positive feedback loop.[142] Although the ISCs or crypts were not specifically isolated for the latter study, the Lgr5$^+$ ISCs were found to have the highest Tert expression, telomerase activity, and telomere length.[143] Therefore, through interaction with Wnt/β-catenin signaling, Tert is likely acting on the cycling ISCs to protect genome integrity and preserve normal function of the intestines (reviewed in Günes and Rudolph).[144]

2.1.5 ISCs can be Isolated and Maintained In Vitro as Organoids in the Presence of Wnt and R-Spondin

The discovery of Lgr5 as an adult intestinal stem cell marker and the availability of the Lgr5-EGFP-ires-CreERT2 reporter mouse model facilitated the development of a new primary culture technique—organoid culture.[62] As mentioned above, $Lgr5^+$ cells expressing high levels of GFP from the Lgr5-EGFP reporter mouse intestine could be distinguished from their immediate daughter cells (GFP^{Low}) and differentiated progeny (GFP^{Neg}) counterparts by FACS, and isolated for in vitro culture. The cells are grown in Matrigel, and supplemented with media containing growth factors, such as Wnt3a, R-spondin1, epidermal growth factor (EGF), fibroblast growth factor 10 (Fgf10), Noggin, and Y27632 (ROCK inhibitor). The fact that Wnt3a and R-spondin1 are obligate components of the medium to maintain ISCs in organoids underscores the importance of R-spondin/Lgr-potentiated Wnt/β-catenin signaling in the intestines.

Single $Lgr5^+$ cells can give rise to self-renewing three-dimensional budding structures that contained crypt-villus regions along with a central lumen akin to the native tissue.[62] The buds of the organoids contain the $Lgr5^+$ stem cells intermingled with Paneth cells, while the regions between the buds contain the differentiated lineages, such as enterocytes, enteroendocrine cells and goblet cells.[145]

The organoids can be propagated long term without the chromosomal abnormalities commonly observed in immortalized tissue culture cell lines, making them a near-physiological in vivo system for studying stem cells without stromal contribution. Furthermore, normal and cancerous intestinal tissue from humans can be maintained as organoids in culture, with the addition of nicotinamide, A83-01 (ALK5 inhibitor) and SB202190 (p38 inhibitor).[146,147] Due to the technique's tractability and utility in studying stem cell populations in a reductionist setting, organoid culture has since been extended to a wide array of tissues from both mouse and humans for assaying the stem potential of putative stem cell populations, with the core growth factors retained across the systems, particularly those required for maintaining potentiated Wnt/β-catenin signaling (reviewed in Fatehullah *et al.*).[148]

2.1.6 ISCs are a Cell-of-Origin of Intestinal Cancer Driven by Aberrant Wnt/β-Catenin Signaling

Wnt/β-catenin signaling was one of the earliest pathways to be associated with cancer, in particular with colorectal cancer (CRC) where aberrant Wnt/β-catenin signaling is associated with more than 93% of CRC cases.[65] Additional mutations in Kras, Tgf-β, PI3-kinase, and p53 signaling pathways promoted colorectal tumorigenesis and carcinogenesis. From this, the notion that CRC progression followed a sequential accumulation of known genetic mutations was conceived and extended to other cancers.[149] The large proportion of human CRCs that harbor mutations in Wnt/β-catenin pathway underscores the requirement for hyperactive Wnt/β-catenin signaling, which is likely acts on the colorectal CSCs.

The tumor suppressor gene *APC* is the most commonly mutated component of the Wnt/β-catenin signaling in CRC, accounting for up to 85% of sporadic CRCs.[65] These mutations can be found in sporadic CRC tumors and in the germlines of patients with the hereditary cancer syndrome familial adenomatous polyposis (FAP).[150,151] These mutations tend to truncate the gene to yield null, hypomorphic, or dominant negative alleles of *APC*, which result in a defective β-catenin destruction complex. As β-catenin molecules accumulate and translocate to the nucleus, Wnt/β-catenin signaling becomes hyperactivated.[152,153] FAP patients inherit one mutated *APC* allele, and subsequent loss of heterozygosity at this gene locus inevitably leads to a significant load of adenomatous polyps in the colon by midteens, some of which eventually progress to malignant adenocarcinoma.[154] In most sporadic CRCs, both *APC* alleles are mutated.[155] Such mutations in *APC*, either inherited or acquired, tend to be one of the earliest events in CRC tumorigenesis.[156] The APC multiple intestinal neoplasia (Min) allele was the first to be modeled in mice.[157] However, for reasons unknown, the APC^{Min} mice developed lesions predominantly in the small intestines, in contrast to human FAP in which the polyps are found in the colon. Mouse models bearing other hypomorphic versions of *APC* also present intestinal lesions with varying severity, which corresponds to the degree of hypomorphism of the *APC* mutant allele.[158–163]

The oncogenic impact of uncurtailed Wnt/β-catenin signaling was unequivocally demonstrated in humans and mouse models with dysregulated Wnt/β-catenin signaling. Upon systemic deletion of *Apc* in adult mice, the intestinal architecture was grossly altered with overproliferation of crypt-like progenitors, causing mortality within 5 days of induction.[51] Similarly, loss of

AXIN2, a Wnt target gene and negative feedback inhibitor of Wnt/β-catenin signaling, is also commonly found in human CRCs.[164] Besides loss of pathway inhibitors, such as *APC* and *AXIN2*, CRC can be initiated via mutations in exon 3 of the *β-catenin* gene (*CTNNB1*), which prevent β-catenin degradation even in the absence of Wnt ligands. The resulting accumulation of β-catenin promotes constitutive formation of Tcf4/β-catenin complex and hyperactivation of the Wnt/β-catenin target gene program.[58] Impressively, reexpressing *APC* in CRCs promotes tumor regression and restores homeostatic crypt behavior despite the presence of oncogenic Kras and absence of p53, indicating that regulation of Wnt/β-catenin signaling is central to CRC therapy.[165]

Mutations in the R-spondin/Lgr signaling axis are also found in CRC. Recurrent gene fusions with *R-spondin2* (*RSPO2*) and *R-spondin3* (*RSPO3*), respectively accounted for 3 and 8% of the CRCs sampled in the Seshagiri *et al.* study.[166] In particular, *RSPO3* was frequently found to be fused in frame with the *Protein tyrosine phosphatase, receptor type K* (*PTPRK*) gene. Tumors with *RSPO2* and *RSPO3* fusions express higher levels of RSPO and LGR4, LGR5, and LGR6 than tumors without the *RSPO* fusions. Moreover, these *RSPO2* and *RSPO3* fusions are sufficient for activating and amplifying Wnt/β-catenin signaling in vitro, indicating escalated levels of Wnt/β-catenin signaling in these tumors. Interestingly, targeting patient-derived xenograft samples that harbor *RSPO3* gene fusions with anti-RSPO3 antibody led to stasis or regression, indicating that RSPO3 is a key driver of these colorectal tumor growths.[167] While *AXIN2*, *MYC*, and *CyclinD1* were only slightly downregulated with anti-RSPO3 treatment, *LGR5* and *ASCL2*, both markers of ISCs, were among the top five most downregulated genes. Cellular differentiation markers were also significantly downregulated, indicating that the anti-RSPO3 treatment significantly reduces Wnt/β-catenin signaling levels and diminishes stem-like properties in the tumor. Similarly, loss-of-function mutations in *RNF43* have been found in 18% of human CRCs,[168] but *ZNRF3* mutations in cancer are less common.[15] Simultaneous knockout of *Rnf43* and *Znrf3*, which increases availability of Frizzled receptors at the cell surface, induces hyperproliferation and intestinal metaplasia in mice.[169] As expected, these Wnt ligand-dependent adenomas are responsive to C59 treatment, which inhibits Porcn activity.[169]

Colorectal tumors capitalize on the homeostatic functions of Wnt/β-catenin signaling to advance tumorigenesis. Inhibition of Wnt/β-catenin signaling in CRC cell lines by expressing the dnTCF4 led to cell cycle arrest and induced differentiation,[58] outcomes similar to in vivo inhibition of β-catenin in xenografts of human CRC tumors.[170] Approximately 70% of human CRCs were found to overexpress *MYC*,[171] a prominent oncogene and a Wnt/β-catenin target gene, which prevents cell cycle arrest.[58,132,133] Deletion of *Myc* in *Apc*-null background prevented tumorigenesis despite the presence of high nuclear β-catenin.[172] Furthermore, loss of a single allele of *Myc* is sufficient to slow down tumorigenesis, although the cancer still progressed.[173] Hence, c-Myc is a key effector of hyperactive Wnt/β-catenin signaling that drives hyperproliferation, apoptosis, and aberrant cell migration in colorectal tumorigenesis.

Additionally, Wnt/β-catenin signaling could protect genomic integrity of colorectal tumors. Similar to observations during homeostasis, β-catenin binds to the Tert promoter in the hyperproliferative cells of the Villin-CreERT2;β-catenin$^{(\Delta ex3)fl/+}$ intestinal epithelium, which will form adenomas.[141] In the intestinal adenomatous lesions of these mice, it is possible that the resulting Wnt-driven increase in Tert expression endows the tumor, including its resident CSC population, with the ability to preserve its telomeres as an essential part of its long-term strategy to ensure survival and growth.[174,175]

Wnt/β-catenin signaling in CRC can also be enhanced by oncogenic Kras signaling, with its oncogenic form Kras^{G12V} found in 50% of CRCs.[65,176] Oncogenic effects of *Apc* inactivation (Apc^{1638N}) are exacerbated by oncogenic Kras^{G12V}, which phosphorylates E-cadherin-bound β-catenin, releasing it from the adherens junction into the cytoplasm.[177] The resulting accumulation of nuclear β-catenin drives expression of Wnt target genes, leading to malignant intestinal adenomas and carcinomas in mice. In CRC cell lines in vitro, oncogenic Ras has also been found to increase Wnt/β-catenin signaling by phosphorylating LRP6 through the Ras-activated MEK pathway. Phosphorylation of LRP6 then activates β-catenin/TCF4 transcriptional activity.[178] Therefore, oncogenic Kras signaling synergizes with Wnt/β-catenin signaling to drive the tumorigenesis program.

Given the emerging plasticity of the intestinal epithelium, are all intestinal cells poised to serve as CRC cell-of-origin upon hyperactivation of Wnt/β-catenin signaling? Or are stem cells the prime tumor-initiating candidates for accumulating the oncogenic mutations due to their longevity? In support of the latter hypothesis, knocking out *APC* in murine Lgr5$^+$ ISCs rapidly generated a lethal burden of intestinal adenomas with high levels of nuclear β-catenin, Ki67, and Myc within 36 days.[64] Similarly, a single allele of stabilized mutant *β-catenin* (*β-catenin*$^{\Delta ex3}$) in Lgr5$^+$ ISCs is sufficient to induce small intestinal adenomas.[179] These tumor-initiating effects can be enhanced by haploinsufficiency for E-cadherin as fewer E-cadherin molecules are present in the membrane to act as sinks for β-catenin.[179] In contrast, differentiated cells cannot form tumors as mutation of *Apc* alone in the differentiated compartment of the intestine is insufficient for tumorigenesis. However, differentiated cells can be reprogrammed to a more stem cell-like state by certain oncogenic events or changes to the local environment, such as colitis-induced inflammation,[180] increased NFκB signaling,[181] or reduced BMP signaling.[182] As a result, these differentiated cells acquire stem cell markers and subsequently drive tumor formation.

Existing CRC tumors rely on resident CSCs to fuel tumor growth. In the heterogeneous colorectal tumor, cells with highest Wnt/β-catenin signaling activity mark CSCs as they have the highest clonogenic and tumorigenic capabilities. Human colorectal tumor cells with high levels of Wnt/β-catenin signaling (Wnt^{HI}) and cells that express high levels of LGR5 protein ($LGR5^{Hi}$) from human colorectal tumor-derived organoids generate tumors in xenografts efficiently.[183,184] Their self-renewal ability was also demonstrated in serially-transplanted xenografts[183] and in in vitro organoids,[184] respectively. Transcriptome profiling of single cells from human colorectal tumors confirmed upregulation of Wnt/β-catenin signaling pathway components in the $LGR5^+$ stem/progenitor subpopulations.[185] Additionally, $LGR5^+$ cells in murine and human colorectal adenomas actively contribute to adenoma growth in vivo.[186,186a] While $LGR5^+$ CSCs are necessary for tumor growth and metastasis, they are dispensable for tumor maintenance.[186a,187] Hence, colorectal CSCs that are responsible for fueling tumor growth are characterized by high Wnt/β-catenin signaling and Lgr5 expression.

Similar to the spatial arrangement in the homeostatic ISC niche, tumor-resident $Lgr5^+$ CSCs are intermingled with Paneth-like cells, reminiscent of the arrangement in the normal crypt,[186] suggesting that Paneth cells also form an essential component of the CSC niche. Also, while it is unclear whether tumor myofibroblasts secreted Wnt ligands, they appear to support the CSCs via secretion of hepatocyte growth factor (Hgf), which could increase β-catenin activity through c-Met signaling.[183] While tumor cells with low Wnt/β-catenin signaling activity (Wnt^{Low}) are nontumorigenic when transplanted on their own, they acquire tumorigenic potential when cotransplanted with tumor myofibroblasts,[183] highlighting the instructive nature of the microenvironment in helping cells acquire stem cell-like characteristics. Similarly, differentiated CRC tumor cells are able to convert into $LGR5^+$ CSCs to promote tumor growth after preexisting $LGR5^+$ CSCs are ablated.[186a]

Currently, clinical studies of several promising therapeutic strategies are underway. CRCs with *RSPO* translocations can be targeted by the anti-RSPO3 antibody mentioned earlier (designated OMP131R10), which is currently in Phase 1 clinical trial (trial identifier NCT02482441), and Porcn inhibitors that deprive the tumors of functional Wnt ligands. Existing small molecule PORCN inhibitors C59, ETC-159, and LGK974 (or WNT974) have shown efficacy in reducing proliferation and stem cell marker expression of patient-derived xenografts of colorectal tumors.[32,34,37] Murine intestinal tumors harboring null mutations in both *Rnf43* and *Znrf3* are also responsive to C59.[169] Encouragingly, these small molecules appear to be effective across multiple cancers. ETC-159 is currently in Phase 1 clinical trial (trial identifier NCT02521844), while LGK974 is currently in Phase 2 (trial identifier NCT02649530). However, both RSPO3 and PORCN-targeted therapeutic approaches will only work on CRCs that rely on paracrine Wnt for growth, in stark contrast to *Apc*-deficient adenomas that are Wnt ligand-independent. One promising therapeutic agent for this class of CRC tumors could be RPI-724, a small molecule that disrupts the binding of β-catenin and its activator (CBP), thereby inhibiting Wnt/β-catenin signaling-driven transcription.[188] RPI-724 was shown to be effective in human CRC cell lines, Apc^{Min} mice and xenografts,[188] and is currently in Phase 2 clinical trial in combination with bevacizumab in patients with metastatic CRC (trial identifier NCT02413853). Furthermore, additional therapeutic strategies would be required to prevent relapse of the cancer as ablating CSCs with high levels of Wnt/β-catenin signaling in murine colorectal adenomas does not obliterate the tumors,[187] potentially due to reprogramming of other tumor cells by the oncogenic niche. Hence, further investigation of the Wnt-promoting tumor microenvironment will help to develop a therapeutic strategy to complement current tumor-targeted treatments.

2.2 Stomach

The stomach is the main site where ingested food is digested into chime, through the actions of the suite of enzymes and acid produced by specialized epithelial cells in the stomach lining. The chime is then released into the intestines where the nutrients can be absorbed. The mouse stomach comprises two distinct parts: the proximal nonglandular part, and the distal glandular part. The glandular part is further subdivided into corpus (proximal) and pylorus (distal) (Fig. 4A). In contrast, the human stomach lacks the nonglandular part, and has counterparts to the murine corpus and pylorus.

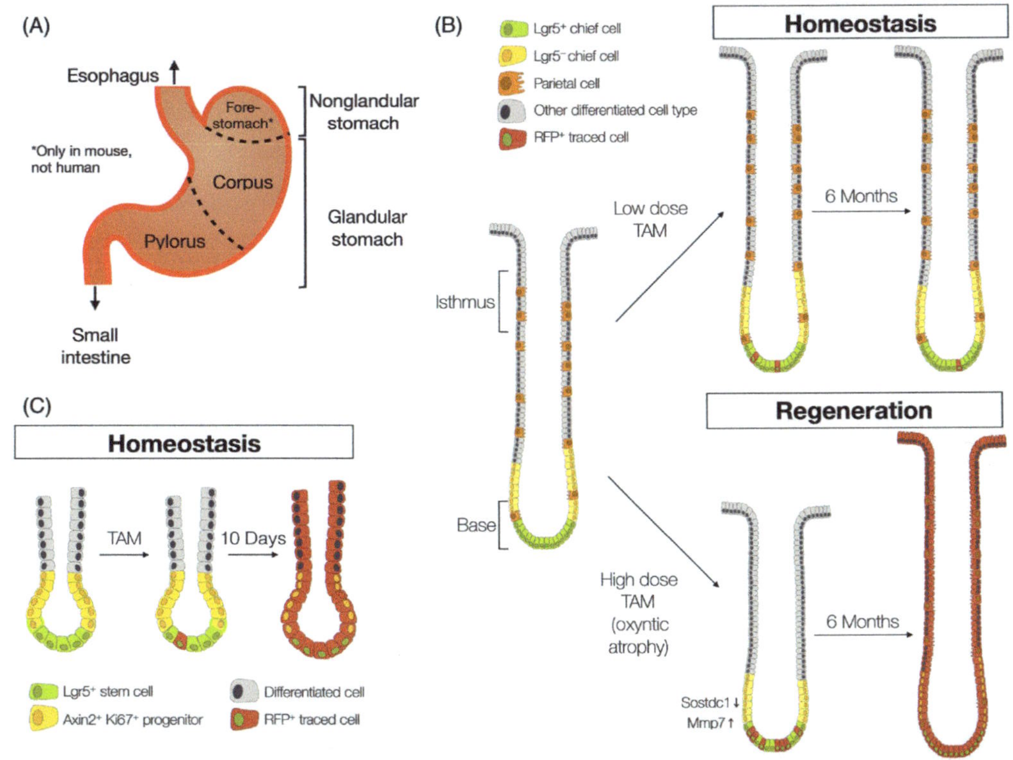

Fig. 4 Gastric stem cell in the corpus and pylorus. (A) Schematic of the murine stomach. (B) Facultative stem cells of the corpus. The chief cells situated at the base of the corpus glands express Lgr5 and do not contribute to regular homeostasis of the corpus, as observed by paucity of $Lgr5^+$ cell-derived RFP^+ traced units after 6 months. However, when oxyntic atrophy (loss of parietal cells) is induced by a high dose of tamoxifen, whole RFP^+ glands comprising all differentiated lineages are observed. The initiation of oxyntic atrophy is marked by reduced Sostdc1, which indicates elevated Wnt/β-catenin signaling. (C) Homeostatic stem cells of the pylorus. $Lgr5^+$ pyloric stem cells are found at the base of the glands, with the proliferative cells in the isthmus region. Only a small subset of $Lgr5^+$ cells is proliferating. When lineage tracing is induced in $Lgr5^+$cells, whole RFP+ pyloric glands are observed after 10 days.

In contrast to the intestines, Wnt/β-catenin signaling-related studies on the homeostatic corpus and pylorus thus far have mostly been limited to stem cells expressing *Lgr5* (and Troy in the corpus), with a lack of mechanistic studies. Hence, this section will mainly discuss the current findings about the corpus and pyloric stem cells in relation to Lgr5, and Wnt/β-catenin signaling in gastric cancer.

2.2.1 Facultative Stem Cells of the Corpus are Wnt-Dependent

The corpus/body of the stomach secretes acid via parietal cells, and the epithelium is protected from the acid and other assaults by a mucous layer produced by the surface mucinous cells. In murine and human corpus, Lgr5 is expressed in basal chief cells (Fig. 4B), which normally function as sources of zymogens to aid digestion.[66] *Troy*, another Wnt/β-catenin target gene, is also expressed in both chief and parietal cells in the corpus,[68] marking a broader population than $Lgr5^+$ chief cells. Both $Lgr5^+$ and $Troy^+$ cells remain quiescent and contribute minimally to homeostasis.[66,68] Nonetheless, these cells are able to generate organoids in vitro with an obligate requirement for Wnt and R-spondin1, an indication of their Wnt-dependent stem potential.[66,68]

However, $Lgr5^+$ and $Troy^+$ cells can be activated by ablation of parietal cells or proliferating cells, respectively, and subsequently produce all differentiated lineages of the gastric epithelium over time[66,68] (Fig. 4B). Interestingly, during early regeneration following parietal cell loss, Wnt signaling is transiently hyperactivated via downregulation of the extracellular Wnt inhibitor Sostdc1. Soon after, $Lgr5^+$ chief cells start to proliferate and display stem cell properties.[66] It is tempting to hypothesize that during homeostasis, Sostdc1 dampens Wnt/β-catenin signaling to a basal level to ensure quiescence of the $Lgr5^+$ cells while retaining their stem potential. If this is true, this would mean that differential levels of Wnt/β-catenin signaling dictate distinct outcomes, indicating an intricately calibrated mechanism.

Since the $Lgr5^+$ and $Troy^+$ cells are not daily stem cells, the day-to-day maintenance of the corpus epithelium must be sustained by another population, possibly the proliferating cells of the isthmus. While Lgr5 is not expressed by the isthmus cells, it has been shown that Wnt5a is expressed in the stromal cells adjacent to the isthmus epithelial cells,[189] suggesting the involvement of Wnt5a-mediated signaling in the isthmus region. However, Wnt5a is likely to be acting through a non-β-catenin-mediated pathway given its known association with the noncanonical Wnt pathways[190] and the absence of Wnt/β-catenin target genes in the isthmus region. Other proposed markers of homeostatic corpus stem cells include Mist1,[189] Sox2,[191] and enhancer of Runx1.[192] However, as these markers do not label isthmus cells exclusively, whether isthmus cells are the stem cells of the homeostatic corpus remains inconclusive.

2.2.2 Wnt-Responding Stem Cells Maintain Pyloric Ppithelium During Homeostasis

The pylorus secretes the hormone gastrin, among many other hormones and enzymes, which stimulates the parietals cells in the corpus to produce acid for digestion. The pylorus comprises gastric glands, which harbor around eight Lgr5$^+$ Axin2$^+$ stem cells at each gland base and Axin2$^+$ Lgr5$^-$ proliferating progenitors in the lower isthmus[193,194] (Fig. 4C). Axin2's expression pattern is reminiscent of Fzd7's, which is the most highly expressed Frizzled in the pylorus.[195]

The Lgr5$^+$ stem cells divide symmetrically,[193] undergo neutral drift,[193] and fuel the turnover of the pyloric epithelium every 7–10 days.[39] Single cells expressing high levels of Lgr5-EGFP generate long-term organoid cultures in the presence of Wnt and R-spondin1,[39] indicating the importance of Wnt/β-catenin signaling in pyloric stem cell maintenance. The requirement for recombinant R-spondin1 in culture media can be substituted with coculturing pyloric organoids with stromal myofibroblasts.[194] In vivo, R-spondin3 secreted by stromal myofibroblasts promotes proliferation of Axin2$^+$ cells, and is required for normal expression of Axin2 and Lgr5.[194] Hence, Wnt-responding stem and progenitor cells of the pylorus are supported by a myofibroblast niche.

2.2.3 Wnt/β-Catenin Signaling and Inflammation Due to Helicobacter pylori Infection Drives Gastric Cancer Originating From Lgr5$^+$ Stem Cells

The current working model of gastric carcinogenesis by Correa[196] begins with *Helicobacter pylori*–induced inflammation, which leads to chronic gastritis, followed by atrophy, intestinal metaplasia, dysplasia, and eventually neoplasia. While environmental factors, such as diet and smoking, are known to correlate with progression in the Correa model, the underlying molecular mechanisms remain poorly understood.

In humans, *H. pylori* infection presents the highest risk factor for gastric cancer. Long-term infection of *H. pylori* leads to chronic gastric inflammation, and approximately 3% of *H. pylori*-infected individuals will later develop gastric cancer.[197] When the lower halves of pyloric glands are colonized by *H. pylori*, there is increased proliferation of Lgr5$^+$ and Axin2$^+$ stem and progenitor cells, driving hyperplasia and chronic inflammation.[198] The virulence factor responsible for driving Lgr5$^+$ stem cell proliferation was determined to be the contact-dependent virulence factor CagA, indicating the need for direct contact between *H. pylori* and the pyloric epithelium.[198] Stromal myofibroblasts, the homeostatic niche of pyloric stem cells, secrete more R-spondin3 upon *H. pylori* infection, thereby inducing higher proliferation in Axin2$^+$ cells,[194] pointing to a direct effect of *H. pylori* infection on Wnt/β-catenin signaling activity.

The Wnt/β-catenin pathway is often hyperactivated in both intestinal and diffuse types of gastric cancer, as evidenced by the presence of nuclear β-catenin.[199] This likely occurs as a result from mutations in Wnt pathway components, such as *APC*, *CTNNB1*, and *RNF43*, which are commonly found in human gastric adenocarcinomas.[67] Human gastric cancer samples have elevated WLS protein expression compared to those from matched normal control gastric tissue.[200] This robust WLS expression was correlated with HER2 expression, increased proliferation and shorter progression-free survival rates, suggesting that Wnt/β-catenin signaling may cooperate with HER2 in promoting tumor progression. Another factor that interacts with the Wnt/β-catenin signaling machinery is the transcription factor Runx3. However, the actual role of Runx3 is unclear as it has been shown to play both activating[201] and repressing[202,203] roles in the modulation of Tcf4/β-catenin complex activity.

However, hyperactivation of Wnt/β-catenin alone only generates preneoplastic lesions in mice,[199] indicating that additional "hits" are required for gastric carcinogenesis. *Helicobacter felis*, a close relative of *H. pylori*, induces cyclooxygenase-2 (Cox-2) and microsomal prostaglandin E synthase-1 (mPGES-1) expression in the stomach.[204] These enzymes synthesize prostaglandins, which are associated with inflammation. Cox-2 and mPGES-1 overexpression synergizes with Wnt1 overexpression to generate dysplastic gastric tumors in mice (known as the Gan mouse model),[199] highlighting the critical role of inflammation in driving the progression of gastric tumors displaying hyperactivated Wnt/β-catenin signaling.

Wnt-responding gastric stem cells can act as the cell-of-origin of gastric cancer. In the murine corpus, the Lgr5$^+$ chief cells can serve as the origin of pseudopyloric metaplasia (SPEM) upon activation of oncogenic Kras^{G12D} and inactivation *Apc* and *Pten*.[66] Importantly, human corpus tumors express elevated levels of LGR5 and AXIN2, underscoring hyperactive Wnt/β-catenin signaling as a feature of corpus tumors in humans. Similarly, the Lgr5$^+$ pyloric stem cells are able to initiate tumors in mice following targeted conditional deletion of the *Apc* tumor suppressor gene, resulting in polyps with hyperactive Wnt/β-catenin signaling.[39] However, the broad action of the Lgr5-CreERT2 driver in multiple Lgr5+ stem cell compartments resulted in rapid lethality due to the resulting tumor burden in the intestines, precluding any progression of the pyloric tumors toward more advanced cancer with time. Subsequently, $Lgr5^+$ pyloric stem cells were formally shown to serve as a cell-of-origin of invasive pyloric tumors when concomitant targeted deletion of *Smad4* and *Pten* tumor suppressor genes in the stem cells generated pyloric adenocarcinomas after 2 months.[69] However, due to technical limitations, a metastatic model of gastric adenocarcinoma is still unavailable.

Many questions regarding stem cells in the homeostatic stomach and gastric cancer remain unsolved. In the corpus, what are the homeostatic stem cells? Are different levels of Wnt/β-catenin signaling required to effect distinct homeostatic and regenerative outcomes? In the pylorus, how are the $Axin2^+Lgr5^+$ and $Axin2^+Lgr5^-$ populations maintained as distinct populations? Do they also experience differential Wnt regulation? Do Wnt/β-catenin signaling and the prostaglandin synthesis pathway interact with each other or do they act independently on the tumor initiating population? The stomach presents an interesting opportunity to examine crosstalk between long-term inflammation and Wnt-driven cancer that can yield insight applicable to other tissues and cancers.

3. SKIN STEM CELLS AND CANCER

The skin is the largest organ of the body and acts as the first line of defense against pathogens and physical environmental assaults. Multiple "miniorgans" are found in the skin, such as hair follicles, interfollicular epidermis (IFE), and appendages, such as sweat and sebaceous glands. Each of these "miniorgans" is maintained by a dedicated population of adult epidermal stem cells. We will examine Wnt/β-catenin signaling in hair follicular stem cells (HFSCs), IFE stem cells and cutaneous cancers in greater detail.

3.1 Hair Follicles

Hair follicles are embedded in the IFE and undergo three phases of hair cycle throughout life—anagen (growth), catagen (destruction), and telogen (rest). During anagen, HFSCs travel from the bulge along the outer root sheath to the matrix, where they proliferate to generate new cells that migrate upward to form the growing hair follicle (Fig. 5A). In catagen, HFSCs migrate along the retracting follicle to return to the bulge where they remain during telogen until the next wave of anagen arrives (Fig. 5B). The repeated cycles of HFSC activation and quiescence make the hair follicle an attractive model for studying stem cell regeneration.

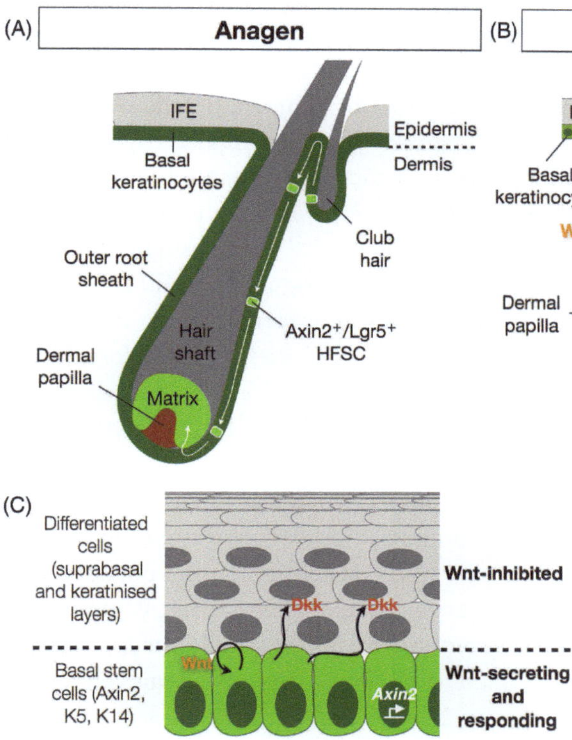

Fig. 5 Skin stem cell in the hair follicle and interfollicular epidermis (IFE). (A) Hair follicular stem cell (HFSC). During anagen, these HFSCs migrate from the outer bulge along the outer root sheath to enter the matrix, where they proliferate and differentiate to generate the hair shaft. (B) During the telogen (resting) phase, the outer bulge stem cells express Axin2 and Lgr5. The outer bulge HFSCs secretes their own Wnts to drive autocrine Wnt/β-catenin signaling. The outer bulge HFSCs prevents the adjacent inner bulge cells from being receptive to Wnt pathway activation by secreting diffusible Dkk protein. (C) IFE stem cell. The IFE stem cell at the base express Axin2, and secrete their own Wnts to set up autocrine Wnt/β-catenin signaling. The IFE stem cells also secrete Dkk proteins, which diffuse to the suprabasal layers and inhibit their response to the Wnt ligands in the environment, thereby restricting Wnt/β-catenin signaling to the basal stem cells.

3.1.1 Wnt/β-Catenin Signaling Induces Anagen in Cooperation With Other Signaling Pathways

High levels of Wnt/β-catenin signaling in late telogen phase have long been known to promote anagen entry, in part promoted by increased levels of R-spondins1–4 in this stage.[205] Constitutive Wnt/β-catenin signaling in basal keratinocytes (which include HFSCs) from development caused de novo formation of hair follicles,[206,207] and increased Wnt/β-catenin signaling during telogen led to precocious anagen entry along with thicker and longer hair follicles.[208] Conversely, inhibiting Wnt/β-catenin signaling from birth prevents hair follicle formation,[209] and abrogating Wnt/β-catenin signaling during anagen halts proliferation in the matrix (where differentiated cells of the hair shaft are produced).[73,210] Similarly, ectopic Dkk1 expression or injection causes the hair follicles to enter catagen prematurely by inducing expression of proapoptotic protein Bax, making Dkk1 a candidate for the molecular mediator of male pattern baldness.[73,211] Surprisingly, sustained proliferation due to constitutive Wnt/β-catenin signaling could also lead to premature hair loss due to stem cell exhaustion, as seen in the murine model in which Wnt1 is overexpressed in basal keratinocytes.[212] Therefore, Wnt/β-catenin signaling during telogen induces hair follicles to enter anagen, and high levels of Wnt/β-catenin signaling promote hair follicle growth.

The telogen-to-anagen transition is actually a combination of two distinct processes—the specification of the hair follicular fate in HFSCs, followed by the proliferation of the committed cells to generate the new hair follicle. As modulating either process would result in the same phenotype in vivo, it has been challenging to determine the exact stage that Wnt/β-catenin signaling acts on. Through a combinatorial in vivo/in vitro approach to uncouple the two processes, Lien *et al.* found that β-catenin-knockout HFSCs retain proliferative capacity, but are unable to specify follicular fate upon grafting.[213] Additionally, plucking β-catenin knockout follicles resulted in sebocyte differentiation instead of new follicle formation,[213] reminiscent of the in vivo β-catenin knockout phenotype.[210] Hence, Wnt/β-catenin signaling is essential for specifying hair follicle fate and preventing sebocyte fate during the telogen-to-anagen transition.

Several Wnt ligands are expressed in the hair follicle during telogen-to-anagen transition: the bulge expresses Wnts 1, 4, 7b; whereas Wnts 6, 7b, 10a, 10b, and R-spondins1–4 are upregulated at the junction of the secondary germ to dermal papilla[74,205,214] (Fig. 5B). The Wnts and R-spondins signal to the proliferative matrix where Wnt/β-catenin signaling is active.

Most notably, genetic studies have revealed the functional relevance of Wnt10a and Wnt10b in the transitioning hair follicle. In agreement with the other studies augmenting Wnt/β-catenin signaling, ectopically expressed Wnt10b increases proliferation at the derma papilla, matrix, and hair shaft, resulting in larger hair follicles.[215] However, the *Wnt10b*-null mouse does not have any hair follicle phenotype,[216] possibly due to redundancy with other Wnt ligands, such as Wnt10a. Wnt10a is associated with odonto-onycho-dermal dysplasia, a rare autosomal recessive disease presenting palmoplantar hyperkeratosis (thickened skin on palms and soles) among a host of other symptoms.[217] Importantly, as Wnt10a and Wnt10b expression is not affected in the *Sonic Hedgehog* (*Shh*)-null mutant,[214] these Wnts likely regulate the expression of *Shh*, which is a Wnt/β-catenin target gene that initiates anagen. In turn, Shh, together with Notch signaling, regulates *Wnt5a* expression in the dermal papilla.[214,218] Without Wnt5a, cysts are generated in place of hair follicles as its target gene *FoxN1*, a critical regulator of keratinocyte differentiation, is not expressed.[218,219]

On the other hand, BMP signaling negatively regulates Wnt/β-catenin signaling during telogen-to-anagen transition. BMPs namely, Bmp2 and Bmp4, are expressed by dermal neighbors comprising fibroblasts, adipocytes, and dermal papillae. These dermis-derived BMP ligands promote BMP signaling in the HFSCs, thereby inhibiting Wnt/β-catenin signaling and anagen entry.[220–224] However, before anagen initiates, the dermal cells will reduce BMP expression and the dermal papillae express the BMP antagonist Noggin. Knocking out Noggin prevents the activation of β-catenin/Lef1-mediated transcription and anagen entry.[225] Hence, it is imperative to inhibit BMP signaling so that the HFSCs are able to respond to Wnt/β-catenin signaling, which then initiates the transition to anagen. Once Wnt/β-catenin signaling kicks off, a suite of β-catenin/Lef1-mediated Wnt target genes, such as *Shh*, *Jagged1* (Notch ligand), *Dlx3* (homeobox transcription factor), *FoxN1*, and keratins are expressed, all of which are required for the proliferation and differentiation of the hair follicular cells during anagen.[206,207,226–229]

3.1.2 Active Wnt/β-Catenin Signaling in HFSC During Telogen Maintains Stemness

For a long time, Wnt/β-catenin signaling was thought to be inactive in the quiescent HFSCs during telogen as there is no proliferation in the bulge during this phase and Wnt/β-catenin signaling has primarily been thought of as a mitogen. However, in vivo lineage tracing approaches employing Axin2-

and Lgr5-Cre drivers have formally proven that telogen HFSCs respond to Wnt/β-catenin signaling. Axin2 and Lgr5 are both expressed in the secondary hair germ during telogen and in the matrix and outer root sheath during anagen. Lineage tracing with Axin2CreERT2 and Lgr5-EGFP-ires-CreERT2 show that Axin2$^+$ and Lgr5$^+$ cells are able to give rise to the differentiated lineages in the entire hair follicle over many hair cycles[74,75] (Fig. 5B). Furthermore, Lgr5-EGFP$^+$ cells can generate new hair follicles when transplanted onto the backs of nude mice, indicative of their stem and transplantation potential.[75]

HFSCs produce their own Wnts to drive autocrine Wnt/β-catenin signaling for maintenance of their stem cell identity, while restricting Wnt/β-catenin signaling response in adjacent daughter cells by secreting the inhibitor Dkk3. During telogen, Wnts 1, 4, 7b are the most highly expressed Wnts ligands, which are maintained throughout the transition to anagen[74] (Fig. 5B). The epithelial origin of Wnts was confirmed when *Wls* was knocked out in basal keratinocytes and outer root sheath cells (including HFSCs), causing a termination of hair growth.[73] Inactivating β-catenin specifically in Axin2$^+$ HFSCs during telogen in adult mice largely abrogates hair growth and reduces proliferation,[73,74] similar to the phenotype observed with pan-basal keratinocyte deletion of β-catenin during telogen.[210] Thus, Wnt-responding HFSCs require β-catenin-mediated Wnt signaling for proliferation and proper hair follicle growth. However, it remains unclear whether Wnt/β-catenin signaling maintains follicular fate during telogen as observed during the transition to anagen,[213] and whether it serves additional functions.

In contrast, another study proposed that during telogen, Wnt/β-catenin target gene expression is repressed by the Tcf3/Tcf4/Groucho complex, thereby rendering Wnt/β-catenin signaling inactive.[213] In the presence of Wnt ligands and nuclear β-catenin, Tcf1/Lef1 activator complex replaces Tcf3/Tcf4/Groucho repressor complex at the promoters of Wnt target genes, thereby effecting their transcription. As *Tcf1* and *Lef1* are Wnt/β-catenin target genes, Wnt/β-catenin signaling is engaged in a positive feedback loop that further enhances its activity. This study proposes that Wnt/β-catenin signaling is only active upon telogen-anagen transition, when β-catenin alleviates transcriptional repression to facilitate activation of follicular fate genes.

Despite the apparent contrast, the findings from the abovementioned studies could be reconciled by examining the relative levels of Wnt/β-catenin signaling activity. It is conceivable that Wnt/β-catenin signaling is present in the HFSCs from telogen through anagen to specify follicular fate and maintain stemness. However, similar to the corpus, it is present at a basal level during telogen, which can only be detected by sensitive reporters, such as the Axin2- and Lgr5-Cre drivers.[74,75] Once the transition into anagen is initiated, positive feedback loops drive Wnt/β-catenin signaling activity to reach the threshold needed to enter anagen and actively drive follicular fate to induce hair follicle growth. Hence, more studies are required to assess the fine-tuning of Wnt/β-catenin signaling in HFSC maintenance and activation.

3.2 Interfollicular Epidermis

The IFE refers to the epidermal layer without the hair follicles and other appendages. It is organized as multiple stratified squamous layers, with the innermost basal cells in contact with the basement membrane, separated from the dermis, and progresses upward toward the increasingly keratinized layers of epidermal cells that forms the outermost layer of the skin (Fig. 5C).

3.2.1 Wnt/β-Catenin Signaling is Required for IFE Homeostasis

Several early studies ablating β-catenin and Lef1 expression in the skin produced hair follicle phenotypes, but not IFE phenotypes,[210,229,230] leading to the view that Wnt/β-catenin signaling is not active in the IFE.

However, some evidence indicate that the IFE is indeed regulated by Wnt/β-catenin signaling. Hyperactivating Wnt/β-catenin signaling often results in thickened IFE. For instance, individuals harboring gain-of-function mutation in *RSPO1* have palmoplantar hyperkeratosis (thickening of the epidermis on the palms and soles) and predisposition to squamous cell carcinoma of the skin.[231] Overexpression of stabilized β-catenin mutant or Wnt1 in keratinocytes also results in a thickened IFE.[206,207,212] Additionally, primary human keratinocyte cultures harboring a gain-of-function β-catenin mutant also has higher colony formation efficiency.[232]

In contrast, abrogating Wnt/β-catenin signaling leads to thinner IFE. Knocking out *Porcn* (and consequently Wnt signaling) in epidermal cells resulted in thinner skin displaying extensive hair-deficient areas and focal dermal hyperplasia,[233] indicating that epidermal cells are an important source of Wnts for skin and hair follicle morphogenesis. Notably, the murine mutants, such as human patients with focal dermal hyperplasia, have thinner, more fragile skin and fewer hair follicles.[233,234] Knocking out *Wls* in basal keratinocytes reduces Wnt/β-catenin signaling levels and also leads to skin thinning with associated diminished proliferation. Importantly, this phenotype can be rescued by transgenic expression of Bmp4, placing BMP signaling downstream of Wnt/β-catenin signaling in the IFE.[235] Other Wnt signaling ablation models, including the *Tcf3/Tcf4* double knockout mouse model, also show thinner IFE and compromised self-renewal ability.[210,236] Surprisingly, constitutive loss of β-catenin in keratinocytes during development leads to epidermal thickening accompanied by loss of hair follicles as developmental multipotent stem cells are directed toward the IFE fate than follicular fate,[210] underscoring the complex, and context-dependent roles of Wnt/β-catenin signaling in the skin.

Wnt/β-catenin signaling acts on IFE stem cells, which reside in the basal keratinocyte layer and express Axin2[70] (Fig. 5C). When Wnt/β-catenin signaling was inhibited in adult mice by conditional deletion of β-catenin in Axin2$^+$ cells, the IFE was thinner and had fewer proliferating cells.[70] This suggests that Wnt/β-catenin signaling maintains the basal stem cells by regulating their proliferation to support a healthy IFE, similar to conclusions from earlier studies with keratinocyte-specific drivers.

Clonal analysis with multicolor lineage tracing in the mouse epidermis from the palm, tail, and ear using Axin2-CreERT2 and other generic inducible Cre drivers showed that like ISCs, the basal stem cells are functionally equivalent. They maintain IFE homeostasis as a population by giving rise to equal numbers of stem and daughter cells during divisions, through a combination of symmetric and asymmetric divisions.[70,237,238] While it is unclear whether Wnt/β-catenin signaling is involved in the decision of division symmetry, a precedent has been set in the mouse embryonic stem cells, where a localized Wnt source is able to direct asymmetric division such that the daughter stem cell proximal to the Wnt source retains stem cell markers while the distal daughter cell tends to differentiate.[239]

Interestingly, like the HFSCs, the Axin2$^+$ basal stem cells act as their own niche by producing their own Wnts, namely, Wnt4 and Wnt10a.[70] To limit Wnt/β-catenin signaling to the stem cells of the basal layer, the basal cells also secrete Dkk3, which diffuses to the suprabasal daughter cells to prevent them from responding to the Wnt ligands.

3.2.2 Multipotent Stem Cells of the IFE, Hair Follicle, and Sebaceous Gland in Adulthood?

Lgr6 was initially reported to be selectively expressed in cells of the isthmus region, above the bulge where Lgr5$^+$ bulge stem cells reside. Lineage tracing analyses showed that these Lgr6$^+$ cells give rise to sebaceous glands, IFE, and hair follicles, constituting a long-term stem cell population that contributes to multiple mini organs of the skin during homeostasis and regeneration over the organism's lifetime.[240] However, recent studies observed additional Lgr6$^+$ populations in the IFE and sebaceous glands.[241,242] These distinct populations of Lgr6$^+$ cells were able to give rise to tracing units in their own respective compartments, exhibiting population asymmetry as previously shown in the skin and other organs.[241] Hence, Lgr6 marks three distinct Wnt-responding adult stem cell populations in the skin.

3.2.3 Cutaneous Cancers

The most frequent cancer in the United States is skin/cutaneous cancer.[243] Of all skin cancers, basal cell carcinoma (BCC) is the most common form of cutaneous cancer. Hedgehog-driven BCCs display active Wnt/β-catenin signaling and human BCCs exhibit nuclear β-catenin localization.[244–246] Strikingly, ectopic expression of Dkk1 in early superficial BCCs blocked tumor formation,[245] and deletion of β-catenin in Hedgehog-responding initiating cells prevented tumor formation.[244] Methylation of promoters of *APC* and *SFRP5* are significantly more frequent in BCCs, which is predicted to increase Wnt/β-catenin signaling activity.[71] Another factor associated with BCC development is reduced Vitamin D receptor levels in the presence of high Wnt/β-catenin signaling activity.[247] If Vitamin D receptor levels are high, then trichofolliculomas (tumors of the sebaceous glands) will develop instead. This echoes the dichotomous cell fates seen in adult murine genetic studies in which Wnt/β-catenin signaling is inhibited by the dominant negative mutant of Lef1,[248] one of the commonly mutated genes in human sebaceous tumors.[249] Hence, Wnt/β-catenin signaling is required for BCC initiation and progression. The pathway is most likely acting on HFSCs and IFE stem cells, both of which have shown to be potential candidates for the cell-of-origin of BCCs.[244,250–252]

Squamous cell carcinoma (SCC) is the second most common form of cutaneous cancer. Murine models of cutaneous SCC can be induced chemically by dimethylbenz(a)anthracene/12-O-tetradecanoylphorbol-13-acetate (DMBA/TPA). The majority of these tumors demonstrate cytoplasmic and nuclear accumulation of β-catenin, and express Wnt/β-catenin target genes, such as *Myc* and *Jun*, indicating active Wnt/β-catenin signaling in SCC.[253] Cytoplasmic and nuclear accumulation of β-catenin is similarly observed in the tumor-propagating CD34$^+$ population and human SCC tumors.[254] Strikingly, these tumors regress upon keratinocyte-specific deletion of β-catenin or ectopic expression of Tcf4, indicating that ectopic expression of Tcf4 reinforces its repressive role and constitutive Wnt/

β-catenin signaling is required for sustained tumor growth via the survival of CSCs. Likewise, promoters of *SFRP1*, *SFRP2*, *SFRP3*, *SFRP4*, and *SFRP5*, a class of Wnt inhibitors, are more significantly methylated in SCC tumors than normal epidermis.[72] In K14-ΔNLef-1 mice in which Wnt/β-catenin signaling in basal keratinocytes is inhibited, DMBA/TPA treatment readily generated sebaceous tumors which contain differentiated sebocytes,[230] demonstrating the dichotomy between hair follicle and sebaceous fates. HFSCs and IFE progenitors have been proposed as the cell of origin for cutaneous SCC (reviewed in Blanpain),[4] both of which are Wnt-responding during homeostasis.[70,74,75]

Hyperactive Wnt/β-catenin signaling has been implicated in a subset of hair follicle tumors, which are densely packed with cysts containing hair follicles, referred to as pilomatricomas (a.k.a. trichomatricomas). Pilomatricomas are benign tumors derived from the hair follicular matrix that often display mutations in the *β-catenin* gene and/or β-catenin expression concentrated in basophilic cells within the tumor.[255] In mice, hyperactivating the Wnt/β-catenin pathway led to epitheloid cysts that progressed to pilomatricomas. The resulting pilomatricomas were addicted to continuous β-catenin expression, highlighting the requirement for constant hyperactive Wnt/β-catenin signaling in these tumors.[206] In human pilomatricomas, the basaloid cells show strong nuclear β-catenin staining,[256] reaffirming active Wnt/β-catenin signaling in the tumors as observed in the mouse model.

Wnt/β-catenin signaling undoubtedly plays an instructive role in lineage determination and adult stem cell maintenance in the skin. The specific role of Wnt/β-catenin signaling is clearly dictated by the particular miniorgan and state of development. During early development, while high levels of Wnt/β-catenin signaling specify a follicular fate, lower levels direct toward sebaceous, and IFE fates. Interestingly, upon establishment of the lineages, the respective adult stem cells require Wnt/β-catenin signaling to maintain homeostasis. However, the exact levels of Wnt/β-catenin signaling required to maintain the stem cells are unclear. Specifically, how do the Wnt/β-catenin signaling activities in the anagen and telogen hair follicles compare to that of the IFE? Is Wnt/β-catenin signaling present at a basal level during mid-telogen before transitioning to much higher levels to induce anagen? If so, what are the underlying mechanics regulating this rheostat? Could potentiation of Wnt/β-catenin by the R-spondin/Lgr signaling axis be involved in this fine-tuning? If a basal level is required for skin-related lineages, is there a common determinant factor among the Wnt-regulated skin lineages?

Addressing these questions would further our understanding of how Wnt/β-catenin signaling achieves such precise regulation of tissue homeostasis and how it is exploited by cutaneous cancers.

4. LIVER STEM CELLS AND CANCER

The liver is the largest internal organ and carries out a variety of functions, including drug metabolism, chemical detoxification, and regulation of glycogen storage. It is organized into hepatic lobules, delineated by the central vein (which brings blood out of the liver), and the portal triad consisting of hepatic artery, hepatic portal vein (both of which transport blood into the liver), and intrahepatic bile duct (Fig. 6A). Blood flows from the hepatic artery and hepatic portal vein to the central vein through sinusoids, which are lined by endothelial cells. While the hepatic epithelial cells surrounding the portal triad and central vein are broadly labeled as hepatocytes, it is clear that the hepatocyte population is heterogeneous and the liver can be divided into multiple zones based on the distinct functions, localizations, and markers of the hepatocytes: (1) pericentral hepatocytes (those surrounding the central vein), (2) periportal hepatocytes (those surrounding the portal triad), and (3) midlobular hepatocytes residing between the first two zones.[257]

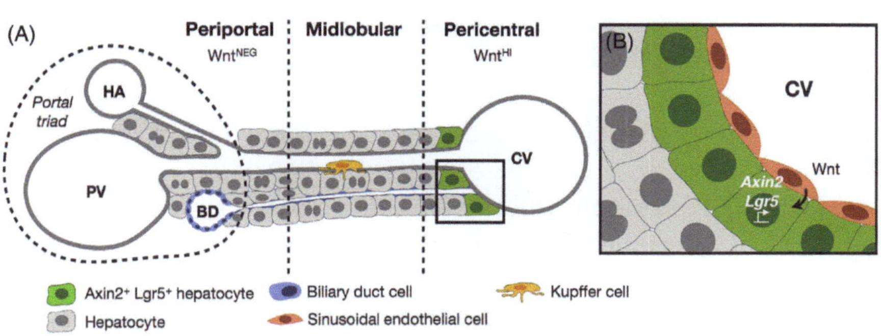

Fig. 6 Liver stem cell. (A) Functional unit of the liver. The portal triad, which consists of portal vein (PV), hepatic artery (HA), and biliary duct (BD), is a region devoid of Wnt/β-catenin signaling; while the region surrounding the central vein (CV) is a site of active Wnt/β-catenin signaling. Lgr4 is expressed in hepatocytes in all three regions. (B) The pericentral hepatocytes immediately surrounding the CV express Lgr5 and Axin2, and are purported to be homeostatic hepatocyte stem cells of the liver. The sinusoid endothelial cells lining the CV express Wnt2 and Wnt9b, forming the niche for the pericentral hepatocytes.

4.1 Wnt/β-Catenin Signaling Regulates Liver Zonation and Hepatocyte Renewal During Homeostasis and Regeneration

R-spondin/Lgr-potentiated Wnt/β-catenin signaling is required for liver zonation. Pericentral hepatocytes around the central vein express Wnt/β-catenin target genes, such as *Axin2*, *Lgr5*, *glutamine synthetase* (*GS*), *transporter-1 of glutamate* (*Glt1*), *leukocyte cell-derived chemotaxin-2* (*Lect2*), *ornithine aminotransferase* (*Oat*), cytochrome proteins, such as *Cyp2e1* and *Cyp1a2*; whereas Wnt/β-catenin signaling is suppressed in the periportal regions.[76,78,258–260]

Modulating Wnt/β-catenin signaling disrupts metabolic zonation in the liver.[78,260–263] Hyperactivation of Wnt/β-catenin signaling by knocking out hepatocyte-specific *Apc* rapidly induces *GS* expression in the entire liver, extending to periportal regions where periportal markers are lost, leading to a fatal metabolic disorder.[260] In contrast, disrupting Wnt/β-catenin signaling via adenoviral Dkk1 administration or knocking out core signaling components, such as *β-catenin*, *Lrp5/Lrp6*, or *Lgr4/Lgr5* abrogates expression of pericentral markers while periportal markers are ectopically expressed in pericentral regions.[78,260–263] These studies underscore the differential regulation by Wnt/β-catenin signaling in the two regions. Mice injected with adenoviral Dkk1 died after a week due to intestinal phenotype[48] but mice with liver-specific knockout of *β-catenin* or *Lrp5/6* were viable despite reduced liver mass with lower proliferation levels.[261–263] The lack of strong phenotype cannot be explained by gamma-catenin (plakoglobin), which can act redundantly with β-catenin, as gamma-catenin was not found in the nucleus, suggesting it is unlikely to have a signal transduction role.[264] Therefore, while Wnt/β-catenin signaling is important in maintaining liver zonation and normal proliferation during homeostasis, it is not essential for survival.

The adult liver has a remarkable regenerative capacity, and can repair itself fully even after two-thirds of the liver has been resected (partial hepatectomy). This is accomplished in part via activation of Wnt/β-catenin signaling, which is thought to regulate the S phase entry of hepatocytes. Hepatocyte-specific knockout of β-catenin or *Lrp5/6* delays liver regeneration due to later S phase entry and reduced proliferation after partial hepatectomy, though the liver mass recovers eventually.[261,263,265] This suggests redundancy in the functions of β-catenin or Wnt signaling with other molecules and pathways in the liver during regeneration. Similarly, affecting the R-spondin/Lgr axis by deleting *Lgr4* and *Lgr5* in the hepatocytes also delays their S phase entry upon partial hepatectomy.[78] Conversely, S phase entry and proliferation rates are accelerated upon partial hepatectomy when Wnt/β-catenin signaling level is increased through expressing stabilized β-catenin mutant in hepatocytes, hydrodynamic delivery of *Wnt1* DNA, systemic injection of R-spondin 1, and knocking out both *Znrf3* and *Rnf43* in hepatocytes.[78,266] Hence, Wnt/β-catenin signaling is required for efficient liver regeneration.

The cellular source of hepatocytes during homeostasis has been a long-standing question in the field. Recently, pericentral hepatocytes and parenchymal hepatocytes have been both proposed as the origin of new hepatocytes in the adult liver. The Wnt-responding pericentral hepatocytes express Axin2[76] and Lgr5.[78] The Axin2$^+$ pericentral hepatocytes proliferate more quickly than the rest of the hepatocytes,[76,267] and give rise to progeny that migrate to the periportal regions over time during homeostasis,[76] suggesting that they are capable of acting as hepatocyte progenitors. However, a separate study using Lgr5-CreERT2 knock in lines generated by Kinzel *etal.*[268] found that hepatocytes expressing Lgr5 did not produce significant progeny.[78] Instead, Lgr4$^+$ hepatocytes present throughout the liver without zonal restriction produce new hepatocytes during homeostasis.[78]

Using an independent Lgr5 reporter line (Lgr5-LacZ), Lgr5 expression is not detected during homeostasis, but is induced in small cells near biliary ducts upon injury by carbon tetrachloride administration, which causes acute liver fibrosis.[79] Lineage tracing with a third independent Lgr5-CreERT2 mouse model demonstrated that these cells give rise to biliary and hepatocyte daughter cells, indicating bipotency.[79] These characteristics are reminiscent of the "oval cells," a bipotent population that only appears upon hepatic injury.[269] Single Lgr5$^+$ cells induced upon injury can generate organoids, which require R-spondin1 in the media for survival, once again underscoring the importance of Wnt/β-catenin activity potentiation by R-spondin/Lgr axis.[79,270]

Given the discrepancies amongest the various in vivo studies, it is important to ensure that the mouse models faithfully recapitulate endogenous expression of the genes of interest, and the physiological behaviors of the cells they mark. For instance, the absence of Lgr5-LacZ expression in liver during homeostasis in the Huch *et al.* study[79] is likely to be technical issue with reporter sensitivity, as Lgr5 mRNA is present in pericentral hepatocytes during homeostasis.[76,78] Therefore, complementary approaches are necessary to establish the true identity of the homeostatic hepatocyte progenitor.

Two niche sources of Wnt signals have been proposed. Sinusoidal endothelial cells secrete Wnt2 and Wnt9b during homeostasis, which are required for normal Axin2 and GS expression.[76] In addition, Wnt2 from sinusoidal endothelial cells is also necessary for complete liver regeneration.[271] The other proposed niche is the hepatic macrophage, Kupffer cell. Knocking out *Wls* in Kupffer cells delays liver regeneration after partial hepatectomy, likely due to later S phase entry as demonstrated by reduced Cyclin D1 levels.[261] In contrast, *Wls*-deficient hepatocytes do not display any phenotype during homeostasis and regeneration.[261] Hence, hepatocytes require paracrine sources of Wnts for homeostasis and regeneration.

4.2 Aberrant Wnt/β-Catenin Signaling is a Crucial Pathway in Hepatocellular Carcinoma

Hepatocellular carcinoma (HCC) is the fifth most common cancer and the third most lethal cancer worldwide.[272] HCC is the uncontrolled growth of hepatocytes, fueled by several aberrant signaling pathways including Wnt/β-catenin signaling. In fact, mutations in Wnt pathway components leading to hyperactivation of the pathway are found in 66% of HCCs,[273] and expression levels of Wnt/β-catenin pathway components correlate significantly with disease progression.[274]

Unsupervised clustering of patient HCC samples identified various subclasses of HCC, some of which are associated with hyperactivated Wnt/β-catenin signaling.[77,275] In another study, Wnt/β-catenin signaling was found to be one of the two major signaling networks, with alterations in Wnt/β-catenin signaling pathway accounting for 65% of the HCC cases.[276] Furthermore, Wnt/β-catenin signaling is highly active in the invasive HCC characterized by positive EpCAM and alpha-fetoprotein expression. Therefore, Wnt/β-catenin signaling plays a crucial role in HCC. Active Wnt/β-catenin signaling is also associated with HCC progression as hepatocellular adenomas with activating mutations in *CTNNB1* are more likely to develop into malignant HCC.[277–280] Additionally, nuclear β-catenin localization is correlated with unfavorable prognosis, disease progression, and expression of the proliferation marker KI67,[277,279,280] particularly in later stages of HCC.[281]

Multiple Wnt/β-catenin signaling components are commonly mutated in HCC. Mutations in exon 3 of *CTNNB1*, which prevents degradation of β-catenin, is the Wnt-related mutation most commonly associated with HCC. *CTNNB1* exon 3 mutations are found in up to 44% of human HCCs across various studies,[276,282–284] and correlate with overexpression of known Wnt/β-catenin target genes, including *Lgr5*, *GS*, and *Glt*.[284,285]

In addition, inactivating mutations in Wnt/β-catenin negative regulators *AXIN1*, *AXIN2*, and *APC* were repeatedly observed in HCCs.[276,285–287] Other alterations found to activate Wnt/β-catenin signaling in HCC include overexpression of the FZD7 receptor,[288–290] increased expression of the ligand WNT3,[288] inactivation of the negative regulator GSK-3β,[291] and epigenetic silencing of the inhibitors sFRPS.[292,293] Of the FZD receptors involved, only FZD7 has been shown to induce nuclear/cytoplasmic accumulation of β-catenin in HCC.[288,290,294] Amplifications in *WNT9A*, *RSPO2*, and *FZD6* leading to their overexpression were also found in 21.6%–25.0% of the HCC tumors analyzed.[276]

While hyperactivated Wnt/β-catenin signaling is prevalent in HCC cases, multiple studies have shown that it is insufficient for inducing HCC on its own. Hepatotoxin-induced HCC originates from mature hepatocytes, not biliary progenitors,[295,296] indicating that the genetic and epigenetic changes promoting tumorigenesis occur in hepatocytes. Murine hepatocytes expressing a stabilized form of mutant β-catenin do not develop HCC spontaneously, though Wnt/β-catenin signaling is active as evidenced by nuclear β-catenin localization.[258,266,297,298] Therefore, to generate mouse models of HCC, additional genetic and/or epigenetic changes are required in a genetic background with hyperactivated Wnt/β-catenin signaling. Additional genetic changes include other tumorigenic factors, such as *Ha-Ras* oncogene[299,300] or loss of tumor suppressors, such as *Liver kinase B1* (*Lkb1*).[301] These changes can also be induced chemically in a Wnt-hyperactivated background by the carcinogen diethylnitrosamine[266] or phenobarbital, which activates constitutive androstane receptor (CAR, NR1I3), a xenobiotic nuclear receptor.[302] Nonetheless, one study showed that it is possible to induce HCC by modulating Wnt/β-catenin signaling on its own. Apc in the liver was inactivated via a low, nonlethal dose of adenoviral-Cre, which led to liver tumorigenesis over a long latency period of 8 months, with associated high levels of Wnt/β-catenin signaling.[303] However, it is unclear if Wnt-independent effects of APC inactivation or viral infection could account for the additional "hit." Overall, other genetic or epigenetic factors are required in combination with a sensitized Wnt/β-catenin background for liver tumorigenesis.

In HCC, Wnt/β-catenin signaling is important in the maintenance of CSCs marked by CD133,[304] EpCAM,[305] OV6,[306] and Lgr5.[307] In fact, EpCAM is a Wnt/β-catenin target gene in HCC.[308] Interestingly, tumor-associated EpCAM is able to perform signaling functions by cleaving its intracellular domain that binds to β-catenin and Lef1 to induce transcription of genes with Lef1 binding sites in the promoter, as well as the hepatic CSC gene signature.[309,310]

The EpCAM$^+$ CSCs from patients with advanced cirrhosis and HCC express high amounts of *WLS*, *MYC*, *CyclinD1*, and *WNT3*. Spheroids formed from these EpCAM$^+$ cells are sensitive to IWP-2, a Porcupine inhibitor, indicating that the CSC-like cells in advanced cirrhosis and CSCs in HCC engage in autocrine Wnt signaling,[311] contrasting the paracrine mode of signaling observed during homeostasis.[76,261,271] However, this therapeutic approach is unlikely to be useful in HCCs harboring mutations in downstream components, such as *CTNNB1*, *APC*, and *AXIN1*, which render Wnt/β-catenin signaling in the tumor independent of Wnt ligands. Wnt/β-catenin signaling in HCC can also be potentially targeted by FH535, a small molecule, which inhibits β-catenin activity (both wildtype and mutant) and has shown efficacy in HCC cell lines and isolated hepatic CSCs marked by CD133, CD44, and CD24 expression.[312]

While it is known that Wnt/β-catenin signaling plays a role in HCC and regulates the CSCs, several outstanding questions remain. It is clear that HCC originates from hepatocytes. However, does HCC arise from a specific subpopulation of hepatocytes, such as the pericentral hepatocytes that have high Wnt/β-catenin signaling activity during homeostasis,[76] or the Lgr4$^+$ parenchymal hepatocytes?[78] Strikingly, while hyperactivation of Wnt/β-catenin signaling appears to be sufficient for tumorigenesis in other tissues, such as the intestines, it is dependent on additional factors to induce the onset of HCC. What are the other crucial functions required for HCC tumorigenesis that are not provided by Wnt/β-catenin signaling? Finally, the precise identity of the liver progenitors during homeostasis remains to be elucidated, as well as their roles in the various models of liver injury and regeneration. Studies of Wnt/β-catenin signaling in the liver will undoubtedly gain further momentum in the future, by harnessing the plethora of tools for dissecting the pathway, which will contribute toward translation impact.

5. CONCLUDING REMARKS

Wnt/β-catenin signaling is instrumental in orchestrating organismal development, and mutations in the pathway often lead to congenital disorders. Postdevelopment, the importance of the pathway in ensuring proper adult homeostasis and regeneration is also evidenced by the increasing number of adult stem cell populations found to require Wnt/β-catenin signalling for maintenance. Through a combination of genetic experiments to modulate the Wnt pathway and lineage tracing, we have achieved a deeper understanding of adult tissue stem cells, with the intestines and skin being prime examples. However, we have only begun to scratch the surface for most other epithelial tissues. While numerous adult stem cells have been found to be Wnt-responding (Table 1), further mechanistic studies are required for a deeper understanding of how Wnt/β-catenin signaling regulates stem cell behavior.

In cancer, Wnt/β-catenin signaling is also intricately involved in tumor initiation, progression, and metastasis. Most of these cancers are associated with hyperactivated Wnt/β-catenin signaling, through genetic mutations or epigenetic misregulation of the pathway components. Such changes associated with Wnt/β-catenin signaling are believed to play a role in the earliest stages of oncogenic transformation. These changes can be broadly classified into two categories—those that are Wnt ligand-dependent, and those that are Wnt ligand-independent. Genes implicated in the former category would include the Wnt ligands, R-spondins, soluble inhibitors, and other pathway components needed for posttranslational modifications and exocytosis of the Wnt ligands (e.g., Porcn and Wls). For the Wnt ligand-dependent cancers, a new class of drugs targeting Porcn is garnering great interest in their early clinical trials for treating various epithelial cancers.[313] Combining Porcn-targeted treatment with other emerging treatments including checkpoint inhibitors, like PD-1 and immunotherapies, such as CAR-T is likely to boost therapeutic efficacy. For instance, the Porcn inhibitor LGK974 has been combined with PD-1 in an ongoing clinical trial for multiple epithelial cancers (NCT01351103).

However, for cancers harboring genetic and epigenetic changes downstream of Wnt/Fzd binding, including mutations in *APC* and *CTNNB1*, Wnt/β-catenin signaling is independent of Wnt ligands. Such cancers would therefore be refractory to Porcn modulators, instead requiring drugs that target the signal transduction machinery in the Wnt-responding cell. A comparison of patents related to Wnt modulators and Wnt-related clinical trials showed that although modulators of β-catenin formed the largest class of patents, the proportion of clinical trials targeting β-catenin was small.[314] This small group of clinical trials (NCT02195440, NCT02413853, NCT01606579) mostly involved one molecule, PRI-724, which blocks the binding of β-catenin and its coactivator, thereby inhibiting Wnt/β-catenin signaling transcription.[188] Given the large proportion of Wnt ligand-independent cancers, it is imperative to develop therapies that can target the Wnt/β-catenin transcription machinery.

A longstanding challenge in drugging the Wnt pathway lies in minimizing off target effects, as many homeostatic adult stem cells are Wnt-dependent. Many of these adult stem cells have been identified or proposed as the cell-of-origin of cancers after accumulating a suite of genome and epigenetic alterations. CSCs, which are responsible for propagating the tumor, are also often dependent on Wnt/β-catenin signaling. CSCs often bear characteristics of adult stem cells, such as self-renewal ability, expression of stem markers, such as Lgr5, CD44 and CD133, and elevated Wnt/β-catenin signaling activity. The tumor microenvironment also plays a critical role in maintaining CSCs. Therefore, modulating Wnt/β-catenin signaling in CSCs and their microenvironment could prove to be crucial for developing effective therapeutic strategies.

REFERENCES

1. Simons BD, Clevers H. Strategies for homeostatic stem cell self-renewal in adult tissues. *Cell.* 2011;145(6):851–862.
2. UK CR. World cancer factsheet. 2014:1–4
3. Hanahan D, Weinberg RA. Hallmarks of cancer: The next generation. *Cell.* 2011;144 (5):646–674.
4. Blanpain C. Tracing the cellular origin of cancer. *Nature.* 2013;15(2):126–134.
5. Veeman MT, Axelrod JD, Moon RT. A second canon. *Dev Cell.* 2003;5(3):367–377.
6. Logan CY, Nusse R. The Wnt signaling pathway in development and disease. *Annu Rev Cell Dev Biol.* 2004;20:781–810.
7. Clevers H, Loh KM, Nusse R. An integral program for tissue renewal and regeneration: Wnt signaling and stem cell control. *Science.* 2014;346(6205). 1248012.
8. Willert K, Nusse R. Wnt proteins. *Cold Spring Harb Perspect Biol.* 2012;4(9):a007864.
9. van de Wetering M, Oosterwegel M, Dooijes D, Clevers H. Identification and cloning of TCF-1, a T lymphocyte-specific transcription factor containing a sequence-specific HMG box. *EMBO J.* 1991;10(1):123–132.
10. Molenaar M, van de Wetering M, Oosterwegel M, et al. XTcf-3 transcription factor mediates beta-catenin-induced axis formation in *Xenopus* embryos. *Cell.* 1996;86 (3):391–399.
11. Behrens J, Kries von JP, Kühl M, et al. Functional interaction of beta-catenin with the transcription factor LEF-1. *Nature.* 1996;382(6592):638–642.
12. Cadigan KM, Waterman ML. TCF/LEFs and Wnt signaling in the nucleus. *Cold Spring Harb Perspect Biol.* 2012;4(11):a007906.
13. Lustig B, Jerchow B, Sachs M, Weiler S. Negative feedback loop of Wnt signaling through upregulation of conductin/axin2 in colorectal and liver tumors. *Mol Cell Biol.* 2002;22:1184–1193.
14. van der Flier LG, Sabates Bellver J, Oving I, et al. The intestinal Wnt/TCF signature. *Gastroenterology.* 2007;132(2):628–632.
15. Hao H-X, Jiang X, Cong F. Control of Wnt receptor turnover by R-spondin-ZNRF3/ RNF43 signaling module and its dysregulation in cancer. *Cancers.* 2016;8(6 [a]):54.
16. Koo B-K, Spit M, Jordens I, et al. Tumour suppressor RNF43 is a stem-cell E3 ligase that induces endocytosis of Wnt receptors. *Nature.* 2012;488:665–669.

17. de Lau W, Barker N, Low TY, et al. LGR5 homologues associate with Wnt receptorsand mediate R-spondin signaling. *Nature.* 2011;476(7360):293–297.
18. Soriano P. Generalized lacZ expression with the ROSA26 Cre reporter strain. *Nat Genet.* 1999;21(1):70–71.
19. Muzumdar MD, Tasic B, Miyamichi K, Li L, Luo L. A global double-fluorescent Cre reporter mouse. *genesis.* 2007;45(9):593–605.
20. Snippert HJ, van der Flier LG, Sato T, et al. Intestinal crypt homeostasis results from neutral competition between symmetrically dividing Lgr5 stem cells. *Cell.* 2010;143 (1):134–144.
21. Madisen L, Zwingman TA, Sunkin SM, et al. A robust and high-throughput Cre reporting and characterization system for the whole mouse brain. *Nat Neurosci.* 2009;13(1):133–140.
22. Feil R, Brocard J, Mascrez B, LeMeur M, Metzger D, Chambon P. Ligand-activated site-specific recombination in mice. *ProcNatlAcadSci USA.* 1996;93(20):10887–10890.
23. Willert K, Brown JD, Danenberg E, et al. Wnt proteins are lipid-modified and can act as stem cell growth factors. *Nature.* 2003;423(6938):448–452.
24. van den Heuvel M, Harryman-Samos C, Klingensmith J, Perrimon N, Nusse R. Mutations in the segment polarity genes wingless and porcupine impair secretion of the wingless protein. *EMBOJ.* 1993;12(13):5293–5302.
25. Kadowaki T, Wilder E, Klingensmith J, Zachary K, Perrimon N. The segment polarity gene porcupine encodes a putative multitransmembrane protein involved in Wingless processing. *Genes Dev.* 1996;10(24):3116–3128.
26. Takada R, Satomi Y, Kurata T, et al. Monounsaturated fatty acid modification of Wnt protein: its role in Wnt secretion. *Dev Cell.* 2006;11(6):791–801.
27. Janda CY, Waghray D, Levin AM, Thomas C, Garcia KC. Structural basis of Wnt recognition by Frizzled. *Science.* 2012;337(6090):59–64.
28. Herr P, Basler K. Porcupine-mediated lipidation is required for Wnt recognition by Wls. *Dev Biol.* 2012;361(2):392–402.
29. Bänziger C, Soldini D, Schütt C, Zipperlen P, Hausmann G, Basler K. Wntless, a conserved membrane protein dedicated to the secretion of Wnt proteins from signaling cells. *Cell.* 2006;125(3):509–522.
30. Bartscherer K, Pelte N, Ingelfinger D, Boutros M. Secretion of Wnt ligands requires Evi, a conserved transmembrane protein. *Cell.* 2006;125(3):523–533.
31. Goodman RM, Thombre S, Firtina Z, et al. Sprinter: a novel transmembrane protein required for Wg secretion and signaling. *Development.* 2006;133(24):4901–4911.
32. Duraiswamy AJ, Lee MA, Madan B, et al. Discovery and optimization of a porcupine inhibitor. *J Med Chem.* 2015;58(15):5889–5899.
33. Madan B, Virshup DM. Targeting Wnts at the source: new mechanisms, new biomarkers, new drugs. *Mol CancerTher.* 2015;14(5):1087–1094.
34. Madan B, Ke Z, Harmston N, et al. Wnt addiction of genetically defined cancers reversed by PORCN inhibition. *Oncogene.* 2015;35(17):2197–2207.
35. Chen B, Dodge ME, Tang W, et al. Small molecule–mediated disruption of Wnt-dependent signaling in tissue regeneration and cancer. *NatChemBiol.* 2009;5(2):100–107.
36. Proffitt KD, Madan B, Ke Z, et al. Pharmacological inhibition of the Wnt acyltransferase PORCN prevents growth of WNT-driven mammary cancer. *Cancer Research.* 2013;73 (2):502–507.
37. Liu J, Pan S, Hsieh MH, et al. Targeting Wnt-driven cancer through the inhibition of porcupine by LGK974. *Proc Natl Acad Sci USA.* 2013;110(50):20224–20229.
38. Leblond CP, Stevens CE. The constant renewal of the intestinal epithelium in the albino rat. *Anat Rec.* 1948;100(3):357–377.
39. Barker N, Huch M, Kujala P, et al. Lgr5(+ve) stem cells drive self-renewal in the stomach and build long-lived gastric units in vitro. *Stem Cell.* 2010;6(1):25–36.

40. Clevers H. The intestinal crypt, A prototype stem cell compartment. *Cell.* 2013;154 (2):274–284.
41. Barker N, Bartfeld S, Clevers H. Tissue-resident adult stem cell populations of rapidly self-renewing organs. *Stem Cell.* 2010;7(6):656–670.
42. Bretscher A, Weber K. Villin: the major microfilament-associated protein of the intestinal microvillus. *Proc Natl Acad Sci USA.* 1979;76:2321–2325.
43. Marjou El F, Janssen KP, Hung-Junn Chang B, et al. Tissue-specific and inducible Cre-mediated recombination in the gut epithelium. *Genesis.* 2004;39(3):186–193.
44. Batlle E, Henderson JT, Beghtel H, et al. Beta-catenin and TCF mediate cell positioning in the intestinal epithelium by controlling the expression of EphB/ephrinB. *Cell.* 2002;111(2):251–263.
45. Fevr T, Robine S, Louvard D, Huelsken J. Wnt/β-catenin is essential for intestinal homeostasis and maintenance of intestinal stem cells. *Mol Cell Biol.* 2007;27 (21):7551–7559.
46. Ireland H, Kemp R, Houghton C, et al. Inducible cre-mediated control of gene expression in the murine gastrointestinal tract: effect of loss of β-catenin. *Gastroenterology.* 2004;126(5):1236–1246.
47. Pinto D, Gregorieff A, Begthel H, Clevers H. Canonical Wnt signals are essential for homeostasis of the intestinal epithelium. *Genes Dev.* 2003;17(14):1709–1713.
48. Kuhnert F, Davis CR, Wang HT, Chu P. Essential requirement for Wnt signaling in proliferation of adult small intestine and colon revealed by adenoviral expression of Dickkopf-1. In: Vol; 2004
49. van Es JH, Haegebarth A, Kujala P, et al. A critical role for the Wnt effector Tcf4 in adult intestinal homeostatic self-renewal. *Mol Cell Biol.* 2012;32(10):1918–1927.
50. Korinek V, Barker N, Moerer P, et al. Depletion of epithelial stem-cell compartments in the small intestine of mice lacking Tcf-4. *Nat Genet.* 1998;19(4):379–383.
51. Sansom OJ, Reed KR, Hayes AJ, et al. Loss of Apc in vivo immediately perturbs Wnt signaling, differentiation, and migration. *Genes Dev.* 2004;18(12):1385–1390.
52. Wong MH, Rubinfeld B, Gordon JI. Effects of forced expression of an NH2-terminal truncated beta-Catenin on mouse intestinal epithelial homeostasis. *J Cell Biol.* 1998;141 (3):765–777.
53. Wong MH, Huelsken J, Birchmeier W, Gordon JI. Selection of multipotent stem cells during morphogenesis of small intestinal crypts of Lieberkuhn is perturbed by stimulation of Lef-1/beta-catenin signaling. *J Biol Chem.* 2002;277(18):15843–15850.
54. Kim KA, Zhao J, Andarmani S, Kakitani M. R-spondin proteins: a novel link to β-catenin activation. *Cell.* 2005;5(1):23–26.
55. Yan KS, Chia LA, Li X, et al. The intestinal stem cell markers Bmi1 and Lgr5 identify two functionally distinct populations. *Proc Natl Acad Sci USA.* 2012;109(2):466–471.
56. Bhanja P, Saha S, Kabarriti R, et al. Protective role of R-spondin1, an intestinal stem cell growth factor, against radiation-induced gastrointestinal syndrome in mice. *PLoS One.* 2009;4(11):e8014.
57. Takashima S, Kadowaki M, Aoyama K, et al. The Wnt agonist R-spondin1 regulates systemic graft-versus-host disease by protecting intestinal stem cells. *J Exp Med.* 2011;208(2):285–294.
58. van de Wetering M, Sancho E, Verweij C, de Lau W. The β-catenin/TCF-4 complex imposes a crypt progenitor phenotype on colorectal cancer cells. *Cell.* 2002;111:241–250.
59. Barker N, van Es JH, Kuipers J, et al. Identification of stem cells in small intestine and colon by marker gene Lgr5. *Nature.* 2007;449(7165):1003–1007.
60. Muñoz J, Stange DE, Schepers AG, et al. The Lgr5 intestinal stem cell signature: robust expression of proposed quiescent "+4" cell markers. *EMBO J.* 2012;31 (14):3079–3091.

61. Cheng H, Leblond CP. Origin, differentiation and renewal of the four main epithelial cell types in the mouse small intestine V. Unitarian theory of the origin of the four epithelial cell types. *AmJAnat.* 1974;141(4):537–561.
62. Sato T, Vries RG, Snippert HJ, et al. Single Lgr5 stem cells build crypt-villus structures in vitro without a mesenchymal niche. *Nature.* 2009;459(7244):262–265.
63. van der Flier LG, van Gijn ME, Hatzis P, et al. Transcription factor achaete scute-like 2 controls intestinal stem cell fate. *Cell.* 2009;136(5):903–912.
64. Barker N, Ridgway RA, van Es JH, et al. Crypt stem cells as the cells-of-origin of intestinal cancer. *Nature.* 2009;457(7229):608–612.
65. Network TCGA.Comprehensive molecular characterization of human colon and rectal cancer. *Nature.* 2012;487(7407):330–337.
66. Leushacke M, Tan SH, Wong A, et al. Lgr5-expressing chief cells drive epithelial regeneration and cancer in the oxyntic stomach. *Nat CellBiol.* 2017;19(7):774–786.
67. Network TCGAG... Comprehensive molecular characterization of gastric adenocarcinoma. *Nature.* 2014;513(7517):202–209.
68. Stange DE, Koo B-K, Huch M, et al. Differentiated Troy+ chief cells act as reserve stem cells to generate all lineages of the stomach epithelium. *Cell.* 2013;155(2):357–368.
69. Li X-B, Yang G, Zhu L, et al. Gastric Lgr5+ stem cells are the cellular origin of invasive intestinal-type gastric cancer in mice. *CellRes.* 2016;1–12.
70. Lim X, Tan SH, Chye Koh WL, et al. Interfollicular epidermal stem cells self-renew via autocrine Wnt signaling. *Science.* 2013;342(6163):1226–1230.
71. Brinkhuizen T, van den Hurk K, Winnepenninckx VRJL, et al. Epigenetic changes in basal cell carcinoma affect SHH and WNT signaling components. *PLoS One.* 2012;7 (12):e51710.
72. Liang J, Kang X, Halifu Y, et al. Secreted frizzled-related protein promotors are hypermethylated in cutaneous squamous carcinoma compared with normal epidermis. *BMC Cancer.* 2015;15:641.
73. Choi YS, Zhang Y, Xu M, et al. Distinct functions for Wnt/β-catenin in hair follicle stem cell proliferation and survival and interfollicular epidermal homeostasis. *Cell Stem Cell.* 2013;13(6):720–733.
74. Lim X, Tan SH, Yu KL, Lim SBH, Nusse R. Axin2marks quiescent hair follicle bulge stem cells that are maintained by autocrine Wnt/β-catenin signaling. *Proc Natl Acad Sci USA.* 2016;113(11):E1498–E1505. 201601599.
75. Jaks V, Barker N, Kasper M, et al. Lgr5 marks cycling, yet long-lived, hair follicle stem cells. *Nat Genet.* 2008;40(11):1291–1299.
76. Wang B, Zhao L, Fish M, Logan CY, Nusse R. Self-renewing diploid Axin2+ cells fuel homeostatic renewal of the liver. *Nature.* 2015;524:180–185.
77. Hoshida Y, Nijman SMB, Kobayashi M, et al. Integrative transcriptome analysis reveals common molecular subclasses of human hepatocellular carcinoma. *CancerRes.* 2009;69 (18):7385–7392.
78. Planas-Paz L, Orsini V, Boulter L, et al. The RSPO–LGR4/5–ZNRF3/RNF43 module controls liver zonation and size. *Nat CellBiol.* 2016;18(5):467–479.
79. Huch M, Dorrell C, Boj SF, et al. In vitro expansion of single Lgr5. *Nature.* 2013;494 (7436):247–250.
80. van Amerongen R, Bowman AN, Nusse R. Developmental stage and time dictate the fate of Wnt/β-catenin-responsive stem cells in the mammary gland. *Cell Stem Cell.* 2012;11(3):387–400.
81. de Visser KE, Ciampricotti M, Michalak EM, et al. Developmental stage-specific contribution of LGR5(+) cells to basal and luminal epithelial lineages in the postnatal mammary gland. *JPathol.* 2012;228(3):300–309.
82. Plaks V, Brenot A, Lawson DA, et al. Lgr5-expressing cells are sufficient and necessary for postnatal mammary gland organogenesis. *CellRep.* 2013;3(1):70–78.

83. Wang D, Cai C, Dong X, et al. Identification of multipotent mammary stem cells by protein C receptor expression. *Nature.* 2015;517(7532):81–84.
84. van Amerongen R, Fuerer C, Mizutani M, Nusse R. Developmental biology. *Dev Biol.* 2012;369(1):101–114.
85. Koboldt DC, Fulton RS, McLellan MD, et al. Comprehensive molecular portraits of human breast tumours. *Nature.* 2012;490(7418):61–70.
86. Ng A, Tan S, Singh G, et al. Lgr5 marks stem/progenitor cells in ovary and tubal epithelia. *Nature.* 2014;16(8):745–757.
87. Jan TA, Chai R, Sayyid ZN, et al. Tympanic border cells are Wnt-responsive and can act as progenitors for postnatal mouse cochlear cells. *Development.* 2013;140(6):1196–1206.
88. Hirata-Tominaga K, Nakamura T, Okumura N, et al. Corneal endothelial cell fate is maintained by LGR5 through the regulation of hedgehog and Wnt pathway. *Stem Cells.* 2013;31(7):1396–1407.
89. Takeda N, Jain R, Li D, Li L, Lu MM, Epstein JA. Lgr5 identifies progenitor cells capable of taste bud regeneration after injury. *PLoS One.* 2013;8(6):e66314.
90. Ren W, Lewandowski BC, Watson J, et al. Single Lgr5- or Lgr6-expressing taste stem/progenitor cells generate taste bud cells ex vivo. *Proc Natl Acad Sci USA.* 2014;111 (46):16401–16406.
91. Yee KK, Li Y, Redding KM, Iwatsuki K, Margolskee RF, Jiang P. Lgr5-EGFP marks taste bud stem/progenitor cells in posterior tongue. *Stem Cells.* 2013;31(5):992–1000.
92. Lehoczky JA, Tabin CJ. Lgr6 marks nail stem cells and is required for digit tip regeneration. *Proc Natl Acad Sci USA.* 2015;112(43):13249–13254.
93. Rinkevich Y, Montoro DT, Contreras-Trujillo H, et al. In vivo clonal analysis reveals lineage-restricted progenitor characteristics in mammalian kidney development, maintenance, and regeneration. *Cell Rep.* 2014;7(4):1270–1283.
94. Barker N, Rookmaaker MB, Kujala P, et al. Lgr5(+ve) stem/progenitor cells contribute to nephron formation during kidney development. *Cell Rep.* 2012;2(3):540–552.
95. Huch M, Bonfanti P, Boj SF, et al. Unlimited in vitro expansion of adult bi-potent pancreas progenitors through the Lgr5/R-spondin axis. *EMBO J.* 2013;32 (20):2708–2721.
96. Wang B-E, Wang X, Long JE, Eastham-Anderson J, Firestein R, Junttila MR. Castration-resistant Lgr5+ cells are long-lived stem cells required for prostatic regeneration. *Stem Cell Rep.* 2015;4(5):768–779.
97. van der Flier LG, Haegebarth A, Stange DE, van de Wetering M, Clevers H. OLFM4 is a robust marker for stem cells in human intestine and marks a subset of colorectal cancer cells. *YGAST.* 2009;137(1):15–17.
98. Tian H, Biehs B, Warming S, et al. A reserve stem cell population in small intestine renders Lgr5-positive cells dispensable. *Nature.* 2011;478(7368):255–259.
99. Sangiorgi E, Capecchi MR. Bmi1 is expressed in vivo in intestinal stem cells. *Nat Genet.* 2008;40(7):915–920.
100. Takeda N, Jain R, LeBoeuf MR, Wang Q, Lu MM, Epstein JA. Interconversion between intestinal stem cell populations in distinct niches. *Science.* 2011;334 (6061):1420–1424.
101. Powell AE, Wang Y, Li Y, et al. The Pan-ErbB negative regulator Lrig1 is an intestinal stem cell marker that functions as a tumor suppressor. *Cell.* 2012;149(1):146–158.
102. Montgomery RK, Carlone DL, Richmond CA, et al. Mouse telomerase reverse transcriptase (mTert) expression marks slowly cycling intestinal stem cells. *Proc Natl Acad Sci USA.* 2011;108(1):179–184.
103. Potten CS, Kovacs L, Hamilton E. Continuous labelling studies on mouse skin and intestine. *Cell Tissue Kinet.* 1974;7(3):271–283.
104. Potten CS. Extreme sensitivity of some intestinal crypt cells to X and gamma irradiation. *Nature.* 1977;269(5628):518–521.

105. Potten CS, Owen G, Booth D. Intestinal stem cells protect their genome by selective segregation of template DNA strands. *J Cell Sci.* 2002;115(Pt 11):2381–2388.
106. Itzkovitz S, Lyubimova A, Blat IC, et al. Technical Report. *Nature.* 2011;14 (1):106–114.
107. Buczacki SJA, Zecchini HI, Nicholson AM, et al. Intestinal label-retaining cells are secretory precursors expressing Lgr5. *Nature.* 2013;495(7439):65–69.
108. Tetteh PW, Basak O, Farin HF, et al. Replacement of lost Lgr5-positive stem cells through plasticity of their enterocyte-lineage daughters. *Stem Cell.* 2016;18 (2):203–213.
109. van Es JH, Sato T, van de Wetering M, et al. ncb2581. *Nature.* 2012;14(10):1099–1104.
109a. Jadhav U, Saxena M, O'Neill NK, et al. Dynamic reorganization of chromatin accessibility signatures during dedifferentiation of secretory precursors into $Lgr5^+$ intestinal stem cells. *Cell Stem Cell.* 2017;21(1):65–77.
109b. Yan KS, Gevaert O, Zhang GXY, et al. Intestinal enteroendocrine lineage cells possess homeostatic and injury-inducible stem cell activity. *Cell Stem Cell.* 2017; 21(1):78–90.
110. Metcalfe C, Kljavin NM, Ybarra R, de Sauvage FJ. Stem cells are indispensable for radiation-induced intestinal regeneration. *Stem Cell.* 2014;14(2):149–159.
111. Baker A-M, Cereser B, Melton S, et al. Quantification of crypt and stem cell evolution in the normal and neoplastic human colon. *Cell Rep.* 2014;8(4):940–947.
112. Grün D, Lyubimova A, Kester L, et al. Single-cell messenger RNA sequencing reveals rare intestinal cell types. *Nature.* 2015;525:251–255.
113. Ritsma L, Ellenbroek SIJ, Zomer A, et al. Intestinal crypt homeostasis revealed at single-stem-cell level by in vivo live imaging. *Nature.* 2014;507(7492):362–365.
113a. Kim TH, Saadatpour A, Guo G, et al. Single-cell transcript profiles reveal multilineage priming in early progenitors derived from Lgr5+ intestinal stem cells. *Cell Reports.* 2016;16(8):2053–2060.
114. Sato T, van Es JH, Snippert HJ, et al. Paneth cells constitute the niche for Lgr5 stem cells in intestinal crypts. *Nature.* 2011;469(7330):415–418.
115. van Es JH, Jay P, Gregorieff A, et al. Wnt signalling induces maturation of Paneth cells in intestinal crypts. *Nature.* 2005;7(4):381–386.
116. Gregorieff A, Pinto D, Begthel H, Destr e O, Kielman M, Clevers H. Expression pattern of Wnt signaling components in the adult intestine. *Gastroenterology.* 2005;129 (2):626–638.
117. Farin HF, Jordens I, Mosa MH, et al. Visualization of a short-range Wnt gradient in the intestinal stem-cell niche. *Nature.* 2016;530(7590):340–343.
118. Flanagan DJ, Phesse TJ, Barker N, et al. Frizzled7 functions as a Wnt receptor in intestinal epithelial Lgr5(+) stem cells. *Stem Cell Rep.* 2015;4(5):759–767.
119. Sasaki N, Sachs N, Wiebrands K, et al. Reg4 +deep crypt secretory cells function as epithelial niche for Lgr5+ stem cells in colon. *Proc Natl Acad Sci USA.* 2016;113(37): E5399–E5407.
120. Durand A, Donahue B, Peignon G. Functional intestinal stem cells after Paneth cell ablation induced by the loss of transcription factor Math1 (Atoh1). *Proc Natl Acad Sci USA.* 2012;109(23):8965–8970.
121. Kim T-H, Escudero S, Shivdasani RA. Intact function of Lgr5 receptor-expressing intestinal stem cells in the absence of Paneth cells. *Proc Natl Acad Sci USA.* 2012;109 (10):3932–3937.
122. Kabiri Z, Greicius G, Madan B, et al. Stroma provides an intestinal stem cell niche in the absence of epithelial Wnts. *Development.* 2014;141(11):2206–2215.
123. Valenta T, Degirmenci B, Moor AE, et al. Wnt ligands secreted by subepithelial mesenchymal cells are essential for the survival of intestinal stem cells and gut homeostasis. *Cell Rep.* 2016;15(5):911–918.

124. San Roman AK, Jayewickreme CD, Murtaugh LC, Shivdasani RA. Wnt secretion from epithelial cells and subepithelial myofibroblasts is not required in the mouse intestinal stem cell niche in vivo. *Stem Cell Rep.* 2014;2(2):127–134.
124a. Stzepourginski I, Nigro G, Jacob J, et al. CD34$^+$ mesenchymal cells are a major component of the intestinal stem cells niche at homeostasis and after injury. *Proc Natl Acad Sci USA.* 2017;114(4):E506–E513.
125. Schuijers J, Junker JP, Mokry M, et al. Ascl2 acts as an R-spondin/Wnt-responsive switch to control stemness in intestinal crypts. *Stem Cell.* 2015;16(2):158–170.
126. Schuijers J, Mokry M, Hatzis P, Cuppen E, Clevers H. Wnt-induced transcriptional activation is exclusively mediated by TCF/LEF. *EMBO J.* 2014;33(2):146–156.
127. Reed KR, Tunster SJ, Young M, Carrico A, John RM, Clarke AR. Entopic over-expression of Ascl2 does not accelerate tumourigenesis in ApcMin mice. *Gut.* 2012;61 (10):1435–1438.
128. Qi Z, Li Y, Zhao B, et al. BMP restricts stemness of intestinal Lgr5+ cells by directly suppressing their signature genes. *Nat Comms.* 2016;7:1–14.
129. He XC, Zhang J, Tong W-G, et al. BMP signaling inhibits intestinal stem cell self-renewal through suppression of Wnt/β-catenin signaling. *Nat Genet.* 2004;36 (10):1117–1121.
130. Auclair BA, Benoit YD, Rivard N, Mishina Y, Perreault N. Bone morphogenetic protein signaling is essential for terminal differentiation of the intestinal secretory cell lineage. *YGAST.* 2007;133(3):887–896.
131. Yan KS, Janda CY, Chang J, et al. Non-equivalence of Wnt and R-spondin ligands during Lgr5. *Nature.* 2017;545(7653):238–242.
132. Gartel AL, Ye X, Goufman E, et al. Myc represses the p21(WAF1/CIP1) promoter and interacts with Sp1/Sp3. *Proc Natl Acad Sci USA.* 2001;98(8):4510–4515.
133. Claassen GF, Hann SR. A role for transcriptional repression of p21CIP1 by c-Myc in overcoming transforming growth factor β-induced cell-cycle arrest. *Proc Natl Acad Sci USA.* 2000;97:9498–9503.
134. Baldin V, Lukas J, Marcote MJ, Pagano M, Draetta G. Cyclin D1 is a nuclear protein required for cell cycle progression in G1. *Genes Dev.* 1993;7(5):812–821.
135. Muncan V, Sansom OJ, Tertoolen L, et al. Rapid loss of intestinal crypts upon conditional deletion of the Wnt/Tcf-4 target gene c-Myc. *Mol Cell Biol.* 2006;26 (22):8418–8426.
136. Welcker M, Singer J, Loeb KR, et al. Multisite phosphorylation by Cdk2 and GSK3 controls cyclin E degradation. *Mol Cell.* 2003;12(2):381–392.
137. Welcker M, Orian A, Jin J, et al. The Fbw7 tumor suppressor regulates glycogen synthase kinase 3 phosphorylation-dependent c-Myc protein degradation. *Proc Natl Acad Sci USA.* 2004;101(24):9085–9090.
138. Diehl JA, Cheng M, Roussel MF, Sherr CJ. Glycogen synthase kinase-3beta regulates cyclin D1 proteolysis and subcellular localization. *Genes Dev.* 1998;12(22):3499–3511.
139. Niehrs C, Acebron SP. Mitotic and mitogenic Wnt signalling. *EMBO J.* 2012;31 (12):2705–2713.
140. Marshman E, Booth C, Potten CS. The intestinal epithelial stem cell. *BioEssays.* 2002;24(1):91–98.
141. Hoffmeyer K, Raggioli A, Rudloff S, Anton R, Hierholzer A, Del Valle I, Hein K, Vogt R, Kemler R. Wnt/β-catenin signaling regulates telomerase in stem cells and cancer cells. *Cancer.* 2012;336(6088):1549–1554.
142. Park J-I, Venteicher AS, Hong JY, et al. Telomerase modulates Wnt signalling by association with target gene chromatin. *Nature.* 2009;460(7251):66–72.
143. Schepers AG, Vries R, van den Born M, van de Wetering M, Clevers H. Lgr5 intestinal stem cells have high telomerase activity and randomly segregate their chromosomes. *EMBO J.* 2011;30(6):1104–1109.

144. Günes C, Rudolph KL. The role of telomeres in stem cells and cancer. *Cell.* 2013;152 (3):390–393.
145. Sato T, Clevers H. Growing self-organizing mini-guts from a single intestinal stem cell: mechanism and applications. *Science.* 2013;340(6137):1190–1194.
146. Sato T, Stange DE, Ferrante M, et al. Long-term expansion of epithelial organoids from human colon, adenoma, adenocarcinoma, and Barrett's epithelium. *Gastroenterology.* 2011;141(5):1762–1772.
147. Jung P, Sato T, Merlos-Suárez A, et al. Isolation and in vitro expansion of human colonic stem cells. *Nat Med.* 2011;17(10):1225–1227.
148. Fatehullah A, Tan SH, Barker N. Organoids as an in vitro model of human development and disease. *Nature.* 2016;18(3):246–254.
149. Fearon ER, Vogelstein B. A genetic model for colorectal tumorigenesis. *Cell.* 1990;61 (5):759–767.
150. Nishisho I, Nakamura Y, Miyoshi Y, et al. Mutations of chromosome 5q21 genes in FAP and colorectal cancer patients. *Science.* 1991;253(5020):665–669.
151. Kinzler KW, Nilbert MC, Vogelstein B, et al. Identification of a gene located at chromosome 5q21 that is mutated in colorectal cancers. *Science.* 1991;251 (4999):1366–1370.
152. Korinek V, Barker N, Morin PJ, et al. Constitutive transcriptional activation by a beta-catenin-Tcf complex in $APC^{-/-}$ colon carcinoma. *Science.* 1997;275(5307):1784–1787.
153. Morin PJ, Sparks AB, Korinek V, et al. Activation of β-catenin-Tcf signaling in colon cancer by mutations in βcCatenin or APC. *Science.* 1997;275(5307):1787–1790.
154. Kinzler KW, Vogelstein B. Lessons from hereditary colorectal cancer. *Cell.* 1996;87:159–170.
155. Rowan AJ, Lamlum H, Ilyas M, et al. APC mutations in sporadic colorectal tumors: a mutational "hotspot" and interdependence of the 'two hits'. *Proc Natl Acad Sci USA.* 2000;97(7):3352–3357.
156. Powell SM, Zilz N, Beazer-Barclay Y, et al. APC mutations occur early during colorectal tumorigenesis. *Nature.* 1992;359(6392):235–237.
157. Su LK, Kinzler KW, Vogelstein B, et al. Multiple intestinal neoplasia caused by a mutation in the murine homolog of the APC gene. *Science.* 1992;256(5057):668–670.
158. Cheung AF, Carter AM, Kostova KK, et al. Complete deletion of Apc results in severe polyposis in mice. *Oncogene.* 2009;29(12):1857–1864.
159. Oshima M, Oshima H, Kitagawa K, Kobayashi M, Itakura C, Taketo M. Loss of Apc heterozygosity and abnormal tissue building in nascent intestinal polyps in mice carrying a truncated Apc gene. *Proc Natl Acad Sci USA.* 1995;92(10):4482–4486.
160. Pollard P, Deheragoda M, Segditsas S, et al. The Apc 1322T mouse develops severe polyposis associated with submaximal nuclear beta-catenin expression. *YGAST.* 2009;136(7). 2204–2213.e1-13.
161. Gaspar C, Franken P, Molenaar L, et al. A targeted constitutive mutation in the Apc tumor suppressor gene underlies mammary but not intestinal tumorigenesis. *PLoS Genet.* 2009;5(7):1–13.
162. Smits R, Kielman MF, Breukel C, et al. Apc1638T: a mouse model delineating critical domains of the adenomatous polyposis coli protein involved in tumorigenesis and development. *Genes Dev.* 1999;13(10):1309–1321.
163. Robanus-Maandag EC, Koelink PJ, Breukel C, et al. A new conditional Apc-mutant mouse model for colorectal cancer. *Carcinogenesis.* 2010;31(5):946–952.
164. Liu W, Dong X, Mai M, et al. Mutations in AXIN2 cause colorectal cancer with defective mismatch repair by activating beta-catenin/TCF signalling. *Nat Genet.* 2000;26(2):146–147.
165. Dow LE, O'Rourke KP, Simon J, et al. Apc restoration promotes cellular differentiation and reestablishes crypt homeostasis in colorectal cancer. *Cell.* 2015;161(7):1539–1552.

166. Seshagiri S, Stawiski EW, Durinck S, et al. Recurrent R-spondin fusions in colon cancer. *Nature.* 2012;488(7413):660–664.
167. Storm EE, Durinck S, De Sousa E, Melo F, et al. Targeting PTPRK-RSPO3 colon tumours promotes differentiation and loss of stem-cell function. *Nature.* 2016;529(7584):97–100.
168. Giannakis M, Hodis E, Jasmine Mu X, et al. RNF43 is frequently mutated in colorectal and endometrial cancers. *Nat Genet.* 2014;46(12):1264–1266.
169. Koo B-K, van Es JH, van den Born M, Clevers H. Porcupine inhibitor suppresses paracrine Wnt-driven growth of Rnf43;Znrf3-mutant neoplasia. *Proc Natl Acad Sci USA.* 2015;112(24):7548–7550.
170. Scholer-Dahirel A, Schlabach MR, Loo A, Bagdasarian L, Meyer R, Guo R, Woolfenden S, Yu KK, Markovits J, Killary K, Sonkin D, Yao YM, Warmuth M, Sellers WR, Schlegel R, Stegmeier F, Mosher RE, McLaughlin ME. Maintenance of adenomatous polyposis coli (APC)-mutant colorectal cancer is dependent on Wnt/β-catenin signaling. *Proc Natl Acad Sci USA.* 2011;108(41):17135–17140.
171. Augenlicht LH, Wadler S, Corner G, Richards C. Low-level c-myc amplification in human colonic carcinoma cell lines and tumors: a frequent, p53-independent mutation associated with improved outcome in a randomized multi-institutional trial. *Cancer Res.* 1997;57(9):1769–1775.
172. Sansom OJ, Meniel VS, Muncan V, et al. Myc deletion rescues Apc deficiency in the small intestine. *Nature.* 2007;446(7136):676–679.
173. Athineos D, Sansom OJ. Myc heterozygosity attenuates the phenotypes of APC deficiency in the small intestine. *Oncogene.* 2010;29(17):2585–2590.
174. Armanios M, Greider CW. Telomerase and cancer stem cells. *Cold Spring Harb Symp Quant Biol.* 2005;70:205–208.
175. Santos Franco S, Raveh-Amit H, Kobolák J, Alqahtani MH, Mobasheri A, Dinnyes A. The crossroads between cancer stem cells and aging. *BMC Cancer.* 2015;15(Suppl 1):S1.
176. Campbell SL, Khosravi-Far R, Rossman KL, Clark GJ, Der CJ. Increasing complexity of Ras signaling. *Oncogene.* 1998;17(11):1395–1413.
177. Janssen KP, Alberici P, Fsihi H, et al. APC and oncogenic KRAS are synergistic in enhancing Wnt signaling in intestinal tumor formation and progression. *Gastroenterology.* 2006;131(4):1096–1109.
178. Lemieux E, Cagnol S, Beaudry K, Carrier J, Rivard N. Oncogenic KRAS signalling promotes the Wnt/β-catenin pathway through LRP6 in colorectal cancer. *Oncogene.* 2014;34(38):4914–4927.
179. Huels DJ, Ridgway RA, Radulescu S, et al. E-cadherin can limit the transforming properties of activating beta-catenin mutations. *EMBO J.* 2015;34(18):2321–2333.
180. Westphalen CB, Asfaha S, Hayakawa Y, et al. Long-lived intestinal tuft cells serve as colon cancer-initiating cells. *J Clin Invest.* 2014;124(3):1283–1295.
181. Schwitalla S, Fingerle AA, Cammareri P, et al. Intestinal tumorigenesis initiated by dedifferentiation and acquisition of stem-cell-like properties. *Cell.* 2013;152(1–2):25–38.
182. Davis H, Irshad S, Bansal M, et al. Aberrant epithelial GREM1 expression initiates colonic tumorigenesis from cells outside the stem cell niche. *Nat Med.* 2014;21(1):62–70.
183. Vermeulen L, De Sousa E, Melo F, van der Heijden M, et al. Wnt activity defines colon cancer stem cells and is regulated by the microenvironment. *Nature.* 2010;12(5):468–476.
184. Kemper K, Prasetyanti PR, de Lau W, Rodermond H, Clevers H, Medema JP. Monoclonal antibodies against Lgr5 identify human colorectal cancer stem cells. *Stem Cells.* 2012;30(11):2378–2386.
185. Dalerba P, Kalisky T, Sahoo D, et al. Single-cell dissection of transcriptional heterogeneity in human colon tumors. *Nat Biotechnol.* 2011;29(12):1120–1127.

186. Schepers AG, Snippert HJ, Stange DE, et al. Lineage tracing reveals Lgr5+ stem cell activity in mouse intestinal adenomas. *Science.* 2012;337(6095):730–735.

186a. Shimokawa M, Ohta Y, Nishikori S, et al. Visualization and targeting of LGR5+ human colon cancer stem cells. *Nature.* 2017;545(7653):187–192.

187. De Sousa E, Melo F, Kurtova AV, Harnoss JM, et al. A distinct role for Lgr5. *Nature.* 2017;543(7647):676–680.

188. Emami KH, Nguyen C, Ma H, et al. A small molecule inhibitor of β-catenin/cyclic AMP response element-binding protein transcription. *Proc Natl Acad Sci USA.* 2004;101 (34):12682–12687.

189. Hayakawa Y, Ariyama H, Stancikova J, et al. Mist1 expressing gastric stem cells maintain the normal and neoplastic gastric epithelium and are supported by a perivascular stem cell niche. *Cancer Cell.* 2015;1–16.

190. Mikels AJ, Nusse R. Purified Wnt5a protein activates or inhibits β-catenin–TCF signaling depending on receptor context. *PLoS Biol.* 2006;4(4):e115.

191. Arnold K, Sarkar A, Yram MA, et al. Adult stem and progenitor cells are important for tissue regeneration and survival of mice. *Stem Cell.* 2011;9(4):317–329.

192. Matsuo J, Kimura S, Yamamura A, et al. Identification of stem cells in the epithelium of the stomach corpus and antrum of mice. *Gastroenterology.* 2017;152(1). 218–231.e14.

193. Leushacke M, Ng A, Galle J, Loeffler M, Barker N. Lgr5(+) gastric stem cells divide symmetrically to effect epithelial homeostasis in the pylorus. *Cell Rep.* 2013;5 (2):349–356.

194. Sigal M, Logan CY, Kapalczynska M, et al. Stromal R-spondin orchestrates gastric epithelial stem cells and gland homeostasis. *Nature.* 2017;548(7668):451–455.

195. Flanagan DJ, Barker N, Nowell C, et al. Loss of the Wnt receptor Frizzled7 in the gastric epithelium is deleterious and triggers rapid repopulation in vivo. *Dis Model Mech.* 2017;10(8):971–980. dmm.029876.

196. Correa P. Human gastric carcinogenesis: a multistep and multifactorial process—First American Cancer Society Award Lecture on Cancer Epidemiology and Prevention. *Cancer Res.* 1992;52(24):6735–6740.

197. Uemura N, Okamoto S, Yamamoto S, et al. Helicobacter pylori infection and the development of gastric cancer. *N Engl J Med.* 2001;345(11):784–789.

198. Sigal M, Rothenberg ME, Logan CY, et al. Helicobacter pylori activates and expands Lgr5(+) stem cells through direct colonization of the gastric glands. *Gastroenterology.* 2015;148(7):1392–404.e21.

199. Oshima H, Matsunaga A, Fujimura T, Tsukamoto T, Taketo MM, Oshima M. Carcinogenesis in mouse stomach by simultaneous activation of the Wnt signaling and prostaglandin E2 pathway. *YGAST.* 2006;131(4):1086–1095.

200. Stewart J, James J, McCluggage GW, et al. Analysis of wntless (WLS) expression in gastric, ovarian, and breast cancers reveals a strong association with HER2 overexpression. *Mod Pathol.* 2014;28(3):428–436.

201. Ju X, Ishikawa T-O, Naka K, Ito K, Ito Y, Oshima M. Context-dependent activation of Wnt signaling by tumor suppressor RUNX3 in gastric cancer cells. *Cancer Sci.* 2014;105 (4):418–424.

202. Ito K, Lim AC., Salto-Tellez M, et al. RUNX3 attenuates beta-catenin/T cell factors in intestinal tumorigenesis. *Cancer Cell.* 2008;14(3):226–237.

203. Ito K, Chuang LSH, Ito T, et al. Loss of Runx3 is a key event in inducing precancerous state of the stomach. *YGAST.* 2011;140(5). 1536–46.e1538.

204. Oshima H, Oshima M, Inaba K, Taketo MM. Hyperplastic gastric tumors induced by activated macrophages in COX-2/mPGES-1 transgenic mice. *EMBO J.* 2004;23 (7):1669–1678.

205. Li N, Liu S, Zhang H-S, et al. Exogenous R-spondin1 induces precocious telogen-to-anagen transition in mouse hair follicles. *IJMS.* 2016;17(4):582.

206. Gat U, DasGupta R, Degenstein L, Fuchs E. De Novo hair follicle morphogenesis and hair tumors in mice expressing a truncated beta-catenin in skin. *Cell.* 1998;95 (5):605–614.
207. Celso CL. Transient activation of β-catenin signalling in adult mouse epidermis is sufficient to induce new hair follicles but continuous activation is required to maintain hair follicle tumours. *Development.* 2004;131(8):1787–1799.
208. Van Mater D, Kolligs FT, Dlugosz AA, Fearon ER. Transient activation of beta -catenin signaling in cutaneous keratinocytes is sufficient to trigger the active growth phase of the hair cycle in mice. *Genes Dev.* 2003;17(10):1219–1224.
209. Andl T, Reddy ST, Gaddapara T, Millar SE. WNT signals are required for the initiation of hair follicle development. *Dev Cell.* 2002;2(5):643–653.
210. Huelsken J, Vogel R, Erdmann B, Cotsarelis G, Birchmeier W. Beta-catenin controls hair follicle morphogenesis and stem cell differentiation in the skin. *Cell.* 2001;105 (4):533–545.
211. Kwack MH, Kim MK, Kim JC, Sung YK. Dickkopf 1 promotes regression of hair follicles. *J Invest Dermatol.* 2012;132(6):1554–1560.
212. Castilho RM, Squarize CH, Chodosh LA, Williams BO, Gutkind JS. mTOR mediates Wnt-induced epidermal stem cell exhaustion and aging. *Stem Cell.* 2009;5(3):279–289.
213. Lien W-H, Polak L, Lin M, Lay K, Zheng D, Fuchs E. In vivo transcriptional governance of hair follicle stem cells by canonical Wnt regulators. *Nature.* 2014;16(2):1–15.
214. Reddy S, Andl T, Bagasra A, et al. Characterization of Wnt gene expression in developing and postnatal hair follicles and identification of Wnt5a as a target of Sonic hedgehog in hair follicle morphogenesis. *Mech Dev.* 2001;107(1–2):69–82.
215. Lei M, Guo H, Qiu W, et al. Modulating hair follicle size with Wnt10b/DKK1during hair regeneration. *Exp Dermatol.* 2014;23(6):407–413.
216. Stevens JR, Miranda-Carboni GA, Singer MA, Brugger SM, Lyons KM, Lane TF. Wnt10b deficiency results in age-dependent loss of bone mass and progressive reduction of mesenchymal progenitor cells. *J Bone Miner Res.* 2010;25(10):2138–2147.
217. Bohring A, Stamm T, Spaich C, et al. WNT10A mutations are a frequent cause of a broad spectrum of ectodermal dysplasias with sex-biased manifestation pattern in heterozygotes. *Am J Hum Genet.* 2009;85(1):97–105.
218. Hu B, Lefort K, Qiu W, et al. Control of hair follicle cell fate by underlying mesenchyme through a CSL-Wnt5a-FoxN1 regulatory axis. *Genes Dev.* 2010;24(14):1519–1532.
219. Li J, Baxter RM, Weiner L, Goetinck PF, Calautti E, Brissette JL. Foxn1 promotes keratinocyte differentiation by regulating the activity of protein kinase C. *Differentiation.* 2007;75(8):694–701.
220. Plikus MV, Mayer JA, la Cruz de D, et al. Cyclic dermal BMP signalling regulates stem cell activation during hair regeneration. *Nature.* 2008;451(7176):340–344.
221. Kobielak K, Pasolli HA, Alonso L, Polak L, Fuchs E. Defining BMP functions in the hair follicle by conditional ablation of BMP receptor IA. *J Cell Biol.* 2003;163 (3):609–623.
222. Kobielak K, Stokes N, la Cruz de J, Polak L, Fuchs E. Loss of a quiescent niche but not follicle stem cells in the absence of bone morphogenetic protein signaling. *Proc Natl Acad Sci USA.* 2007;104(24):10063–10068.
223. Andl T, Ahn K, Kairo A, et al. Epithelial Bmpr1a regulates differentiation and proliferation in postnatal hair follicles and is essential for tooth development. *Development.* 2004;131(10):2257–2268.
224. Yuhki M, Yamada M, Kawano M, et al. BMPR1A signaling is necessary for hair follicle cycling and hair shaft differentiation in mice. *Development.* 2004;131(8):1825–1833.
225. Jamora C, DasGupta R, Kocieniewski P, Fuchs E. Links between signal transduction, transcription and adhesion in epithelial bud development. *Nature.* 2003;422 (6929):317–322.

226. Estrach S, Ambler CA, Celso Lo CL, Hozumi K, Watt FM. Jagged 1 is a β-catenin target gene required for ectopic hair follicle formation in adult epidermis. *Development.* 2006;133(22):4427–4438.
227. Hwang J, Mehrani T, Millar SE, Morasso MI. Dlx3 is a crucial regulator of hair follicle differentiation and cycling. *Development.* 2008;135(18):3149–3159.
228. Balciunaite G, Keller M, Balciunaite E, Piali L, Zuklys S, Mathieu YD, Gill J, Boyd R, Sussman DJ, Holländer GA. Wnt glycoproteins regulate the expression of FoxN1, the gene defective in nude mice. *Nat Immunol.* 2002;3(11):1102–1108.
229. Merrill BJ, Gat U, DasGupta R, Fuchs E. Tcf3 and Lef1 regulate lineage differentiation of multipotent stem cells in skin. *Genes Dev.* 2001;15(13):1688–1705.
230. Niemann C, Owens DM, Schettina P, Watt FM. Dual role of inactivating Lef1 mutations in epidermis: tumor promotion and specification of tumor type. *Cancer Res.* 2007;67(7):2916–2921.
231. Parma P, Radi O, Vidal V, et al. R-spondin1 is essential in sex determination, skin differentiation and malignancy. *Nat Genet.* 2006;38(11):1304–1309.
232. Zhu AJ, Watt FM. β-Catenin signalling modulates proliferative potential of human epidermal keratinocytes independently of intercellular adhesion. *Development.* 1999;126(10):2285–2298.
233. Wang X, Reid Sutton V, Omar Peraza-Llanes J, et al. Mutations in X-linked PORCN, a putative regulator of Wnt signaling, cause focal dermal hypoplasia. *Nat Genet.* 2007;39 (7):836–838.
234. Grzeschik K-H, Bornholdt D, Oeffner F, et al. Deficiency of PORCN, a regulator of Wnt signaling, is associated with focal dermal hypoplasia. *Nat Genet.* 2007;39 (7):833–835.
235. Zhu X-J, Liu Y, Dai Z-M, et al. BMP-FGF signaling axis mediates Wnt-induced epidermal stratification in developing mammalian skin. *PLoS Genet.* 2014;10(10): e1004687.
236. Nguyen DX, Chiang AC, Zhang XHF, et al. WNT/TCF signaling through LEF1 and HOXB9 mediates lung adenocarcinoma metastasis. *Cell.* 2009;138(1):51–62.
237. Clayton E, Doup DP, Klein AM, Winton DJ, Simons BD, Jones PH. A single type of progenitor cell maintains normal epidermis. *Nature.* 2007;446(7132):185–189.
238. Doupe DP, Klein AM, Simons BD, Jones PH. The ordered architecture of murine ear epidermis is maintained by progenitor cells with random fate. *Dev Cell.* 2010;18 (2):317–323.
239. Habib SJ, Chen B-C, Tsai F-C, et al. A localized Wnt signal orients asymmetric stem cell division in vitro. *Science.* 2013;339(6126):1445–1448.
240. Snippert HJ, Haegebarth A, Kasper M, et al. Lgr6 marks stem cells in the hair follicle that generate all cell lineages of the skin. *Science.* 2010;327(5971):1385–1389.
241. Füllgrabe A, Joost S, Are A, et al. Stem Cell Reports. *Stem Cell Rep.* 2015;5 (5):843–855.
242. Liao X-H, Nguyen H. Epidermal expression of Lgr6 is dependent on nerve endings and Schwann cells. *Exp Dermatol.* 2014;23(3):195–198.
243. Guy GPJ, Machlin SR, Ekwueme DU, Yabroff KR. Prevalence and costs of skin cancer treatment in the U.S., 2002–2006 and 2007–2011. *Am J Prev Med.* 2015;48(2):183–187.
244. Youssef KK, Lapouge GL, Bouvr e K, et al. Adult interfollicular tumour-initiating cells are reprogrammed into an embryonic hair follicle progenitor-like fate during basal cell carcinoma initiation. *Nature.* 2012;14(12):1282–1294.
245. Yang SH, Andl T, Grachtchouk V, et al. Pathological responses to oncogenic Hedgehog signaling in skin are dependent on canonical Wnt/beta-catenin signaling. *Nat Genet.* 2008;40(9):1130–1135.
246. Salto-Tellez M, Peh BK, Ito K, et al. RUNX3 protein is overexpressed in human basal cell carcinomas. *Oncogene.* 2006;25(58):7646–7649.

247. Pálmer HG, Anjos-Afonso F, Carmeliet G, Takeda H, Watt FM. The vitamin D receptor is a Wnt effector that controls hair follicle differentiation and specifies tumor type in adult epidermis. *PLoS One.* 2008;3(1):e1483.
248. Niemann C, Owens DM, Hülsken J, Birchmeier W. Expression of ΔNLef1 in mouse epidermis results in differentiation of hair follicles into squamous epidermal cysts and formation of skin tumours. *Development.* 2002;129(1):95–109.
249. Takeda H, Lyle S, Lazar AJF, Zouboulis CC, Smyth I, Watt FM. Human sebaceous tumors harbor inactivating mutations in LEF1. *Nat Med.* 2006;12(4):395–397.
250. Adolphe C, Hetherington R, Ellis T, Wainwright B. Patched1 functions as a gatekeeper by promoting cell cycle progression. *Cancer Res.* 2006;66(4):2081–2088.
251. Wang GY, Wang J, Mancianti M-L, Epstein Jr EH. Basal cell carcinomas arise from hair follicle stem cells in Ptch1. *Cancer Cell.* 2011;19(1):114–124.
252. Peterson SC, Eberl M, Vagnozzi AN, et al. Basal cell carcinoma preferentially arises from stem cells within hair follicle and mechanosensory niches. *Stem Cell.* 2015;16 (4):400–412.
253. Malanchi I, Peinado H, Kassen D, et al. Cutaneous cancer stem cell maintenance is dependent on β-catenin signalling. *Nature.* 2008;452(7187):650–653.
254. Bhatia N, Spiegelman VS. Activation of Wnt/beta-catenin/Tcf signaling in mouse skin carcinogenesis. *Mol Carcinog.* 2005;42(4):213–221.
255. Kajino Y, Yamaguchi A, Hashimoto N, Matsuura A, Sato N, Kikuchi K. β-Catenin gene mutation in human hair follicle-related tumors. *Pathol Int.* 2001;51(7):543–548.
256. Kim YS, Shin DH, Choi JS, Kim K-H. The immunohistochemical patterns of the β-catenin expression in pilomatricoma. *Ann Dermatol.* 2010;22(3):284–289.
257. Jungermann K, Kietzmann T. Zonation of parenchymal and nonparenchymal metabolism in liver. *Annu Rev Nutr.* 1996;16:179–203.
258. Cadoret A, Ovejero C, Terris B, et al. New targets of beta-catenin signaling in the liver are involved in the glutamine metabolism. *Oncogene.* 2002;21(54):8293–8301.
259. Ovejero C, Cavard C, Perianin A, et al. Identification of the leukocyte cell-derived chemotaxin 2 as a direct target gene of beta-catenin in the liver. *Hepatology.* 2004;40 (1):167–176.
260. Benhamouche S, Decaens T, Godard C, et al. Apc tumor suppressor gene is the "Zonation-Keeper" of mouse liver. *Dev Cell.* 2006;10(6):759–770.
261. Yang J, Mowry LE, Nejak-Bowen KN, et al. Beta-catenin signaling in murine liver zonation and regeneration: a Wnt-Wnt situation. *Hepatology.* 2014;60(3):964–976.
262. Sekine S, Lan BY., Bedolli M, Feng S, Hebrok M. Liver-specific loss of beta-catenin blocks glutamine synthesis pathway activity and cytochrome p450 expression in mice. *Hepatology.* 2006;43(4):817–825.
263. Tan X, Behari J, Cieply B, Michalopoulos GK, Monga SPS. Conditional deletion of beta-catenin reveals its role in liver growth and regeneration. *YGAST.* 2006;131 (5):1561–1572.
264. Wickline ED, Awuah PK, Behari J, Ross M, Stolz DB, Monga SPS. Hepatocyte gamma-catenin compensates for conditionally deleted beta-catenin at adherens junctions. *J Hepatol.* 2011;55(6):1256–1262.
265. Sekine S, Gutierrez PJA, Lan BY., Feng S, Hebrok M. Liver-specific loss of beta-catenin results in delayed hepatocyte proliferation after partial hepatectomy. *Hepatology.* 2007;45(2):361–368.
266. Nejak-Bowen KN, Thompson MD, Singh S, et al. Accelerated liver regeneration and hepatocarcinogenesis in mice overexpressing serine-45 mutant beta-catenin. *Hepatology.* 2010;51(5):1603–1613.
267. Magami Y, Azuma T, Inokuchi H, et al. Cell proliferation and renewal of normal hepatocytes and bile duct cells in adult mouse liver. *Liver.* 2002;22(5):419–425.

268. Kinzel B, Pikiolek M, Orsini V, et al. Developmental Biology. *Dev Biol.* 2014;390 (2):181–190.
269. Oh SH, Hatch HM, Petersen BE. Hepatic oval "stem" cell in liver regeneration. *Semin Cell Dev Biol.* 2002;13(6):405–409.
270. Huch M, Gehart H, van Boxtel R, et al. Long-term culture of genome-stable bipotent stem cells from adult human liver. *Cell.* 2015;160(1–2):299–312.
271. Ding B-S, Nolan DJ, Butler JM, et al. Inductive angiocrine signals from sinusoidal endothelium are required for liver regeneration. *Nature.* 2010;468(7321):310–315.
272. Ferlay J, Soerjomataram I, Dikshit R, et al. Cancer incidence and mortality worldwide: Sources, methods and major patterns in GLOBOCAN 2012. *Int J Cancer.* 2014;136(5): E359–E386.
273. Totoki Y, Tatsuno K, Covington KR, et al. Trans-ancestry mutational landscape of hepatocellular carcinoma genomes. *Nat Genet.* 2014;46(12):1267–1273.
274. Kim J-Y, Lee H-Y, Park K-K, Choi Y-K, Nam J-S, Hong I-S. CWP232228 targets liver cancer stem cells through Wnt/β-catenin signaling: a novel therapeutic approach for liver cancer treatment. *Oncotarget.* 2016;7(15):20395–20409.
275. Boyault S, Rickman DS, de Reynies A, et al. Transcriptome classification of HCC is related to gene alterations and to new therapeutic targets. *Hepatology.* 2007;45(1):42–52.
276. Kan Z, Zheng H, Liu X, et al. Whole-genome sequencing identifies recurrent mutations in hepatocellular carcinoma. *Genome Res.* 2013;23(9):1422–1433.
277. Nhieu JT, Renard CA, Wei Y, Cherqui D, Zafrani ES, Buendia MA. Nuclear accumulation of mutated beta-catenin in hepatocellular carcinoma is associated with increased cell proliferation. *Am J Pathol.* 1999;155(3):703–710.
278. Nault J-C, Bioulac-Sage P, Zucman-Rossi J. Hepatocellular benign tumors-from molecular classification to personalized clinical care. *YGAST.* 2013;144(5):888–902.
279. Inagawa S, Itabashi M, Adachi S, et al. Expression and prognostic roles of beta-catenin in hepatocellular carcinoma: correlation with tumor progression and postoperative survival. *Clin Cancer Res.* 2002;8(2):450–456.
280. Kondo Y, Kanai Y, Sakamoto M, et al. Beta-catenin accumulation and mutation of exon 3 of the beta-catenin gene in hepatocellular carcinoma. *Jpn J Cancer Res.* 1999;90 (12):1301–1309.
281. Mao TL, Chu JS, Jeng YM, Lai PL, Hsu HC. Expression of mutant nuclear beta-catenin correlates with non-invasive hepatocellular carcinoma, absence of portal vein spread, and good prognosis. *J Pathol.* 2001;193(1):95–101.
282. Woo HG, Kim SS, Cho H, et al. Profiling of exome mutations associated with progression of HBV-related hepatocellular carcinoma. *PLoS One.* 2014;9(12):e115152.
283. La Coste de A, Romagnolo B, Billuart P, et al. Somatic mutations of the beta-catenin gene are frequent in mouse and human hepatocellular carcinomas. *Proc Natl Acad Sci USA.* 1998;95(15):8847–8851.
284. Cieply B, Zeng G, Proverbs-Singh T, Geller DA, Monga SPS. Unique phenotype of hepatocellular cancers with exon-3 mutations in beta-catenin gene. *Hepatology.* 2009;49(3):821–831.
285. Zucman-Rossi J, Benhamouche S, Godard C, et al. Differential effects of inactivated Axin1 and activated beta-catenin mutations in human hepatocellular carcinomas. *Oncogene.* 2007;26(5):774–780.
286. Taniguchi K, Roberts LR, Aderca IN, et al. Mutational spectrum of beta-catenin, AXIN1, and AXIN2 in hepatocellular carcinomas and hepatoblastomas. *Oncogene.* 2002;21(31):4863–4871.
287. Satoh S, Daigo Y, Furukawa Y, et al. AXIN1 mutations in hepatocellular carcinomas, and growth suppression in cancer cells by virus-mediated transfer of AXIN1. *Nat Genet.* 2000;24(3):245–250.

288. Kim M, Lee HC, Tsedensodnom O, et al. Functional interaction between Wnt3 and Frizzled-7 leads to activation of the Wnt/beta-catenin signaling pathway in hepatocellular carcinoma cells. *J Hepatol.* 2008;48(5):780–791.
289. Nambotin SB, Lefrançois L, Sainsily X, et al. Pharmacological inhibition of Frizzled-7 displays anti-tumor properties in hepatocellular carcinoma. *J Hepatol.* 2011;54 (2):288–299.
290. Merle P, la Monte de S, Kim M, et al. Functional consequences of frizzled-7 receptor overexpression in human hepatocellular carcinoma. *YGAST.* 2004;127(4):1110–1122.
291. Ban KC, Singh H, Krishnan R, Seow HF. GSK-3beta phosphorylation and alteration of beta-catenin in hepatocellular carcinoma. *Cancer Lett.* 2003;199(2):201–208.
292. Takagi H, Sasaki S, Suzuki H, et al. Frequent epigenetic inactivation of SFRP genes in hepatocellular carcinoma. *J Gastroenterol.* 2008;43(5):378–389.
293. Shih Y-L, Shyu R-Y, Hsieh C-B, et al. Promoter methylation of the secreted frizzled-related protein 1 gene SFRP1 is frequent in hepatocellular carcinoma. *Cancer.* 2006;107 (3):579–590.
294. Bengochea A, de Souza MM, Lefrancois L, et al. Common dysregulation of Wnt/ Frizzled receptor elements in human hepatocellular carcinoma. *Br J Cancer.* 2008;99 (1):143–150.
295. Tummala KS, Brandt M, Teijeiro A, et al. Hepatocellular carcinomas originate predominantly from hepatocytes and benign lesions from hepatic progenitor cells. *Cell Rep.* 2017;19(3):584–600.
296. Shin S, Wangensteen KJ, Teta-Bissett M, et al. Genetic lineage tracing analysis of the cell of origin of hepatotoxin-induced liver tumors in mice. *Hepatology.* 2016;64 (4):1163–1177.
297. Harada N, Miyoshi H, Murai N, et al. Lack of tumorigenesis in the mouse liver after adenovirus-mediated expression of a dominant stable mutant of beta-catenin. *Cancer Res.* 2002;62(7):1971–1977.
298. Tan X, Apte U, Micsenyi A, et al. Epidermal growth factor receptor: a novel target of the Wnt/beta-catenin pathway in liver. *YGAST.* 2005;129(1):285–302.
299. Bauer-Hofmann R, Klimek F, Buchmann A, Müller O, Bannasch P, Schwarz M. Role of mutations at codon 61 of the c-Ha-Ras gene during diethylnitrosamine-induced hepatocarcinogenesis in C3H/He mice. *Mol Carcinog.* 1992;6(1):60–67.
300. Harada N, Oshima H, Katoh M, Tamai Y, Oshima M, Taketo MM. Hepatocarcinogenesis in mice with beta-catenin and Ha-ras gene mutations. *Cancer Res.* 2004;64(1):48–54.
301. Miyoshi H, Deguchi A, Nakau M, et al. Hepatocellular carcinoma development induced by conditional β-catenin activation in Lkb1+/-mice. *Cancer Sci.* 2009;100 (11):2046–2053.
302. Dong B, Lee J-S, Park Y-Y, et al. Activating CAR and β-catenin induces uncontrolled liver growth and tumorigenesis. *Nat Commun.* 2015;6:5944.
303. Colnot S, Decaens T, Niwa-Kawakita M, Godard C, Hamard G, Kahn A, Giovannini M, Perret C. Liver-targeted disruption of Apc in mice activates β-catenin signaling and leads to hepatocellular carcinomas. *Proc Natl Acad Sci USA.* 2004;101 (49):17216–17221.
304. Ma S, Chan K-W, Hu L, et al. Identification and characterization of tumorigenic liver cancer stem/progenitor cells. *YGAST.* 2007;132(7):2542–2556.
305. Yamashita T, Ji J, Budhu A, et al. EpCAM-positive hepatocellular carcinoma cells are tumor-initiating cells with stem/progenitor cell features. *YGAST.* 2009;136 (3):1012–1024.
306. Yang W, Yan H-X, Chen L, et al. Wnt/beta-catenin signaling contributes to activation of normal and tumorigenic liver progenitor cells. *Cancer Research.* 2008;68 (11):4287–4295.

307. Lei Z-J, Wang J, Xiao H-L, et al. Lysine-specific demethylase 1 promotes the stemness and chemoresistance of Lgr5(+) liver cancer initiating cells by suppressing negative regulators of beta-catenin signaling. *Oncogene.* 2015;34(24):3188–3198.
308. Yamashita T, Budhu A, Forgues M, Wang XW. Activation of hepatic stem cell marker EpCAM by Wnt-beta-catenin signaling in hepatocellular carcinoma. *Cancer Res.* 2007;67(22):10831–10839.
309. Maetzel D, Denzel S, Mack B, et al. Nuclear signalling by tumour-associated antigen EpCAM. *Nat Cell Biol.* 2009;11(2):162–171.
310. Mani SKK, Zhang H, Diab A, et al. EpCAM-regulated intramembrane proteolysis induces a cancer stem cell-like gene signature in hepatitis B virus-infected hepatocytes. *J Hepatol.* 2016;65(5):888–898.
311. Khosla R, Rastogi A, Ramakrishna G, et al. EpCAM+ liver cancer stem-like cells exhibiting autocrine Wnt signaling potentially originate in cirrhotic patients. *Stem Cells Transl Med.* 2017;6(3):807–818.
312. Gedaly R, Galuppo R, Daily MF, et al. Targeting the Wnt/β-catenin signaling pathway in liver cancer stem cells and hepatocellular carcinoma cell lines with FH535. *PLoS One.* 2014;9(6):e99272.
313. Pai SG, Carneiro BA, Mota JM, et al. Wnt/beta-catenin pathway: modulating anticancer immune response. *J Hematol Oncol.* 2017;10(101):1–12.
314. Lu B, Green B, Farr J, Lopes F, Van Raay T. Wnt drug discovery: weaving through the screens, patents and clinical trials. *Cancers.* 2016;8(9):82.

CHAPTER THREE

Wnt Signaling in the Central Nervous System: New Insights in Health and Disease

Carolina A. Oliva*, Carla Montecinos-Oliva*,†, Nibaldo C. Inestrosa*,‡,§,1

* Center for Aging and Regeneration (CARE-UC), Pontifical Catholic University of Chile, Santiago, Chile
† Interdisciplinary Institute for Neuroscience (IINS), University of Bordeaux, Bordeaux, France
‡ Center for Healthy Brain Ageing, University of New South Wales, Sydney, NSW, Australia
§ Center of Excellence in Biomedicine of Magallanes (CEBIMA), University of Magallanes, Punta Arenas, Chile
[1] Corresponding author. E-mail address: ninestrosa@bio.puc.cl

Contents

Progress in Molecular Biology and Translational Science, Volume 153
ISSN 1877-1173
http://dx.doi.org/10.1016/bs.pmbts.2017.11.018

Abstract

Since its discovery, Wnt signaling has been shown to be one of the most crucial morphogens in development and during the maturation of central nervous system. Its action is relevant during the establishment and maintenance of synaptic structure and neuronal function. In this chapter, we will discuss the most recent evidence on these aspects, and we will explore the evidence that involves Wnt signaling on other less known functions, such as in adult neurogenesis, in the generation of oscillatory neural rhythms, and in adult behavior. The dysfunction of Wnt signaling at different levels will be also discussed, in particular in those aspects that have been found to be linked with several neurodegenerative diseases and neurological disorders. Finally, we will address the possibility of Wnt signaling manipulation to treat those pathophysiological aspects.

1. INTRODUCTION

Wnt was first discovered as an integration site for a mouse mammary tumor virus, and it was called protooncogene *integrase-1* (*int-1*). The *int-1* sequence was probed and found to be identical to its homolog in *Drosophila*, the called *Wingless* gene, one of the first genes described to be involved in segmentation and appendage patterning in the fly. Later, new techniques allowed the knock out of the so-called *Wingless/int-1* (*Wnt*) gene, establishing that it is a fundamental gene that participates in the correct formation of the neural tube and is involved in many steps during the development of several brain regions.[1,2] Since its discovery, Wnt and its downstream signaling has been ascribed to fundamental processes, including embryonic development, neurogenesis, synaptogenesis, and adult neural plasticity.[3,4] However, Wnt has been also implicated in several dysfunctions and diseases, such as different types of cancer,[5] in some metabolic disorders, such as diabetes mellitus, and also in neurodegenerative diseases,[6,7] which reflects its relevance in fundamental biological processes that may be highly conserved across species.[8] In most of these pathologies, Wnt signaling is unproperly regulated, revealing the existence of a narrow window where timing and expression level of Wnt components are critical for the correct Wnt activity. This chapter focuses on Wnt signaling in postnatal development and function of the central nervous system (CNS), with an emphasis in the recent evidence that relate Wnt with several neurological disorders.

The Wnt ligands are secreted glycoproteins and to date 19 members of the Wnt family have been described in mammals. Wnt ligands have been classified in two major groups, canonical ligands and noncanonical ligands, based on the intracellular signaling cascade that is activated.[9] However, cumulative evidence indicates that this classification is not a property of the Wnt ligand itself. Several other aspects must be considered to classify the pathway involved in signaling. One of the most important aspects is the number of receptors involved that allow an enormous possibility of interactions, which are difficult to predict. The typical Wnt receptor is Frizzled (Fz), a seven-transmembrane receptor that contains a cysteine-rich domain (CRD) which forms the Wnt-binding site.[10] Ten Fz receptors have been described in mice and humans, all of them are ortholog pairs and share ~94% identical amino acid.[11] They bind Wnt ligands with high affinity, resulting in very specific interactions.[12] In addition to Fz, other proteins act as coreceptors increasing the possibilities of interaction, such as the single transmembrane low-density lipoprotein receptor–related protein 5/6 (LRP5/6)[13–15] and the single-pass transmembrane receptor tyrosine kinases (RTKs) Ror1, Ror2, and Ryk.[9,16–19] The vast possibilities of interactions among ligands, receptors, and coreceptors, makes the cellular response difficult to predict, increasing the complexity of Wnt intracellular signaling activation. There are also several steps where signaling activity can be regulated. For example, the receptors can be phosphorylated from the intracellular side. The activation of Wnt signaling can also be intrinsically regulated by endogenous antagonists, such as by the endogenous secreted antagonists Cerberus[20] and Dickkopf-related protein-1 (Dkk1), which binds LRP5/6 to disrupt the interaction of Wnt/Fz;[14,21] by the inhibitor soluble Fz-related protein (sFRP) which directly binds Wnt because of its similarity to Fz[22,23] or by the Wnt inhibitory factor-1 (WIF-1).[24] Other molecules have been described as endogenous Wnt activators, such as Norrin and R-spondin.[25] Therefore, dysregulation at any of these levels can be crucial during the signal transduction process.

Once the Wnt/receptor/coreceptor binding is established, a specific sequence of intracellular downstream responses is elicited, which makes the cellular response complex and diverse.[26] The evidence indicates that the activation of a canonical or noncanonical pathway in a cell by a given Wnt ligand may depend on the context of how that Wnt couples to the receptor and coreceptor.[27,28] Multiple Wnt ligands can compete to bind to specific receptors.[28] However, the induction of the canonical or noncanonical pathway depends on the particular coreceptor activated.[28] The activation of a specific coreceptor by a Wnt ligand involves a common mechanism that requires the recruitment of the proteins disheveled (Dvl), Axin, a scaffold protein, and glycogen synthase kinase-3β (GSK-3β).[28] It is of consensus that the signaling through the LRP5/6 coreceptor triggers the canonical Wnt pathway, whereas signaling through the receptor tyrosine kinase-like orphan receptor 1 and 2 (Ror1/2) coreceptor triggers the noncanonical pathway. Therefore, the combinatorial binding of Wnt ligands to specific receptors and coreceptors can activate different intracellular signaling mechanisms, resulting in consequences at different levels. We will explore those possibilities in the text further.

1.1 The Canonical Wnt Pathways

1.1.1 *β-Catenin–Dependent Pathway*

The first Wnt signaling pathway to be identified was the "canonical pathway." This pathway is also called the Wnt/β-catenin pathway since the induction of some embryonic structures in *Xenopus*[1] and the transformation of mouse mammary cells[29] are both strongly correlated with an increase in β-catenin/T-cell factor (TCF) protein levels.[30] Evolutionary conservation of the Wnt-related genes is proof of its relevance; they are highly conserved in all animals described to date and together with BMP conforms the Cartesian coordinate system that determine the body axis patterning.[31] Activation of the Wnt/β-catenin pathway has been prominently involved in regulating cell differentiation and proliferation,[31–33] and controlling growth and cell fate specifications during development.[34] Since deregulation of β-catenin/TCF protein levels is a relevant feature in a variety of cancers,[35] this aspect of Wnt signaling is, currently, the most investigated. Although it is usually referred as the canonical Wnt/β-catenin pathway, β-catenin is a hub for many different others signaling processes besides Wnt signaling.

The activation of canonical pathway starts with the binding of Wnt ligand to its receptor. There are several Wnt ligands classified as canonical ligands (i.e., Wnt1, Wnt3a, Wnt7a/b, and Wnt8).[9] In steady-state or in the absence of Wnt stimulation, cytoplasmic levels of β-catenin are low because of sequential phosphorylation by the enzymes casein kinase-1α (CK-1α) and GSK-3β, and is targeted for ubiquitination and proteasome degradation.[4,36,37] Upon Wnt binding to both the receptor Fz and the coreceptor LRP5/6 to form a Wnt-receptor complex that recruits the protein Dvl to the membrane. This complex oligomerizes in the plasma membrane forming the signalosome (Fig. 1), which serves as a platform for the recruitment and allocation of the "β-catenin destruction complex."[38] The destruction complex is formed by

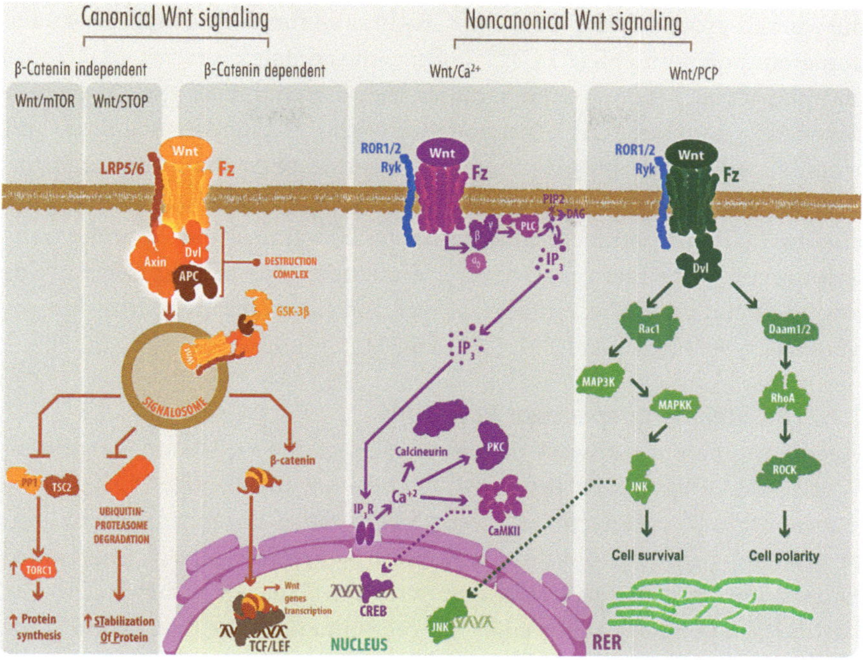

Fig. 1 Wnt signalling cascades. Two main classifications, canonical (GSK-3β dependent) and noncanonical (GSK-3β independent). In every case, Wnt binds to a Frizzled (Fz) receptor, from there, the intracellular cascades differ. In the canonical signalling, the Wnt-Frizzled-LRP5/6 complex is endocytosed into a signalosome and GSK-3β binds to the complex, like this disassembling the destruction complex formed by Axin, APC, CSK1, and GSK-3β (not shown for simplification). Inactivation of the destruction complex can lead to increased cytoplasmic β-catenin levels, followed by its translocation to the nucleus and increased transduction of Wnt target genes. This is the canonical Wnt-β-catenin–dependent pathway. But also, the Wnt canonical pathway can act on a β-catenin independent way, which diverges from the signalosome leading to the inhibition proteasome and *ST*abilization *O*f *P*roteins (Wnt/STOP). Also, can cause the inhibition of TCS2, which will lead to activation of TORC1 and an increase in protein synthesis (Wnt/mTOR). Noncanonical pathways are completely independent on GSK-3β action. The coreceptor can be either Ror1/2 or Ryk and it has not been studied whether it requires signalosome formation, or not. In the Wnt/Ca^{2+} signaling, Frizzled receptor activation leads to G-protein and PLC activation, which will cleave PIP_2 into DAG and IP_3, which will activate IP_3 receptors in the RER allowing Ca^{2+} release into the cytoplasm. The increase in Ca^{2+} concentration will activate different proteins, such as kinases CaMKII and PKC or the phosphatase calcineurin. CaMKII activation, into successive steps *(dashed line)*, will lead to the activation of the transcription factor CREB and consequent gene transcription. Finally, activation of the noncanonical Wnt/PCP pathway acting mainly on small GTPases, such as RhoA and Rac1 which will activate kinases, such as ROCK, MAPK, and JNK to cause cytoskeleton rearrangements and cell survival. Figure is not to scale and in order to simplify, not all molecules involved are shown.

the scaffold protein Axin, GSK-3β, CK-1α, and the tumor suppressor adenomatous polyposis coli (APC).[39,40] Once the complex is recruited, CK-1α phosphorylates LRP5/6 which causes the disassembly of the "β-catenin destruction complex." As a consequence of this, β-catenin is stabilized and accumulated in the cytoplasm and can translocate to the nucleus. In the nucleus, β-catenin will displace the Groucho transcriptional repressor and bind to TCF/lymphoid enhancer factor (LEF)[8] to finally activate the transcription of Wnt target genes. Several Wnt target genes are expressed in this process, including *c-Myc*, *cyclin D1*, *Axin2*, and *Ca^{2+}/calmodulin-dependent protein kinase type IV (CamKIV)*[4,33,41,42] (Fig. 1).

1.1.2 β-Catenin–Independent Pathways

Recently, different pathways that are parallel to the well-known canonical Wnt signaling have been described.[43] These are the Wnt/mTOR (mammalian target of rapamycin) and Wnt/STOP (Wnt-dependent stabilization of proteins) pathways. Like the canonical β-catenin pathway (mentioned earlier), the two new pathways start with the Wnt ligand binding to the Fz–LRP5/6 complex, followed by the recruitment of the destruction complex to the membrane, including the recruitment of GSK-3β (Fig. 1). From this point, two differences characterize this pathway from the traditional canonical pathway: the cascades are completely independent on β-catenin action. In the first case, Wnt-LRP5/6 is the signal for GSK-3β to be recruited. This fact avoids the phosphorylation of tuberous sclerosis protein 2 (TSC2, also known as Tuberin) mediated by GSK-3β, a negative regulator of mTOR complex 1 (TORC1), upregulating mTOR signaling. The final result is the upregulation of protein translation and modulation of cytoskeleton dynamics. As a consequence, the mTOR pathway activated by Wnt-LRP5/6, increases the synthesis of proteins involved in cell cycle, such as cyclin D1, coupling cell growth, and the progression into cell cycle[44] (Fig. 1).

In the second case, the newly described Wnt/STOP pathway works in a different way. GSK-3β can phosphorylate several other proteins, besides β-catenin. When the activation of Wnt-LRP5/6 recruits GSK-3β, this avoids several proteins to be ubiquitinated and be send for degradation, which stabilize them prolonging its half-life. This fact gives the Wnt/STOP pathway its name. This signaling is dominant during the mitotic processes, and stabilizes proteins to preserve their integrity during cell division[45] (Fig. 1).

Including the signal through Wnt/β-catenin, both new signaling pathways (Wnt/mTOR and Wnt/STOP) work synchronized with the cell growth program. These Wnt-related pathways increase gene expression through the transcriptional activity of β-catenin, increase protein synthesis through Wnt/mTOR, and reduce protein degradation through Wnt/STOP. These pathways have been described for dividing cells; however, because neurons are postmitotic cells, there is no evidence that these pathways are functioning in the same manner in neurons. It is well known that under standard conditions, mTOR is downstream of any signaling acting through the phosphatidylinositol-4,5-biphosphate 3-kinase (PI3K)/protein kinase B (Akt) pathway, and is responsible for initiating the translation of synaptic proteins in spines and dendrites, especially during plasticity phenomena.[46,47] However, whether mTOR signaling through PI3K/Akt is coupled to Wnt-LRP5/6 remains to be elucidated. In the case of Wnt/STOP signaling, less information is available regarding activity in neurons, but we can predict that it would be relevant in severe neuronal injury, hypoxia, and stem cell proliferation or differentiation, where the protein content must be preserved. Both issues remain to be elucidated and open a new possibility for research in the field of neurobiology.

1.2 The Noncanonical Wnt Pathways

The noncanonical Wnt pathways unite different intracellular signaling cascades that are triggered by a Wnt ligand binding to a Fz receptor and a growing variety of coreceptors. There are several Wnt ligands classified as noncanonical (i.e., Wnt4, Wnt5a, Wnt11).[9] One of the main differences between the canonical and noncanonical Wnt pathways is that noncanonical Wnt signaling does not include GSK-3β or β-catenin as intermediary molecules. The noncanonical pathways are subdivided in two, the Wnt/Planar Cell Polarity (PCP) pathway, also known as Wnt/c-Jun N-terminal Kinase (JNK), and the Wnt/Ca^{2+} pathway.

1.2.1 The Wnt/PCP Pathway

The Wnt/PCP signaling pathway was the first noncanonical Wnt pathway identified initially in *Drosophila*.[48,49] In this pathway, the binding of Wnt to the Fz receptor on the membrane surface is followed by the activation of Rho and Rac small GTPases which in turn stimulate Rho-associated protein kinase (ROCK) and JNK, respectively (Fig. 1). The downstream effect of this pathway is the regulation of the cytoskeletal organization, cell motility, and gene expression by JNK-dependent transcription factors, for example, activating transcription factor 2 (ATF2), with concomitant activation of its target genes.[50] In mammals, Wnt/PCP regulates the development of left–right asymmetry in a coordinate manner.[51,52] Importantly, deregulation of the PCP pathway leads to failed convergent extension, which has been related to incomplete neural tube closure and spina bifida.[50]

If we consider all the types and shapes cells can encompass, neurons are the most polarized cells in any animal. Pre- and postsynaptic regions within buttons and dendritic spines, such as axon terminals, and somatodendritic arborization, respectively, are neuronal features that need to be carefully developed and maintained. Wnt/JNK signaling is fundamental in the development and maintenance of this functional polarization.[53–55]

1.2.2 The Wnt/Ca^{2+} Pathway

In neurons, the calcium ion (Ca^{2+}) has been ascribed to several functions. Among them, it exerts a fast synaptic response in a controlled spatiotemporal manner.[56] Postsynaptic increases in Ca^{2+} have been directly linked to increased potentiation in hippocampal neurons[57,58] and cortical inhibitory neurons.[59] Ca^{2+} is fundamental for synaptic function but can also cause unwanted effects, such as synaptotoxicity. Therefore, any pathway that regulates intracellular Ca^{2+} concentration is crucial in neurons.

The Wnt/Ca^{2+} pathway is mostly a G-protein–dependent signaling pathway[60,61] Receptor–ligand interaction leads to activation of phospholipase C (PLC) in the plasma membrane, which cleaves PIP_2 with a consequent increase in intracellular inositol-1,4,5-thriphosphate [Ins (1,4,5) P_3] (IP_3) and diacylglycerol (DAG). IP_3 then acts as a second messenger and binds to IP_3 receptors (IP_3R) in the rugose endoplasmic reticulum (RER), causing diffusion of Ca^{2+} from the lumen of the RER to the cytoplasm, thereby increasing intracellular Ca^{2+} concentration and activating CaMKII, the main downstream effector. DAG acts in parallel and also favors Ca^{2+} release from the RER and PKC activation. Then, both CaMKII and PKC can activate different transcription factors, such as cAMP response element-binding portein-1 (CREB) and nuclear factor kappa-light chain-enhancer of activate B cells (NFκB), allowing the transcription of several genes that are classified as Wnt-target genes, many of which have a significant relevance in the CNS.[62] Once activated by Ca^{2+}, CaMKII activation persists long after Ca^{2+} transients end, for this reason CaMKII is quintessential in long term potentiation (LTP) (Fig. 1).

1.3 Cross-Talk Between Wnt Pathways

The canonical and noncanonical Wnt pathways described can cross-talk and negatively regulate each other. For instance, inducible overexpression of

Wnt5a, a commonly known noncanonical ligand, causes an increase in Wnt/β-catenin signaling activation in the forming meninges.[63] Interestingly, on the skin of the same mice, overexpression of Wnt5a caused inhibition of Wnt β-catenin signaling.[63] Additionally, it has been reported that Wnt5a activates Wnt noncanonical pathways and in parallel, inhibits the canonical Wnt/β-catenin pathway in HeLaS3 cells by inhibiting phosphorylation of the LRP6 coreceptor.[64] This further shows that the final effect of one specific Wnt ligand is going to be determined by the cellular environment in which it acts. Additionally, Wnt5a acting noncanonically can genetically interact with Wnt1 acting canonically to control different stages in the neurogenesis of dopaminergic neurons, both in an antagonist or cooperative manner.[65] Even more, authors showed that it is possible to replace Wnt1 by Wnt3a and have similar effects,[65] demonstrating the existence, in vivo, of cross-talk between canonical and noncanonical pathways.

Of note, it has been shown that in the same cell, Wnt3a and Wnt5a ligands can trigger phosphorylation of completely different sets of intracellular receptors.[28] Strikingly, Wnt ligands can compete for the Fz receptor at the cell surface and cause inhibition of the reciprocal pathway,[28] therefore providing another control point to determine the specificity of signaling. Regarding to Wnt/Ca^{2+} signaling, several ligands that are considered to be "canonical" can mobilize Ca^{2+}, such as Wnt3a, Wnt4, Wnt7a, Wnt9b, Wnt10b, and Wnt11.[66,67] Strikingly, evidence from cell lines, demonstrated that Wnt5a not only has the highest response to Ca^{2+} but also translocate β-catenin into the nucleus.[67] What has been proposed is that, an initial Ca^{2+} wave caused by Wnt ligands facilitate depolarization of the nuclear envelope and in this way allows β-catenin to enter de nucleus.[67] This evidence accounts for the high variability of responses of Wnt ligands and how they can facilitate the action of other Wnt ligands.

2. WNT EXPRESSION IN THE POSTNATAL AND ADULT CNS

In rodents, the presence of Wnt ligands and other components of Wnt signaling have been described in major subdivisions of the cerebral cortex, in the olfactory bulb (OB) and OB related areas, in the hippocampus, neocortex, and thalamus.[68–71] The expression of different Wnt proteins is particularly high in those brain areas where the neurons are undergoing continuous renovation, such as in the OB and dentate gyrus (DG) zone of the

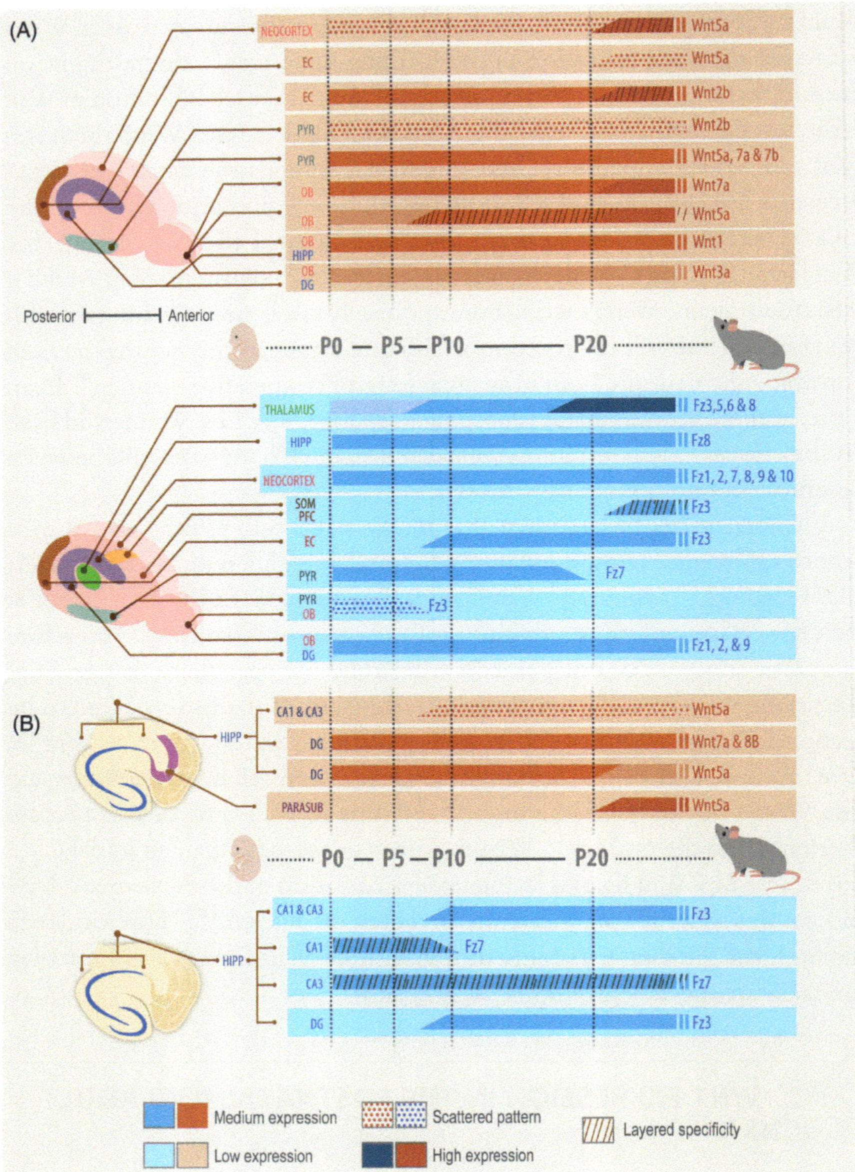

Fig. 2 Postnatal gene expression of Wnt ligands and Fz receptors. Diagram represents the postnatal gene expression of Wnt and Fz, from P0 up to P20. Most of the data represented here are based on the study of Shimogori *et al.* and other related published papers.[68] (A) *Upper panel*, expression of major class of Wnt ligands. Wnt5a is highly expressed along the brain, with either full or layer specificity or scattered pattern in the major brain subdivisions, such as OB, Pyr, and EC. Other like Wnt2b is mainly present in

hippocampus,[68] both of which are regions where adult neurogenesis has been identified (Fig. 2). Wnt protein expression is also prevalent during synaptogenesis and dendrite development in the hippocampus.[69] Interestingly, in areas, such as the OB and the cerebellum, the presence of Wnt proteins, such as Wnt7a and Wnt5a are detected in the postsynaptic cells and in other target cells, while Fz receptors and endogenous inhibitors are detected in the presynaptic cells. It is believed that in this way, Wnt ligand allocated in the postsynaptic or "recipient" cell would act as a synaptogenic factor, transforming the axonal growth cone into a presynaptic structure with increased levels of synaptic proteins that is ready to establish physical contact. This could be especially relevant in structures that are continuously being renovated through life. It is noteworthy that Fz receptors are also present on the target cells, which indicates that Wnt proteins are also exerting an autocrine effect as well.

As mentioned earlier, in the hippocampus Wnt signaling appears to be robust in structures where neurogenesis, axonal growth, and synaptogenesis occur. Specific Wnts, Fz receptors, and sFRPs are expressed in the lining of the DG and in the synapses between DG mossy fibers and hippocampal CA3 region neurons. A scattered presence of Wnt5a remains during adulthood and decreases in the DG, while Wnt7 remains present in the DG and CA3. Notably, Wnt5a is highly expressed at postnatal day 20 (P20) in the pre-subiculum and parasubiculum, but is scattered in the entorhinal cortex (EC); while Wnt2, which has a diminished expression in hippocampus, is highly expressed in the EC (Fig. 2A and B).

◀ archicortex and paleocortex, full and with layer specificity in EC and scattered in Pyr, while Wnt7a is fully expressed in olfactory related areas (OB and Pyr). Wnt1 is fully expressed in OB and hippocampus, while Wnt3a expression is in neurogenic-related areas (OB and DG). *Lower panel*, expression of the major class of Fz receptors. Thalamus expresses several Fr, including Fz3, 5, 6, and 8, all of them with an increased expression by time, starting at P0. Fz3 is also expressed in other cortices and olfactory related areas are time- and pattern-specific. Fz8 is also expressed in hippocampus and fully expressed in neocortex. Fz7 is also fully expressed in Pyr but decaying by P20 and after. (B) *Upper panel*, specific expression pattern of Wnt in subregions of hippocampus. Wnt5a, the most abundant, present a nonhomogenous temporal expression, starting at DG, to CA3 and CA1, and finishing at parasubiculum. Wnt7a and Wnt8b are fully expressed in DG. *Lower panel*, specific expression pattern of Fz in subregions of hippocampus. Fz3 has the same expression pattern starting at P7 in DG, CA3 and CA1. Instead, Fz7 starts from P0 in layered pattern in CA3 and CA1. *EC*, entorhinal cortex; *DG*, dentate gyrus; *Hipp*, hippocampus; *OB*, olfactory bulb; *Parasub*, parasubiculum; *PFC*, prefrontal cortex; *Pyr*, pyriform cortex; SOM, somatosensory cortex.

Wnt signaling components are prevalent in the thalamus during postnatal life.[68,72] The thalamus is the relay station for most of the sensory information to the cortex. The main period of thalamic maturation occurs postnatally, when inputs coming from the sensory organs are being established. Recently, it has been demonstrated that β-catenin controls thalamic gene expression,[73] specifically for genes that encode voltage-gated ion channels, neurotransmitter receptors, synaptic vesicle proteins, and synaptic structural proteins, suggesting that canonical Wnt signaling can regulate neuronal excitability in the thalamus. Ca^{2+}, potassium ion (K^{+}), and chloride ion (Cl^{-}) conductances, most of which directly propagate, can inhibit or modify electric signals. Thus, by regulating the expression of conductance-related expression genes, Wnt/β-catenin signaling can determine the degree of excitability in thalamic neurons and modulate thalamocortical circuit activity.[72,73] This means that Wnt signaling might be involved in thalamocortical connectivity derived from the emergent sensory information, based on experience, which is continuously processed during the lifespan of an animal. The expression of Wnt ligands and receptors in brain areas, including those that are actively participating in neurogenesis or implicated in sensory processing, suggest that Wnt pathways are involved not only in structural functions but also in higher superior cognitive processes.

3. WNT FUNCTION IN THE DEVELOPING AND MATURE CNS

In this and the following sections, we will describe the recent advances in deciphering the role of Wnt signaling in mature neurons. We include data from mature in vitro neurons [12 days in vitro (DIV) or more] and from ex vivo and in vivo analyses. Unless clearly stated, the accumulated data referenced here were performed in murine models. Table 1 shows a complete and comprehensive list of the reported effects for different Wnt ligands.

3.1 Role of Wnt Signaling in the Presynaptic Terminal

Wnt ligands and other neurotrophic factors acting as guidance cues [i.e., brain-derived neurotrophic factor (BDNF)] are crucial for the establishment of neuronal polarity[108,109] and synaptic connections[110,111] during the development and maturation of neurons, including processes, such as branching, axonal outgrowth, pruning, etc. Research on this subject is comprehensive,

Table 1 Reported Roles of Wnt Ligands in Mature Neurons or Adult Central Nervous System (CNS).

Ligand	Effects	Cell Type/Tissue/Model	Pathway	References
Wnt1				
	Favors neurogenesis	Adult midbrain SC (in vivo)/progenitor neurons (in vitro)	Canonical (β-catenin)	74,75
	Protects against oxidative stress	Hippocampal neurons (in vitro)/rats (in vivo)	Canonical (β-catenin-Akt)	76
	Protects against neurodegeneration	Dopaminergic neurons/substantia nigra	Canonical (β-catenin)	77
	Necessary for neurogenesis	Midbrain dopaminergic neurons (in vitro)	Canonical (β-catenin)	65
Wnt2				
	Mediate antidepressant effect	Ventral hippocampus	Canonical (β-catenin)	78,79
	Dendritogenesis and spinogenesis	Cortical neurons (in vitro)	nd.	80
Wnt3				
	Mediates antidepressant effect	Ventral hippocampus	Canonical (β-catenin)	78
	Promotes neurogenesis	Dentate gyrus	Canonical (β-catenin)	81,82
Wnt3a				
	Increases basoon clusters	Hippocampus (in vivo)/hippocampal neurons (in vitro)	Canonical (β-catenin)	71,83
	Increases vGlut1 clusters	Hippocampal neurons (in vitro)	Canonical (β-catenin)	83
	Increases synaptic vesicle release	Hippocampal neurons (in vitro)	Canonical (β-catenin)	66,83
	Regulates oscillatory activity	mECx-CA1-CA3 (ex vivo)	nd.	84
	Favors neurogenesis	Midbrain dopaminergic neurons (in vitro)	nd.	65
	Induces exosome secretion	Microglia (in vitro)	Noncanonical (nd.)	85
	Counteracts proinflammatory lipopolysaccharides	Microglia (in vitro)	nd.	86

(Continued)

Table 1 Reported Roles of Wnt Ligands in Mature Neurons or Adult Central Nervous System (CNS).—cont'd.

Ligand	Effects	Cell Type/Tissue/Model	Pathway	References
	Regulates inhibition of mPTP opening	Isolated mitochondria from hippocampal slices (in vitro)	Canonical (β-catenin) mitochondrial GSK-3β	86,87
Wnt5a				
	Increases clustering of PSD95	Hippocampal neurons (in vitro)	Noncanonical (JNK)	88
	Increases clustering of SynGAP	Hippocampal neurons (in vitro)	Noncanonical (nd.)	89
	Stimulates synaptic differentiation	Hippocampal neurons (in vitro)	Noncanonical (CaMKII)	90
	Increases AMPA and NMDA currents	Hippocampal neurons (in vitro)	Noncanonical (CaMKII)	90–92
	Increases fEPSP transmission	Hippocampal neurons (in vitro)	Noncanonical (CaMKII)	90,92
	Protects against Aβ oligomers synaptoxicity	Hippocampal neurons (in vitro)	Noncanonical (nd.)	92
	Modulates NO production	Hippocampal neurons (in vitro)	Noncanonical (Ror2)	91,93
	Increases dendritic spine density	Hippocampal neurons (in vitro)	Noncanonical (Ca^{2+}, nd.)	90,94
	Favors dendritic maintenance/ plasticity	Hippocampal neurons (in vitro)	Noncanonical (CaMKII)	95
	Recycling of functional $GABA_A$ receptors	Hippocampal neurons (in vitro)	Noncanonical (CaMKII)	96
	Long-term information storage	Hippocampus (in vivo)	nd.	97
	Prevents memory impairment	Rats (in vivo)	nd.	98
	Regulates oscillatory activity	mECx-CA1-CA3 (ex vivo)	nd.	84
	Favors neurogenesis	Midbrain dopaminergic neurons (in vitro)	nd.	65
	Differentiation, axonal growth, and repulsion	Midbrain dopaminergic neurons (in vitro)	nd.	99
Wnt7a				
	Promotes dendritic spine growth	Hippocampal neurons (in vitro)/ hippocampus (in vivo)	Non-canonical (CaMKII)	100

	Induces synapse formation	Hippocampal neurons (in vitro)	Canonical (β-catenin-cyclin D1)	100,101
	Increases clustering of PSD95	Hippocampal neurons (in vitro)	Canonical (β-catenin-NGN2)	100,102
	Vesicle cycle	Hippocampal neurons (in vitro)	Noncanonical (CaMKII)	103
	Modulates α7-nAChR expression/localization	Hippocampal neurons (in vitro)	Noncanonical (CaMKII)	103
	Increases clustering of vGlut	Hippocampal neurons (in vitro)	Noncanonical (CaMKII)	100
	Increases clustering of Synaptophysin	Hippocampal neurons (in vitro)	nd.	97
	Long-term information storage	Hippocampus (in vivo)	nd.	97
	Induces clustering of bassoon (Wnt7a/b)	Hippocampus (in vivo)	nd.	71
	Modulates excitatory neurotransmitter release	Hippocampus (ex vivo)	Canonical (β-catenin)	104
	Increases clustering of synapsin 1	Cerebellar neurons	nd.	105
	Induces stem cell proliferation	Ventral midbrain and dentate gyrus	Canonical (β-catenin)	102,106
	Induces neural stem cell maturation	Dentate gyrus	Canonical (β-catenin)	102
Wnt7b				
	Promotes dendritic growth	Hippocampal neurons	Noncanonical (Rac-JNK)	54
	Induces clustering of VAMP2	Mossy fibers	nd.	107
	Induces clustering of Bassoon	Mossy fibers	nd.	107

Akt, Ak strain transforming; CA1, cornu ammonis 1; CA3, cornu ammonis 3; CaMKII, calcium calmodulin dependent kinase II; JNK, c-Jun N-terminal kinase; mECx, medial entorhinal cortex; mPTP, mitochondrial permeability transition pore; α7-nAChR, α7-nicotinic acetylcholine receptor; nd., nondetermined, authors did not study the specific signaling pathway by which it acts; NGN2, neurogenin 2; PSD95, postsynaptic density 95; SC, stem cells; VAMP2, Rac: Ras-related C3 botulinum toxin substrate 1; vGlut1, vesicular glutamate transporter 1.

Those ligands that are not mentioned have not been found referenced for studies in mature neurons or CNS, in the literature. Wnt3 refers to work in which no particular isoform of the ligand (a or b was used).

but the role of Wnt signaling on the presynaptic sites of mature neurons and mature synaptic connections has not been extensively studied.

First, it is necessary to clarify that in presynaptic terminals, the complete machinery for Wnt signaling is present. Ligands and Fz receptors can be found on either side of central synapses. Fz5 is found in hippocampal neurons in vitro and in the hippocampus in vivo, and Fz5 expression levels are increased in mature neurons and toward adulthood.[101,107] Fz5 colocalizes with pre- and postsynaptic proteins. In fact, gain-of-function experiments of Fz5 causes an increase in presynaptic sites and in synaptic density contacts in hippocampal neurons at 9–11 DIV, when the neurons were not completely mature.[101] Instead, Fz1 is found exclusively at presynaptic sites in mature hippocampal neurons (14 DIV) and its expression increases from birth to adulthood, as detected in brain lysates and in hippocampal neurons.[83] Gain-of-function experiments performed to overexpress Fz1, causes an increase in the clustering of the presynaptic protein Bassoon. Therefore, a ligand like Wnt3a binds to the Fz1 receptor in order to act in presynaptic sites.[83] Other ligands can also bind to presynaptic Fz receptors. Acting through canonical Wnt signaling, Wnt7a disassembles the β-catenin destruction complex in presynaptic terminals,[103,112] also releasing APC from the complex.[103] Free APC translocate to the membrane and recruits presynaptic terminal proteins, one of which is α7-nicotinic acetylcholine receptor (α7-nAChR) which gets stabilized in the plasma membrane.[103] There, α7-nAChRs act as modulators of presynaptic activity. Later, it was shown that effectively, Wnt7a stimulates synaptic vesicle cycle inducing synaptic release and recycling of vesicles.[112] Wnt7a increases miniature excitatory postsynaptic currents (mEPSCs)[100,112] and reduces paired-pulse facilitation. It was later proposed that Wnt7a exerted these effects through coupling and activation of Fz5 receptor,[101] and it requires Dvl to regulate the normal recycling rates.[107,113] Indeed, double null mutant Wnt7a/Dvl significant reduces mEPSCs.[107] Wnt7a also increases clustering of presynaptic proteins, such as synaptophysin, synaptotagmin, and SV-2.[112] In fact, in Wnt7a mutant neurons there is a delay in synaptic maturation,[113] confirming the relevance of Wnt7a in presynaptic assembly and function. The isoform Wnt7b also has a role on regulating presynaptic architecture, inducing clustering of VAMP2 and Bassoon,[107] the last one being a common feature for Wnt7a/b.

Wnt3a also induces an increment in vesicle recycling.[112] Wnt3a increases the frequency of synaptic vesicle release, measured by the increment in the frequency of mEPSCs and by the unloading of FM1-43.[66] These effects are dependent on rapid presynaptic Ca^{2+} influx, causing the increase in excitatory synaptic transmission, independently of postsynaptic rearrangements,[66,112] demonstrating that Wnt3a preferentially acts on presynaptic sites.

In vivo studies suggest that experience also modifies synaptic architecture in the adult mouse hippocampal stratum lucidum.[71] It was shown that after exposure to an enriched-environment, there was an increase in endogenous Wnt7a/b mRNA and protein levels in CA3 pyramidal neurons. By applying recombinant Wnt ligands, it was clear that Wnt3a and Wnt7a but not Wnt5a caused a significant increase in presynaptic Bassoon puncta, with Wnt7a having a much greater effect than Wnt3a.[71,83] This translates into an increased complexity of mossy fiber terminals and increase synapse numbers. Local application of sFRP1 to the CA3 reverted this effect, thus demonstrating that Wnt signaling is both necessary and sufficient to regulate connectivity in adult hippocampus.[71] It was also found that Wnt3a increases vesicular glutamate transporter 1 (vGlut1) clustering[83] in cultured hippocampal neurons. Interestingly, it seems that at a presynaptic level, the actions of Wnt3a and Wnt7a overlap, but Wnt7a plays relevant roles both at pre- and postsynaptic sites, that will be explained in the next section. Altogether these studies confirm a role of Wnt signaling, mainly canonical Wnt signaling, in the maintenance and plasticity of synaptic structures into the presynaptic terminal. For a detailed summary on the described roles of Wnt signaling in presynaptic architecture and activity, see Table 1.

3.2 Role of Wnt Signaling in the Postsynaptic Region

As mentioned earlier, the role of Wnt ligands on synaptic functioning has been demonstrated in several studies (63, 73, and 74). Wnt signaling orchestrates the synaptic architecture of postsynaptic density proteins. For instance, in cultured hippocampal neurons Wnt5a increases the amount of postsynaptic density protein 95 (PSD95), a scaffolding protein, by clustering soluble PSD95 into new dendritic spines. Similarly, Wnt5a increases the size of synaptic clusters Ras GTPase activating protein 1 (SynGAP), another scaffolding protein, in hippocampal neurons.[88,89] Interestingly, the short-time scale in which this occurs, 15–30 min after Wnt5a treatment, suggests that these changes involves a reaccommodation of existing proteins and do not dependent on gene expression. The Wnt5a-dependent increased number of PSD95 clusters depends on JNK activity, not CaMKII.[88] Noteworthy, Wnt7a also increases the density of PSD95 clusters, but after 3 h of treatment.[100] The clustering of PSD95 correlates with the increase of synaptophysin (presynaptic protein) cluster density, indicating that new excitatory synaptic connections have been established, while there was no change in inhibitory synapses.[100] Interestingly, this effect of Wnt7a is associated to the activity of CaMKII (Wnt/Ca^{2+} pathway) and not through the expected canonical β-catenin–dependent pathway, as shown before.[103,112] This emphasized the fact that Wnt actions are cell-context dependent.

In direct relation with the remodeling capacity of postsynaptic proteins, Wnt ligands modulate synaptic architecture, dendritogenesis, spinogenesis, and synaptogenesis. Several studies have revealed a role for Wnt ligands in arborization and dendritic spine formation on excitatory neurons. In particular, Wnt2 positively modulates dendrite growth and dendritic spine formation in cortical neurons through a BDNF-mediated mechanism.[80] Interestingly, BDNF in a positive feedback loop, causes an increase in Wnt2 mRNA expression with no need for increased neuronal activity; therefore this would be a basal mechanism for dendrite- and spinogenesis.[80]

Another Wnt ligand that has a role in dendrito- and spinogenesis is Wnt5a. This is the most abundant ligand in the hippocampus, reason why most of the research on Wnt5a has been accomplished in hippocampal neurons (Fig. 2). Wnt5a causes a rapid increase in dendritic spine density[90,94] and more recently has been found to be essential in the dendritic maintenance of hippocampal neurons.[95] On the other hand, it has been found that the Wnt coreceptor Ryk (component of the noncanonical pathway, Fig. 1) is a negative regulator of dendritic arborization. In vitro overexpression of Ryk causes a decrease in dendritogenesis.[114] Accordingly, in vivo experiments showed that Ryk haploinsufficient ($Ryk^{+/-}$) animals have excessive dendritic growth and augmented complexity in somatosensory cortex.[114] Both evidences suggest that Wnt signaling, through Ryk activity, acts by restricting dendritic arborization.[114] Despite the specific ligand involved in this action is unknown, the authors suggest that Wnt5a may be mediating the response.[114] This would be an important mechanism to control synaptic plasticity. Additionally, Wnt3a also interacts with Ryk to induce neurite outgrowth in explants of dorsal root ganglia.[18] The extracellular domain of Ryk physically interacts with Fz and Wnt3a, while its intracellular domain interacts with Dvl.[18] Therefore, Ryk constitutes a link between Wnt and Dvl, and is fundamental for the neurite outgrowth and axonal guidance induced by Wnt3a.[18] Wnt7a also regulates spine morphogenesis and growth through another mechanism that involves the rapid activation of CaMKII in spines.[100] Overall the data is robust to support a positive role of Wnt in the formation, maintenance, and functioning of synaptic activity.

The evidence presented support the role of Wnt ligands in postsynaptic density and dendritic spine remodeling, and in dendritogenesis and synaptogenesis. Electrophysiological studies are fundamental to determine the functional implications and relevance of the changes mentioned earlier. In Section 4, we will discuss the evidence in support of the idea that Wnt participates in basal neuronal activities as well as in activity-dependent changes.

3.3 Adult Neurogenesis in the Hippocampus is Modulated by Wnt Signaling

There is a plethora of cues and signaling pathways that regulate the maintenance of stem cells and their differentiation (i.e., Wnt, Notch, sonic hedgehog, etc.). The balance between undifferentiated and differentiated cells is fundamental in neurogenesis. The role of Wnt signaling in development and neurogenesis in embryonic stages was early elucidated in the field.[115] However, its role during adulthood came only recently. It is important to clarify that the actual role and importance of newly generated neurons and their capacity to integrate into the existing circuitry of the brain, are still a matter of debate. Here, and when discussing the role of Wnt signaling in neurological and neuropsychiatric disorders (Section 5), we will address evidence that supports a functional role for adult neurogenesis.

As shown in Table 1, different Wnt ligands play a role in adult neurogenesis. Two brain areas are main niches for adult neurogenesis: the subventricular zone of the lateral ventricle and the subgranular zone (SGZ) of the DG in the hippocampus. Wnt3 and other Wnt signaling components (i.e., Fz 1, Dvl 2, GSK-3β, and β-catenin) are expressed in the hippocampal formation,[68] surrounding the SGZ, particularly in astrocytes.[81,82] Moreover, Wnt3 expression in hippocampal astrocytes declines during aging which correlates with a decrease in the neurogenic rate,[82] suggesting a key role of Wnt components in this neurogenic niche. Notably, overexpression of Wnt3 caused an increase in adult hippocampal progenitors (AHPs) conversion into neurons both in vitro and in vivo in 15-week-old rats.[81] Moreover, lentiviral injection in rat DG of a dominant negative (dn) of Wnt1 protein, a protein from a mutated gene that causes antagonist effects, caused a marked (eight-fold) decrease in neurogenesis in 8–9-week-old rats.[81,116] Wnt3a increased the expression levels of neurogenic differentiation 1 (Neurod1) through a β-catenin–dependent pathway during early neurogenesis in the DG of mice.[117] In addition, the same animals were tested for a spatial memory tests, showing significant differences for long-term retention but not for short-term spatial memory.[116] This demonstrate that Wnt-dependent adult neurogenesis is fundamental for long-term memory formation.

Loss of Wnt7a causes a significant reduction in the amount of isolated Neural Stem Cells (NSCs), as it has been demonstrated in 6-weeks-old Wnt7a$^{-/-}$ mice. Moreover, these animals showed less number of newborn cells and a reduction in dendritic development of DG neurons.[102] This was further corroborated by in vivo studies.[102] Thus, Wnt7a is involved in two crucial steps of neurogenesis in the adult mouse brain. First, it contributes to the proliferation of neural stem cells through the expression of cyclin D1,[102] a Wnt target gene. Second, Wnt7a promotes the differentiation of precursor cells into differentiated neurons through neurogenin 2 (Neurog2) expression.[102] Decreased expression of neurogenin 1 (Neurog1), a known Wnt target gene, has also been linked to low Wnt/β-catenin signaling and consequent minimal adult neurogenesis in DG.[118] In cases were GSK-3β is hyperactivated, this causes a reduction in the amount of β-catenin, resulting in a defective Wnt signaling activity. As a consequence, the Neurog1 expression is altered modifying the fate specification of NSCs by reducing neuronal differentiation and favoring commitment to glial cells.[118]

Altogether, these evidences support the fact that Wnt signaling plays a significant role in adult hippocampal neurogenesis and that newborn neurons can result in memory consolidation changes.

4. WNT EFFECT IN NEURONAL PROCESSING

4.1 Wnt Signaling in Basal Neuronal Activities

An increasing number of studies have demonstrated that Wnt is present or constitutively released during basal synaptic processing. This means that the endogenous Wnt signaling pathways have a significant role in basal synaptic neurotransmission. Much of this evidence was obtained using sFRP, a physiological Wnt scavenger, acting as an endogenous inhibitor regulating the availability of Wnt in the media. The number of dendritic branches in cultured hippocampal neurons is significantly reduced in the presence of sFRP, suggesting that Wnt is being released in this preparation and is participating in the remodeling of the dendritic arborization.[54] Furthermore, electrophysiological recordings of mouse brain slices demonstrated that the presence of sFRP reduces the amplitude of evoked field excitatory postsynaptic potentials (fEPSPs) in hippocampal slices, suggesting that Wnt is being released and controlling the basal tone of synaptic activity.[90] More specifically, the evidence indicates that the application of sFRP reduces the basal N-methyl-D-aspartate receptor (NMDAR) currents but not the α-amino-3-hydroxy-5-methyl-4-isoxazolepropionic acid receptor (AMPAR) currents, suggesting that Wnt ligands are being released and specifically regulating NMDARs activity.[119] Moreover, using PKC or JNK inhibitors to block downstream signaling, the basal NMDAR synaptic neurotransmission is reduced, indicating that Wnt is using some noncanonical pathway components to maintain glutamatergic neurotransmission.[119] Studies on hippocampal cultures demonstrated that the increase in the number and length of dendritic branches depends on β-catenin availability[120] and this effect is reduced by Dkk1. This study also demonstrated that the neurons but not the astrocytes secrete measurable amounts of Wnt, although this study fails in determine what kind of Wnt ligand is being released.[120]

Thus, it seems likely that Wnt molecules are actively preserving a functional and structural connectivity even during basal activity. Animals maintained in an enriched environment increase the number of synapses of mossy fibers. This effect is reduced by lowering the levels of available Wnt7a/b through the local administration of sFRP.[71] sFRP can also reduce the number of synapses in animals housed in nonenriched environment. This suggests that even in basal condition, Wnt signaling is relevant to generate structural plasticity.[71] Using antibodies either against Wnt3a, Wnt 5a or Wnt7a ligands and sFRP, our laboratory recently demonstrated a reduction in the hippocampal firing rate and a reduction in the generation of spontaneous oscillations in the EC; both Wnt3a and Wnt5a were able to increase the firing rate, indicating that endogenous Wnts were released and participated in the maintenance of relevant rhythmic activity in neuronal circuits.[84]

4.2 Wnt Signaling in Activity-Dependent Processes

Dendritic spines are the structural place where the experience dependent-induced changes happen.[121–123] The establishment of the synaptic contacts occur during early development,[122] while during postnatal development and adolescence, the refinement of the circuits and the definitive contacts are defined by environment and activity-dependent processes, all of which will lead to spine maintenance or elimination.[121] Earlier, we discussed evidence demonstrating that Wnt participates during basal activity. Further we summarize the data showing that Wnt signaling is responsible for structural dendritic changes during activity-dependent plasticity.

Sustained depolarization by high K^+ in hippocampal neuronal cultures activates the secretion of Wnt by the same neurons. The increment of this secreted Wnt in turn stimulate the same neurons to accumulate β-catenin and it is this availability that generate the increase in number and length of dendritic branches.[120] Notably, this study demonstrated that this stimulus induce the release of Wnt from neurons, not from astrocytes, and the effect in dendritic arborization is reduced by Dkk1.[120] Another evidence demonstrated that Ca^{2+}-entry-induced by depolarization causes an increment in dendritic arborization.[124] This effect was through CaMKIV activity over CREB transcription factor, but the final effector was unknown.[124] Now we know that the effector is Wnt2 through a mechanism that involves NMDARs, Ca^{2+}-entry-Ca^{2+}-release and the consequent activation of several intracellular proteins (CaMKs).[69] Thus, CaMKIV activates CREB, which in turn induces the transcription of Wnt2 and its secretion, stimulating dendritic outgrowth.[69] Indeed, adding Wnt2 can itself stimulate dendrite development.[80] The dendritic arborization induced by high K^+ can be inhibited by WIF (a Wnt inhibitor),[69] confirming that sustained activity induces synthesis of more Wnt to potentiate the cellular response.

Wnt signaling has been reported to be relevant during protocols to elicit long-term plasticity, such as LTP. Both theta burst stimulation (TBS) or high frequency stimulation (HFS)[125] are protocols to induce LTP, and both can increase the expression of Wnt signaling components and Wnt release in an activity-dependent manner. LTP induced by tetanic stimulation activates Wnt/β-catenin/TCF pathway, confirmed by the accumulation of β-catenin in the nucleus and the increment of several Wnt-target genes.[126] Moreover, LTP also induces changes in mRNA levels and the immunostaining pattern of several Wnt-related proteins, such as Wnt3a, Fz4, β-catenin, and Dvl3.[126] Chen *et al.* found that tetanic stimulation reduces size and number of Wnt3a puncta. This process is dependent on NMDAR activation, which suggest that Wnt3a is being released in an activity-dependent process.[126] The same group determined that lithium, an activator of canonical Wnt/β-catenin, increases LTP. Conversely, the blockage of Wnt3a reduces the magnitude of LTP, which suggest that Wnt3a could be endogenously released to keep active the Wnt/β-catenin pathway.[126] Moreover, sustained neuronal activity reduces the expression of sFRP in DG.[127] As a consequence, the neural progenitor proliferation and neuronal development regulated by activity increases. Therefore, neuronal activity would act as a deregulatory mechanism, essential for the activity-dependent neuronal maturation, releasing the tonic inhibition by sFRP.[127]

4.3 Wnt Signaling Functions: In Vivo Evidence

Until now, the evidence has revealed that Wnt components play a significant role in generating and maintaining the basal activity of brain circuits. Moreover, sustained depolarization activates Wnt signaling which increases synaptic transmission and facilitates synaptic plasticity in hippocampal brain slices and in cultured neurons.[7,128] Recently, other aspects have involved Wnt signaling in more physiological aspects, such as during active behavior. An example of this occurs during the establishment of fear memory in the circuits corresponding to the amygdala, specifically in the basolateral amygdala. During the process of fear memory consolidation in the amygdala the mRNA levels of β-catenin transiently increases. On the other hand, the specific deletion of β-catenin negatively affects fear memory consolidation in adult mice.[129] A stereotaxic injection of the Wnt antagonist Dkk1 into the basolateral amygdala interferes with the long-term memory consolidation without affecting short-term memory.[130] Surprisingly, the continuous administration of Wnt1 during fear memory formation also interferes with the long-term memory consolidation in the amygdala.[130] This means that avoiding the transient increase and/or decrease in the Wnt1 mRNA levels that occurs immediately after fear conditioning, is critical for fear memory formation.[130] Several other genes related to Wnt signaling (15 out of 17) were immediately downregulated following fear conditioning. This strongly indicates that Wnts are highly correlated with new memory formation.[130]

Another example involving Wnt signaling in more physiological aspects, occurs during consolidation and recall of spatial memory. In this case, the levels of Wnt7a/b and Wnt5a, but not Wnt3, increase in hippocampus.[97] Interestingly, the level of Wnt ligands are correlated with the quality of training. Animals trained to locate a hidden platform in a water maze showed higher levels of Wnt7a concentrated in the granular layer of DG. Moreover, these animals have better memory retention 30 days after the training completion, compared to the animals trained in a visible platform.[97]

A third example involves memory consolidation in an object recognition task, where the activation of Wnt/β-catenin pathway is also required.[131] Recognition of familiar objects is eliminated by intrahippocampal administration of Dkk1. It also causes the reduction of β-catenin, Cyclin D1, c-myc, and Wnt7 protein levels, suggesting that canonical Wnt signaling is actively involved in hippocampal consolidation memory.[131] Additional evidence has demonstrated severe deficiency in working memory performance in older APC heterozygous knockout ($APC^{+/-}$) mice.[132] This evidence, that appears to be contradictory because it implies an increment in the levels of Wnt/β-catenin, suggest that the expression of the different components of Wnt signaling may be maintained in a strict equilibrium; where too low or too high protein expression causes disequilibrium. In support, is the evidence that, by controlling β-catenin stability, APC negatively regulate gene expression through TCF.[133] Therefore, in the $APC^{+/-}$ mice less APC would alter the gene expression dependent of TCF, what at the end could be related to the age-dependent working memory deficits observed.[132] Despite that mice that overexpress TCF also exhibit abnormal behavior in test related to contextual and fear conditioning, the phenotype is not similar to the $APC^{+/-}$ mice. Still these evidences support the hypothesis that Wnt signaling is determinant for the correct higher order functions of the brain, however more studies are required to elucidate this conundrum.

The mechanisms that involve long-lasting changes dependent on Wnt signaling are highly conserved across species. For example in *Drosophila*, the mechanism for long-term olfactory memory formation requires *armadillo* (a homolog of β-catenin), *Wingless* (a Wnt ligand), and *arrow* (a LRP5/6 homolog),[134] which suggest that the activation of Wnt signaling is key to induce long-lasting changes relevant for animal behavior.

The effect of Wnt5a has been indirectly tested in other functional studies, such as in spatial memory in in vivo adult mice.[135] Direct application into the CA1 hippocampal region of the formyl-hexapeptide 5 (Foxy5), a molecule that mimic the effect of Wnt5a, or Wnt-activating small molecule potentiator-1 (WASP1), which is an enhancer of Wnt5a, causes an improvement of animal spatial memory performance. Moreover, hippocampal slices obtained from the same animals showed an enhancement in LTP in HFS protocols.[135] Altogether, the evidence supports the idea that Wnt/β-catenin pathway is involved in long-term memory consolidation and that activation of Wnt signaling in response to a sensory stimulus induces functional long-lasting changes to improve the animal's performance.

5. WNT SIGNALING DYSFUNCTION IN NEURODEGENERATION AND MENTAL DISORDERS

Currently, the general idea that Wnt signaling plays a continuous role in development and maintenance in the CNS is well accepted. Recent advances in psychiatric disorder etiology have repeatedly found a relationship with Wnt signaling malfunction. During the past few decades, increasing data have suggested that the progression of a neurological disease is the result of severe neuroanatomical abnormalities. However, more evidence shows that changes in synaptic connectivity, such as an imbalance between excitation and inhibition may result in a cascade of multiple effects at the neural network level which in many cases precedes cell degeneration or cell death.[136–139] Excitatory/inhibitory balance critically regulates cortical network function, neural network activity, and behaviors associated with psychiatric disorders.[139] Multiple lines of evidence implicate excitatory and inhibitory synaptic circuits in the cortex, striatum, and hippocampus as key cellular substrates for the pathogenesis in these disorders. In fact, major neurological diseases, such as autism, schizophrenia (SZ), bipolar disorder (BP), and major neurodegenerative disorders including Alzheimer disease (AD), Huntington's disease (HD), and Parkinson's disease (PD) share the characteristics of progressive neuronal loss. In this section, we review the most recent studies related to the progression of the most common neurological diseases in which the Wnt signaling pathway and its components are relevant.

5.1 Wnt in Alzheimer's Disease

An important issue in the study of Alzheimer therapies is the realization that we do not clearly understand what causes the disease. Two main hypotheses have been presented in the last few decades: the tau hypothesis and the Aβ oligomers hypothesis; we will not delve into the fundaments, since they have been addressed by others.[140,141] Aβ has been targeted in clinical trials on human patients, with no favorable results. So far, drugs can only temporarily ameliorate symptoms and extend the cognitive function and independence of patients but not cure AD. Among the FDA-approved drugs for treating AD, there are two main types: cholinesterase inhibitors and NMDA receptor antagonists. The most recent FDA-approved drug to treat some of the symptoms of AD was donepezil, a cholinesterase inhibitor, approved in 2002,[142] but none of the inhibitors by itself, or in combination, are an effective treatment for AD, for this reason, new approaches to treat AD are needed.

Several studies have linked a decrease in neurogenic rate to neurodegenerative diseases. In this regard, and considering the extensive evidence supporting a role of Wnt signaling in adult hippocampal neurogenesis (Section 3.3), there is a new-found interest in studying Wnt signaling as a cause of and/or treatment for AD. Several components of the Wnt pathway are altered in AD.[143] For example, it has been found a broad region of susceptibility for late-onset AD in chromosome 12 that contains the *LRP6* gene.[144] Moreover, this susceptibility depends on the apolipoprotein E (*APOE*)-ε4 allele status, as has been demonstrated by genome-wide analysis studies (GWAS).[144] Mutations in *APOE* gene have been widely accepted as a risk factor for sporadic and late onset AD. APOE is one out of several secreted lipoproteins that mediate the interaction between ligand-receptor presentation/endocytosis through the LDLR family, such as LRP1.[145] Therefore, it is hypothesized that LRP5/6 could be involved in APOE processing of lipoproteins. A particular *LRP6* allele found in AD patients shows reduced activation of β-catenin signaling, supporting the involvement of Wnt signaling in the AD.[144]

Initial evidence associates these findings with a defective Wnt/β-catenin pathway.[144] β-catenin levels in microglia of AD postmortem tissue was elevated, which is consistent with an inflammatory response of microglia. Additionally, levels of Dkk1 an antagonist of the pathway, were increased in postmortem brains of AD patients[146] and in mouse models.[147] Dkk1 is rapidly overexpressed after Aβ oligomers treatments in acute hippocampal slices.[148] In fact, Dkk1 action is necessary for the synaptotoxic effects of Aβ oligomers in hippocampal neurons and hippocampal slices.[148] Inducible overexpression of Dkk1 caused synaptic failures similar to Aβ oligomers.[149] Likewise, overexpression of Dkk1 resulted in excitatory synapse loss, impaired LTP, enhanced long-term depression, and hippocampus-associated learning and memory deficits.[149] Interestingly, when overexpression of Dkk1 was ceased or when an anti-Dkk1 antibody was added, the phenotype returned to control levels for every parameter measured.[148,149] On the other hand, GSK-3β but not GSK-3α, has been extensively linked to AD,[150,151] and new research suggests a direct role of GSK-3β inhibition in recovering cognitive function in mice models of AD.[150,152–154] In any case, the role of GSK-3α cannot be overlooked since there are some reports of different contributions from both GSK-3 isoforms (α and β) in AD pathogenesis.[150] These data support the GSK-3β hypothesis of AD,[155,156] which states that Aβ impairs PI3K/Akt signaling and leads to the activation of GSK-3β, lowering total β-catenin levels. This induces tau hyperphosphorylation, neurofibrillary tangle formation, synaptotoxicity, synaptic loss, and eventually, neuronal death.[155] Even more, there is evidence in Chinese hamster ovary (CHO) cells that a decline in Wnt signaling leads to an increase in amyloid precursor protein (APP) processing and high levels of Aβ oligomers,[157] and thus a direct link between Wnt signaling loss and the onset of AD. Furthermore,

Wnt/β-catenin signaling activation through pharmacological tools generates significant improvements in cognitive tests in transgenic mouse models of AD,[135,158,159] further supporting that Wnt signaling is dysregulated and could be a cause of AD.[144,146] Additionally, activating the Wnt-canonical pathway has shown to be neuroprotective in a double transgenic mice model (APP_{swe}/PS1ΔE9 mice), as shown by recovery of synaptic protein levels, a lower Aβ burden in the cortex and hippocampus, and improved spatial memory.[160] Activation of canonical and noncanonical pathways, rescues LTP and memory impairment in the same transgenic mice model,[135] supporting a role of Wnt signaling in the etiology of AD and proposes Wnt signaling as a plausible target to treat this disease.

5.2 Wnt in Autism Spectrum Disorder

Increasing evidences has revealed a relationship between the Wnt pathway and a broad array of autism spectrum disorders (ASDs),[161] especially for the role of Wnt signaling during development. The earliest evidence comes from the late 1990s. The *Dvl KO* mouse exhibited complex behavioral responses involving deficits in social interaction, reduced mothering behavior, and abnormalities in sensorimotor gating, such as acoustic startle and tactile startle,[162] suggesting that Wnt signaling might be related to ASD etiology. At the same time, the *Wnt2* gene was found to be allocated on chromosome 7q (7q31–33 linked region), a site that has been associated with autism in several familial and twin autism studies.[163] Indeed, mutations in the *Wnt2* gene have been identified that can potentially increase susceptibility to autism; these mutations have been observed in patients but not in healthy controls.[164] However, other study shows no association of Wnt2 with autistic disorder,[165] indicating that there is no consensus on this matter. More investigations are necessary to determine the genetic factors involved in ASDs.

Recently, other insights have been obtained from indirect studies. These studies showed that children who were exposed to valproic acid (VPA) in utero, a drug used in bipolar disorder, exhibited severe symptoms of autism. These children and animals treated with VPA demonstrated repetitive behavior, hyperactivity, and an oversized brain characteristic of autistic patients,[166] all of which are considered part of the autism-like phenotype. Interestingly, at the molecular level VPA increases the mRNA expression and the stabilized form of β-catenin, reduces the mRNA expression of GSK-3β, and enhances the phosphorylated form of GSK-3β.[167] Moreover, VPA also increases the expression of the canonical Wnt ligands, Wnt1 and Wnt2, in the prefrontal cortex and the hippocampus.[167] On the other hand, previous results showed that transgenic mice overexpressing stable forms of β-catenin exhibited macrocephaly.[168] Both evidences that appears to be contrary to the role of Wnt signaling that has been suggested in this review, are indeed a strong demonstration that hyperactivation of the canonical Wnt pathway during prenatal life can cause the phenotypic ASD features, such as an oversized brain and an autistic behavior. How could Wnt signaling be involved in physical and psychological features in autism-like disorders? The answer came recently. In an elegant study, Belinson and coworkers demonstrated that the control of neuronal proliferation and differentiation during embryonic development is under tight control by the canonical Wnt signaling pathway. Using a mutant of the Dvl1 and 3 protein, they found that β-catenin and a group of specific transcription factors (Brn2/Tbr2) are involved in the regulation of embryonic brain size and deep layer neurogenesis.[169] During normal embryonic development, Wnt signaling activates the transcriptional activity of β-catenin, which increases the expression of Brn2 and represses the expression of Tbr2, reducing the proliferation of neural progenitors.[169] Instead, in the Dvl double mutant, less β-catenin causes the repression of Brn2 and the derepression of Tbr2. This causes the premature deep layer neurogenesis of neural progenitors located in specific brain regions during a specific period of embryonic development. This would affect the correct establishment of neural connections in the future prefrontal cortex, which is evidenced by the serious deficits in social behavior that the animal exhibited during adult life.[169] When these animals were treated in utero with an inhibitor of GSK-3β to activate the canonical Wnt pathway, deficits in brain size and social behavior in adulthood were reversed. For the first time, this study links developmental Wnt signaling to adult social behavior.[169] Therefore, the recent evidence indicates that the correct expression and developmental-time of Wnt signaling pathway activation are crucial to determine social behavior in adulthood and that Wnt signaling dysfunction can cause autism like-behavior.

In the most common heritable form of autism, the fragile-X syndrome (FXS), a reduction in the expression of some components of the Wnt signaling pathway have been recently described. In FXS the deficiency of fragile-X linked mental retardation protein (FMRP) is complete or partial and generates different degrees of the autistic-like behavior. Indeed, FMRP works as a negative regulator of protein translation; its deficiency causes dysregulation and upregulation of protein synthesis and their signaling.[170,171]

One of the most striking characteristics of FXS at neuronal level, is the high density of dendritic spines with immature morphology, that is thin and elongated shape, and the absence of mushroom shape of mature neurons.[172,173] At the functional level, the signaling downstream upon the activation of group 1 metabotropic glutamate receptor (gp1 mGluR), is upregulated in FXS. As a result, an exacerbated increase in protein synthesis causes dysfunctional synaptic plasticity. Several studies have demonstrated the relationship between FXS and Wnt signaling. Zhang *et al.* were the first to describe reduced levels of Wnt2 in specific regions of the brain, including the cortex, hippocampus and cerebellum in fragile-X mental retardation 1 gene (Fmr1) KO mice, suggesting that a lack of FMRP affects the translation of Wnt2 mRNA.[174] This was supported with more evidence demonstrating a significant reduction in the expression of Wnt7a and a moderate reduction in β-catenin–dependent signaling in patients with FXS.[175]

Among the proteins that are indirectly involved in FMRP and Wnt signaling, an interesting link was observed between the proteins CaMKIV and PSD-95 proteins. The activation of gp1 mGluR induces an increase in protein translation, including FMRP, through a Ca^{2+}-dependent signaling mechanism that depends on CaMKIV.[176,177] Thus, CaMKIV acts as a transcriptional regulator of FMRP1 during glutamatergic signaling.[176,178] Interestingly, CaMKIV is a gene target of Wnt/β-catenin signaling; β-catenin directly binds to the promoter region of CaMKIV and upregulates its gene expression.[42] These results suggest that, in a system that lacks FMRP, the availability of CaMKIV can be modulated by Wnt signaling. On the other hand, FMRP directly binds to PSD-95 mRNA, a scaffold protein that is involved in signaling and synaptic plasticity.[179] PSD-95 mRNA is highly localized in dendrites both in vitro and in vivo. PSD-95 is important for the assembly of the postsynaptic structure during memory processes, and it has been associated with Wnt signaling. Fz receptors can bind to PSD-95, and together recruit other proteins during the Wnt downstream signaling cascade.[180] Through the Fz9-Gαo complex, Wnt5a activates the intracellular cascade essential for the formation of dendritic spines,[94] inducing the clustering of PSD-95 using a JNK-dependent signaling mechanism.[88] On the other hand, it has been shown that mRNA of PSD-95 and FMRP are rapidly translated in response to the activation of gp1 mGluR.[181] Indeed, in Fmr1 KO mice the absence of PSD-95 expression induced by gp1 mGluR activation suggest that FMRP is required for its dependent expression.[181] During complex spatial navigation tasks, there is a specific increase in PSD-95 in wild-type mice but not in Fmr1 KO mice.[182] This correlates with the number of errors made during a navigation task, where the animals that make more errors have less PSD-95 protein and vice versa, which indicates a reduction in learning in the KO mice.[182] This evidence suggests that FMRP regulates the abundance and stability of PSD-95 mRNA specifically in the hippocampus[179] The effect of Wnt signaling on PSD-95 has been widely documented. Wnt5a promotes the clustering of PSD-95 in the postsynaptic terminal.[88] The protein levels of Wnt7a and PSD-95 in hippocampus increases when the animals are maintained in an enriched environment.[71] PSD-95 increases when the animal is performing learning tasks and reduces its expression in the presence of Dkk-1.[131] Altogether these evidences indicate that modulation of PSD-95 by Wnt signaling is a crucial step in the formation of a functional synaptic structure, and it could be relevant to resolve the problem of the PSD-95 instability in synaptic structures caused by the absence of FMRP. More studies are required to elucidate this possibility.

One of the clearest pieces of evidence regarding the cross-talk between FMRP and Wnt signaling comes from studies on the role of FMRP in neurogenesis. In a very elegant study, Luo *et al.* have demonstrated both in vivo and in vitro that FMRP can regulate neurogenesis in adult neural progenitor cells (aNPCs) located in the DG through the regulation of GSK-3β and β-catenin.[118] The authors found that loss of FMRP caused an increase in GSK-3β and a reduction of β-catenin in aNPCs. In fact, FMRP has a physical association with GSK-3β mRNA and represses its translation. Therefore, when FMRP is absent, the GSK-3β protein translation increases. On the other hand, the transcription factor Neurog-1, which is an early initiator of neuronal differentiation and an inhibitor of glial differentiation, can be regulated by Wnt signaling. In the absence of FMRP, Wnt signaling is reduced, and promoter activity for the early differentiation gene Neurog-1 decreases. Consequently, these cells display high rates of proliferation and a reduction in neuronal differentiation but an increase in glial differentiation. Therefore, the increase in GSK-3β and reduction in β-catenin alters the proliferation rate and the fate destination of NPCs. In summary, FMRP can control the differentiation of aNPCs through the modulation of Wnt/β-catenin signaling, which activates the promoter activity for Neurog-1.[118] No unique molecular mechanism controls neurogenesis. In fact, embryonic neurogenesis and adult neurogenesis appear to be controlled by independent mechanisms, some of which are parallel. Whether the mechanism that includes FMRP-Wnt/β-catenin signaling is involved in multiple types of neurogenesis is unknown. However, it would be more interesting to find a direct correlation between higher-level cognitive tasks and the incorporation of new neurons in the circuit. In fact, in Fmr1 KO mice, fewer numbers of new neurons are incorporated into the DG circuit.[183] Altogether, these data suggest that FMRP controls Wnt signaling to modulate adult neurogenesis and could be relevant for the hippocampal neurogenesis associated with ASD.

Several other molecules that interact to some degree with the Wnt pathway, have been reported to be affected in ASD. For example, tuberous sclerosis syndrome that causes epilepsy, mental retardation, and autism, negatively control the stability of β-catenin by controlling the degradation complex.[184] The tyrosine kinase receptor Met which is a target gene of the canonical Wnt pathway, has been associated with colorectal cancer when overexpressed and with autism when underexpressed.[185,186] Some cadherins that interact with β-catenin in cell-adhesion complexes during the establishment of the synaptic structure have been associated with malformations and deficits related to macrocephaly.[187] Phosphatase and tensin homolog (PTEN), a relevant negative regulator of PI3K/Akt signaling, has also been indicated in ASD due to its roles in development, including morphology, number of dendritic spines, and other autistic features, such as macrocephaly and seizures.[188] A mutation in the PTEN gene causes an increase in Akt and the subsequent inhibition of GSK-3β.[188] A mutation in the APC protein, which is part of the destruction complex for β-catenin degradation, causes an increase in excitability, in dendritic spines density with immature shape, and in autistic-like behavior, such as poor social interaction and repetitive behavior.[189] Finally, the association between β-catenin and cadherins also regulates the recruitment of synaptic vesicles into terminals, and the status of this association has been correlated with ASD.[190]

5.3 Wnt in Bipolar Disorder

Even before knowing its mechanism of action, lithium was used as a clinical mood stabilizer for depression and bipolar disorder (BP). It was and still is the first-line drug in the treatment of BP. Lithium has been effective in the amelioration of the symptoms of mania with demonstrated antidepressant properties,[191] and has significantly reduced the risk of suicide.[192] Lithium is a potent in vivo inhibitor of GSK-3 activity[193–195] and directly or indirectly increases adult hippocampal neurogenesis.[81,154,196,197] The second most used medication for bipolar disorder is VPA and it has been found to inhibit histone deacetylases and GSK-3β.[195] Interestingly, haploid sufficiency of

GSK-3β causes similar effects as lithium treatment in mouse models of AD,[194] proving that lithium mainly acts through GSK-3β inhibition, with lower efficiency in inhibiting GSK-3α. By inhibiting GSK-3β, lithium can mimic the activation of the Wnt/β-catenin–dependent pathway by stabilizing intracellular β-catenin.[198–200] Moreover, transgenic mice overexpressing β-catenin show the same behavior as lithium treated animals,[201] and mice with constitutively active GSK-3 show the same behavior as that observed in an animal model of bipolar disease,[202,203] which strongly suggests that the symptoms of bipolar disease are related to the Wnt signaling cascade.

In 2010, a systematic analysis of the publications and the Genetic Association Database information on bipolar disorder, was published.[204] Considering the variety of approaches analyzed (GWAS, gene expression, epigenetics, pharmacogenomics, mouse models, among others.), the most replicable results are the ones involving GSK-3β.[204] Thus, GSK-3β not only is a crucial protein in the effective pharmacological treatment of bipolar disorder, but would also be part of the etiology of the disease. Since 2010, there have been new and promising investigations on the roles of GSK-3β in patients with bipolar disorder. There are several studies arguing that dysregulation of GSK-3β activity occurs in the brains of bipolar disorder patients.[204] In relation to this, patients with bipolar I disorder (the heritable form of BP) have a decreased hippocampal volume, which is recovered to normal volume after lithium treatment,[205–207] suggesting that the recovery in hippocampal volume is due to an increase in adult neurogenesis. Again, the neurogenic role of the activated Wnt/β-catenin pathway in the hippocampus, plays a pivotal role in bipolar disorder treatment. Analysis of postmortem brain samples showed that total mRNA levels of GSK-3β and β-catenin were significantly decreased in the dorsolateral prefrontal and temporal cortices, but not in the cingulate cortex.[208] In the same study, protein levels of GSK-3β, phosphorylated (p)-GSK-3β-Ser9 (i.e., the inactive form of GSK-3β) and β-catenin, were also decreased in the same areas as the mRNA.[208] This evidence strongly suggest that these proteins are involved in the pathophysiology of BD. More in vivo studies are required to evaluate how could lithium restore their levels in a long-term therapy.

5.4 Wnt in Epilepsy and Seizures

Epilepsy is a chronic neurological disease characterized by abnormal discharges of neuronal activity. Increasing data demonstrate the relevance of Wnt signaling in early and late processes involved in epilepsy. One example is in temporal lobe epilepsy, which involves one-third of all known forms of epilepsy and mainly involves the hippocampus.[209] It has been shown that in early stages of the disease, the first epileptogenic events induce neurogenesis in the hippocampus. This has also been reported in pharmacologically induced animal epileptic models. Interestingly, these newborn cells do not integrate correctly into the circuit and establish aberrant connectivity that contributes to epileptic behavior.[210] Later during the chronic phases of the disease, neurogenesis becomes substantially decreased. A significant proportion of cells that would become interneurons also decreases.[211–213] This could be responsible for the exacerbated hyperexcitability in the brains of adult patients and for the reduced abilities in learning processes.

It has been suggested that initial epileptic insults may convey a dysfunction in Wnt signaling. Patients with hippocampal sclerosis, a neurological condition characterized by severe neuronal loss, gliosis,[214] and temporal lobe epilepsy, also shows an early expression of Dkk1. The expression of Dkk1 in hippocampal sclerosis correlates with neuronal death,[214] as it does in hypoxia/ischemia in vivo and in excitotoxicity-related neuronal death in vitro.[215] The increase of Dkk1 precedes neuronal death in both olfactory neurons and in the hippocampus of rats that developed kainate-induced seizures.[214] High levels of Dkk1 are found in animals that are considered high responders to kainate in relation to the severity of their seizures; conversely, no Dkk1 was found in the low responders. Treatment with lithium protected the animals against neuronal loss but not against the kainate-induced seizures.[214] The administration of the Dkk1 antisense sequence directly into the brain almost completely prevents seizure-induced neuronal death[146,214] but not ictal activity seizures.[214] On the other hand, several inhibitors of GSK-3β have been successfully been shown to increase neurogenesis selectively through the stimulation of Wnt signaling.[216] Altogether these data suggest that the mechanisms that control of neuronal loss and seizure activity are different, and therapies directed against Wnt signaling components would protect against the neuronal loss induced by epilepsy.

Recently, dysfunctions related to Wnt/β-catenin signaling have been related to infantile spasms (IS), an infantile form of epilepsy. This syndrome is characterized by motor spasms during childhood that worsen into chronic seizures and cognitive deficits in adulthood.[217] Genetic evidence indicated that most of the risk genes in this syndrome are related to the Wnt/β-catenin pathway. Therefore, Pirone and coworkers developed a conditional KO of the APC protein (*APC cKO*), which under normal conditions control the protein levels of β-catenin, thus generating an animal with high levels of β-catenin. This cKO show many of the manifestations developed in IS, including motor spams, hyperexcitability with abnormal brain discharges, and adult seizures.[217] These data show an interesting link between dysregulated Wnt/β-catenin signaling and the developmental manifestation of physiological features related to epilepsy and seizure activity. This is a new indication that tight regulation of Wnt signaling is fundamental not only during developmental processes but also throughout life in the physiological processes that depend on it.

5.5 Wnt in Parkinson's Disease

PD is a motor neurodegenerative disorder characterized by the midbrain dopaminergic neuron loss in the *substantia nigra* pars compacta. Surviving neurons will accumulate aggregates of α-synuclein and parkin, which will eventually suffer synaptic loss followed by cell death.[218] As with most neurogenerative or neuropsychiatric diseases, there is no clear origin of PD. Similarly, to Alzheimer's disease, a considerably low proportion of cases are of genetic origin.[218]

There is extensive evidence proving that Wnt ligands are fundamental in neurogenesis, differentiation and survival of midbrain dopaminergic neurons.[7,65,219] Much work has been done on the development and embryonic stages. More recently, there is evidence for the role of Wnt signaling on the maintenance of dopaminergic neurons in the adult brain. Examination of postmortem brain tissue from sporadic PD patients has delivered crucial information. First, GSK-3β levels are increased in the striatum, suggesting low Wnt activity. Second, over 230 genes showed abnormal methylation profiles, and 34 of them are Wnt-related genes. The Wnt and neurogenesis related genes forkhead box C1 (FOXC1), Neurog2, sprouty homolog 1 (SPRY1), and catenin beta 1 (CTNNB1) showed marked hypermethylation and their protein products showed a clear reduction in expression.[220] Aside from the neurogenesis-related Wnt genes, genes coding for the coreceptor LRP5/6 and the sFRPs were also hypermethylated in the brains of PD patients.[220] Overall, this tells us that Wnt signaling is inhibited in PD and may play a significant role as a cause or consequence of the disease.

Research conducted on knock-in Wnt1 mice showed that activation of the Wnt/β-catenin pathway prevents premature neuronal death of dopaminergic neurons in the substantia nigra and ventral tegmental area, but no evidence indicates a generation of new tyrosine hydroxylase positive midbrain neurons.[77] These results contrasts with those described by L'Episcopo and coworkers, where astroglial-derived Wnt1 favored in vitro dopaminergic cell neurogenesis in cells taken from adult midbrain neural progenitor/stem cells, after 1-methyl-4-phenyl-1,2,3,6-tetrahydropyridine (MPTP) injury (a pharmacological model of PD).[74] On the other hand, blocking Wnt action reverted astrocytic neuroprotection against 1-methyl-4-phenylpyridinum (MPP+) and 6-hydroxydopamine (6-OHDA).[74] This was confirmed in vivo, since astrocytes lacking Wnt1 were not able to recover neighboring dopaminergic neurons from injury. Conversely, pharmacological activation of Wnt/β-catenin pathway recovered neurons from MPTP-induced damage,[74,75] as well as the neurogenic potential of dopaminergic neurons.[74] In vivo experiments showed that Wnt signaling deficits can be restored by activating Wnt signaling in astrocytes.[75] It is proposed that Wnt/β-catenin signaling and Wnt1 expression are increased in the midbrain aqueduct periventricular regions in response to cell injury of midbrain dopaminergic neurons.[75]

Inestrosa and Arenas suggested that the Wnt pathway may have a therapeutic application in cell-replacement therapy in PD.[7] It is necessary to establish if the dopaminergic neurons, damaged during PD, are able to react and fully transduce the Wnt-dependent signals. In this regard, new therapies have been focused to induce progenitors cell differentiation to dopaminergic neurons through the activation of several signaling pathways, including Wnt signaling. Indeed, based on this approach, the differentiation of human astrocytes to dopaminergic neurons reached an efficiency of 16% approximately.[221] Altogether, there seems to be a therapeutic potential in inducing Wnt/β-catenin signaling into astrocytes to promote recovery of midbrain dopaminergic neurons.

5.6 Wnt in Schizophrenia

One of the most studied neurological diseases is schizophrenia, affecting 1% of the population around the world. Despite the amount the information, only recently the studies confirm the relevance of the genetic factors over the environmental factors. This is a complex disease with variable phenotypes that can vary within the patient lifespan. The consensus today, recognizes the relevance of developmental factors as a cause of the aberrant neuronal connections observed in patients. Among its characteristics, patients show defects in brain asymmetry connections and activity, and a reduction in hippocampal volume,[222] which may be related to the reduced neuronal size of hippocampal subregions.[223,224] The relationship with Wnt signaling appears to come from the fact that several differences in the expression levels and distribution along several brain regions of some Wnt components, such as Dkk3, have been detected in schizophrenia samples compared to control.[225] Reduction of the staining pattern of β-catenin in postmortem hippocampus of schizophrenia patients compared to controls has been also observed.[226] Moreover, protein levels and activity of GSK-3β has been found to be reduced in prefrontal cortex of patients[227,228] but also its mRNA levels,[229] confirming that Wnt signaling could be highly involved in schizophrenia. The possibility of a role of Wnt from developmental stages should be explored.

A reduced number of newly formed cells in the hippocampus have been detected in this disease.[230] One of the most studied genes involved in the pathology is *disrupted in schizophrenia 1* (*DISC-1*) gene, the major risk gene of the disease. This gene has been related to the integration of newly formed cells into the hippocampal circuit to regulate its morphology and positioning. This role is mediated through the interaction of DISC-1 with GSK-3β; the latter is inhibited to control the amount and stability of β-catenin.[231] In general, downregulation of the DISC-1 gene causes misalignment of cells during neurogenesis and aberrant physiological behavior, such as hyperexcitability.[232] The deficit in proliferation due to DISC-1 knockdown was rescued with GSK-3β inhibitors, demonstrating for the first time the relationship between Wnt signaling and *DISC-1*.[231] An increase in GSK-3β and a reduction in β-catenin (i.e., downregulated Wnt signaling) changes the proliferation rate and fate destination of NPCs. When Wnt signaling is reduced, the promoter activity for the early differentiation gene *Neurog-1* decreases. Consequently, these cells display high rates of proliferation and a reduction in neuronal differentiation, favoring an increase into glial differentiation.[118]

Recently, a cohort study obtained from a population of schizophrenic patients found novel mutations in two genes involved in Wnt signaling, in the *Wnt7b* gene and in *β-catenin* gene.[233] The novel mutation in β-catenin resulted in a damaged protein; in the case of *Wnt7b*, the mutations create new binding sites for micro-RNA (miRNA) regulation. Functional studies are necessary to determine the relevance of these mutations. The study of Wnt signaling at the levels of development, synaptic structure, and function will allow the prediction that Wnt signaling may be involved in the etiology of the disorder. Another line of evidence that indirectly links schizophrenic behavior with Wnt signaling comes from the dopaminergic-signaling pathway. Activation of the D2 receptor (D2R) by dopamine induces hallucinations and psychotic behaviors.[234] D2R activation inhibits the enzyme Akt, which is an inhibitor of GSK-3β. As a result, GSK-3β becomes hyperactivated. Also, D2R bind and sequester β-catenin to deactivate Wnt signaling at the level of TCF/LEF mediated transcription.[235] Antipsychotic medication increases the level of Dvl and the binding of GSK-3β to D2Rs, and as a consequence releases β-catenin to allow its signaling.[195,236] Whether this antipsychotic effect is related to the upregulation of Wnt is manifested during the life of a schizophrenic patient is an interesting question that must be solved to design feasible therapies.

5.7 Wnt in Depression

There are numerous causes for depression. One of the most prevalent is chronic stress that has a causal relationship with depression. Stress causes an abnormal and steady increase in cortisol levels and deregulation in the glucocorticoid homeostasis.[237] At the same time, stress is a well-known cause for decreased neurogenesis. Therefore, it appears logical to think that stress among many other effects, causes depression by downregulating adult hippocampal neurogenesis. This is not only a theoretical hypothesis but also has been empirically proven. One of the main mechanisms of action for currently used antidepressants is the upregulation of adult hippocampal neurogenesis.[78,79,238] It was found that chronic administration of antidepressants, such as fluoxetine, citalopram, venlafaxine, and atomoxetine increased the expression of several Wnt signaling molecule genes (Wnts and Fz receptors, among others) in the hippocampus of depressed mice.[79] Wnt2 was the only Wnt molecule that had increased levels with all antidepressants tested.[79] It was also shown that chronic antidepressant administration increased the inactivation of GSK-3β, consistent with previous findings arguing this as an antidepressant mechanism.[239–241] Antidepressants also caused an increase in Neurod1 expression,[79] which is associated with increased neurogenesis. The antidepressant roles of Wnt2 and Wnt3 were later confirmed by others.[78]

Recent research has focused on the study of miRNAs to uncover their roles in the effective action of antidepressants. Two studies have performed miRNA screening on plasma from patients with major depression that were treated with different antidepressants.[242,243] In the first study, it was found that miR-146a-5p, miR-146b-5p, miR-221-3p, miR-24-3p, and miR-26a-5p were downregulated[242] after a 12-week escitalopram treatment. In the second study, miRNA analysis was performed on blood from patients treated for 8-weeks with different antidepressants (fluoxetine, escitalopram, nortriptyline, or duloxetine) and compared with an 8-week placebo treatment. This study found three downregulated miRNAs, miR-146a-5p, miR-146b-5p, and miR-24-3p in common with the first study.[242] Additionally, and a new one, miR-425-3p, was found.[243] These miRNAs regulate over 30 genes linked to MAPK and the Wnt signaling pathway as confirmed by in-silico and in vitro analyses.[242,243] Thus, it seems evident that the Wnt pathway is crucially involved in effective antidepressant action. It remains to be determined at what level Wnt signaling is acting, knowing the role of Wnt signaling in several aspects in postnatal life.

6. CONCLUDING REMARKS

New and exciting evidence reveals a fundamental role of Wnt signaling in the mature CNS. Wnt modifies the presynaptic compartment, regulates the dynamic of synaptic apparatus, including proteins and synaptic vesicles. At the postsynaptic compartment Wnt modulates the assembly of the postsynaptic structure and dendritic spine morphology and function. Wnt pathway modulates both basal and evoked neuronal activity, enhances synaptic plasticity and improves cognitive function in the adult brain. Wnt activity is required for the normal functioning of neuronal circuits and its inhibition can cause synaptic plasticity impairments and memory deficits. This evidence has allowed to unveil the relationship of Wnt signaling with the etiology of human diseases and its potential therapeutic use in neurological and neuropsychiatric diseases. We also learned that Wnt signaling components are in tight regulation in terms of timing of expression and amount of expression; too much or too low can have disturbing consequences at different levels in the functioning of CNS. As we deepen our understanding on the complexity of the Wnt pathway, and its role in physiological and pathophysiological processes within the CNS, better design, precise, and clean treatments will be performed.

REFERENCES

1. McMahon AP, Moon RT. Ectopic expression of the proto-oncogene int-1 in Xenopus embryos leads to duplication of the embryonic axis. *Cell.* 1989;58(6):1075–1084.
2. McMahon AP, Moon RT. int-1–a proto-oncogene involved in cell signalling. *Development.* 1989;107(Suppl):161–167.

3. Nusse R, Brown A, Papkoff J, et al. A new nomenclature for int-1 and related genes: the Wnt gene family. *Cell.* 1991;64(2):231.
4. Nusse R, Varmus H. Three decades of Wnts: a personal perspective on how a scientific field developed. *EMBO J.* 2012;31(12):2670–2684.
5. Polakis P. Wnt signaling in cancer. *Cold Spring Harb Perspect Biol.* 2012;4(5):a008052.
6. Berwick DC, Harvey K. The importance of Wnt signalling for neurodegeneration in Parkinson's disease. *Biochem Soc Trans.* 2012;40(5):1123–1128.
7. Inestrosa NC, Arenas E. Emerging roles of Wnts in the adult nervous system. *Nat Rev Neurosci.* 2010;11(2):77–86.
8. Clevers H, Nusse R. Wnt/beta-catenin signaling and disease. *Cell.* 2012;149 (6):1192–1205.
9. Gordon MD, Nusse R. Wnt signaling: multiple pathways, multiple receptors, and multiple transcription factors. *J Biol Chem.* 2006;281(32):22429–22433.
10. Bhanot P, Brink M, Samos CH, et al. A new member of the frizzled family from *Drosophila* functions as a Wingless receptor. *Nature.* 1996;382(6588):225–230.
11. Bjarnadóttir TK, Gloriam DE, Hellstrand SH, Kristiansson H, Fredriksson R, Schiöth HB. Comprehensive repertoire and phylogenetic analysis of the G protein-coupled receptors in human and mouse. *Genomics.* 2006;88(3):263–273.
12. Rulifson EJ, Wu CH, Nusse R. Pathway specificity by the bifunctional receptor frizzled is determined by affinity for wingless. *Mol Cell.* 2000;6(1):117–126.
13. Tamai K, Semenov M, Kato Y, et al. LDL-receptor-related proteins in Wnt signal transduction. *Nature.* 2000;407(6803):530–535.
14. Mao B, Wu W, Li Y, et al. LDL-receptor-related protein 6 is a receptor for Dickkopf proteins. *Nature.* 2001;411(6835):321–325.
15. MacDonald BT, Tamai K, He X. Wnt/beta-catenin signaling: components, mechanisms, and diseases. *Dev Cell.* 2009;17(1):9–26.
16. Cadigan KM, Liu YI. Wnt signaling: complexity at the surface. *J Cell Sci.* 2006;119(Pt 3):395–402.
17. Green JL, Kuntz SG, Sternberg PW. Ror receptor tyrosine kinases: orphans no more. *Trends Cell Biol.* 2008;18(11):536–544.
18. Lu W, Yamamoto V, Ortega B, Baltimore D. Mammalian Ryk Is a Wnt coreceptor required for stimulation of neurite outgrowth. *Cell.* 2004;119:97–108.
19. Oishi I, Suzuki H, Onishi N, et al. The receptor tyrosine kinase Ror2 is involved in non-canonical Wnt5a/JNK signalling pathway. *Genes Cells.* 2003;8(7):645–654.
20. Nusse R. Wnts and Hedgehogs: lipid-modified proteins and similarities in signaling mechanisms at the cell surface. *Development.* 2003;130(22):5297–5305.
21. Bafico A, Liu G, Yaniv A, Gazit A, Aaronson SA. Novel mechanism of Wnt signalling inhibition mediated by Dickkopf-1 interaction with LRP6/Arrow. *Nat Cell Biol.* 2001;3 (7):683–686.
22. Rattner A, Hsieh JC, Smallwood PM, et al. A family of secreted proteins contains homology to the cysteine-rich ligand-binding domain of frizzled receptors. *Proc Natl Acad Sci USA.* 1997;94(7):2859–2863.
23. Finch PW, He X, Kelley MJ, et al. Purification and molecular cloning of a secreted. Frizzled-related antagonist of Wnt action. *Proc Natl Acad Sci USA.* 1997;94(13):6770–6775.
24. Hsieh JC, Kodjabachian L, Rebbert ML, et al. A new secreted protein that binds to Wnt proteins and inhibits their activities. *Nature.* 1999;398(6726):431–436.
25. Cruciat C-M, Niehrs C. Secreted and transmembrane Wnt inhibitors and activators. *Cold Spring Harb Perspect Biol.* 2013;5(3):a015081.
26. van Amerongen R, Nusse R. Towards an integrated view of Wnt signaling in development. *Development.* 2009;136(19):3205–3214.
27. Mikels a J, Nusse R. Wnts as ligands: processing, secretion and reception. *Oncogene.* 2006;25(57):7461–7468.

28. Grumolato L, Liu G, Mong P, et al. Canonical and noncanonical Wnts use a common mechanism to activate completely unrelated coreceptors. *Genes Dev.* 2010;24 (22):2517–2530.
29. Wong GT, Gavin BJ, McMahon AP. Differential transformation of mammary epithelial cells by Wnt genes. *Mol Cell Biol.* 1994;14(9):6278–6286.
30. Shimizu H, Julius MA, Giarré M, Zheng Z, Brown AM, Kitajewski J. Transformation by Wnt family proteins correlates with regulation of beta-catenin. *Cell Growth Differ.* 1997;8(12):1349–1358.
31. Niehrs C. On growth and form: a Cartesian coordinate system of Wnt and BMP signaling specifies bilaterian body axes. *Development.* 2010;137(6):845–857.
32. Wray J, Hartmann C. WNTing embryonic stem cells. *Trends Cell Biol.* 2012;22(3):159–168.
33. Toledo EM, Colombres M, Inestrosa NC. Wnt signaling in neuroprotection and stem cell differentiation. *Prog Neurobiol.* 2008;86(3):281–296.
34. Cadigan KM, Nusse R. Wnt signaling: a common theme in animal development. *Genes Dev.* 1997;11(24):3286–3305.
35. Reya T, Clevers H. Wnt signalling in stem cells and cancer. *Nature.* 2005;434 (7035):843–850.
36. Aberle H, Bauer A, Stappert J, Kispert A, Kemler R. β-catenin is a target for the ubiquitin–proteasome pathway. *EMBO J.* 1997;16(13):3797–3804.
37. Liu C, Li Y, Semenov M, et al. Control of beta-catenin phosphorylation/degradation by a dual-kinase mechanism. *Cell.* 2002;108(6):837–847.
38. Bilic J, Huang Y-L, Davidson G, et al. Wnt induces LRP6 signalosomes and promotes dishevelled-dependent LRP6 phosphorylation. *Science.* 2007;316(5831):1619–1622.
39. Gao C, Chen Y-G. Dishevelled: the hub of Wnt signaling. *Cell Signal.* 2010;22 (5):717–727.
40. Cliffe A, Hamada F, Bienz M. A role of dishevelled in relocating Axin to the plasma membrane during wingless signaling. *Curr Biol.* 2003;13(11):960–966.
41. Hodar C, Assar R, Colombres M, et al. Genome-wide identification of new Wnt/beta-catenin target genes in the human genome using CART method. *BMC Genom.* 2010;11:348.
42. Arrázola MS, Varela-Nallar L, Colombres M, et al. Calcium/calmodulin-dependent protein kinase type IV is a target gene of the Wnt/beta-catenin signaling pathway. *J Cell Physiol.* 2009;221(3):658–667.
43. Acebron SP, Niehrs C. β-Catenin-independent roles of Wnt/LRP6 signaling. *Trends Cell Biol.* 2016;26(12):956–967.
44. Inoki K, Ouyang H, Zhu T, et al. TSC2 integrates Wnt and energy signals via a coordinated phosphorylation by AMPK and GSK3 to regulate cell growth. *Cell.* 2006;126(5):955–968.
45. Acebron SP, Karaulanov E, Berger BS, Huang Y-L, Niehrs C. Mitotic Wnt signaling promotes protein stabilization and regulates cell size. *Mol Cell.* 2014;54(4):663–674.
46. Kumar V, Zhang M-X, Swank MW, Kunz J, Wu G-Y. Regulation of dendritic morphogenesis by Ras-PI3K-Akt-mTOR and Ras-MAPK signaling pathways. *J Neurosci.* 2005;25(49):11288–11299.
47. Lee C-C, Huang C-C, Hsu K-S. Insulin promotes dendritic spine and synapse formation by the PI3K/Akt/mTOR and Rac1 signaling pathways. *Neuropharmacology.* 2011;61(4):867–879.
48. Vinson CR, Adler PN. Directional non-cell autonomy and the transmission of polarity information by the frizzled gene of *Drosophila*. *Nature.* 1987;329(6139):549–551.
49. Lawrence PA, Shelton PM. The determination of polarity in the developing insect retina. *J Embryol Exp Morphol.* 1975;33(2):471–486.
50. Simons M, Mlodzik M. Planar cell polarity signaling: from fly development to human disease. *Annu Rev Genet.* 2008;42:517–540.

51. Vladar EK, Antic D, Axelrod JD. Planar cell polarity signaling: the developing cell's compass. *Cold Spring Harb Perspect Biol.* 2009;1(3):a002964.
52. Bayly R, Axelrod JD. Pointing in the right direction: new developments in the field of planar cell polarity. *Nat Rev Genet.* 2011;12(6):385–391.
53. Chen CH, He CW, Liao CP, Pan CL. A Wnt-planar polarity pathway instructs neurite branching by restricting F-actin assembly through endosomal signaling. *PLoS Genet.* 2017;13(4):e1006720.
54. Rosso SB, Sussman D, Wynshaw-Boris A, Salinas PC. Wnt signaling through Dishevelled, Rac and JNK regulates dendritic development. *Nat Neurosci.* 2005;8(1):34–42.
55. Kim MJ, Futai K, Jo J, Hayashi Y, Cho K, Sheng M. Synaptic accumulation of PSD-95 and synaptic function regulated by phosphorylation of serine-295 of PSD-95. *Neuron.* 2007;56(3):488–502.
56. Leybaert L, Sanderson MJ. Intercellular Ca(2+) waves: mechanisms and function. *Physiol Rev.* 2012;92(3):1359–1392.
57. Malenka RC, Kauer JA, Zucker RS, Nicoll RA. Postsynaptic calcium is sufficient for potentiation of hippocampal synaptic transmission. *Science.* 1988;242:81–84.
58. Lee S-JR. Escobedo-Lozoya Y, Szatmari EM, Yasuda R. Activation of CaMKII in single dendritic spines during long-term potentiation. *Nature.* 2009;458 (7236):299–304.
59. Topolnik L. Dendritic calcium mechanisms and long-term potentiation in cortical inhibitory interneurons. *Eur J Neurosci.* 2012;35(4):496–506.
60. Slusarski DC, Corces DG, Moon RT. Interaction of Wnt and a Frizzled homologue triggers phosphatidylinositol signalling. *Lett to Nat.* 1997;390:410–413.
61. Kohn AD, Moon RT. Wnt and Calcium signaling: beta-catenin-independent pathways. *Cell Calcium.* 2005;38(3–4):439–446.
62. Inestrosa NC, Montecinos-Oliva C, Fuenzalida M. Wnt Signaling: role in Alzheimer disease and schizophrenia. *J Neuroimmune Pharmacol.* 2012;7(4):788–807.
63. van Amerongen R, Fuerer C, Mizutani M, Nusse R. Wnt5a can both activate and repress Wnt/β-catenin signaling during mouse embryonic development. *Dev Biol.* 2012;369(1):101–114.
64. Sato A, Yamamoto H, Sakane H, Koyama H, Kikuchi A. Wnt5a regulates distinct signalling pathways by binding to Frizzled2. *EMBO J.* 2010;29(1):41–54.
65. Andersson ER, Salto C, Villaescusa JC, et al. Wnt5a cooperates with canonical Wnts to generate midbrain dopaminergic neurons in vivo and in stem cells. *Proc Natl Acad Sci.* 2013;110(7):E602–E610.
66. Avila ME, Sepúlveda FJ, Burgos CF, et al. Canonical Wnt3a modulates intracellular calcium and enhances excitatory neurotransmission in hippocampal neurons. *J Biol Chem.* 2010;285(24):18939–18947.
67. Thrasivoulou C, Millar M, Ahmed A. Activation of intracellular calcium by multiple Wnt ligands and translocation of β-catenin into the nucleus: a convergent model of Wnt/Ca2+ and Wnt/β-catenin pathways. *J Biol Chem.* 2013;288(50):35651–35659.
68. Shimogori T, VanSant J, Paik E, Grove EA. Members of the Wnt, Fz, and Frp gene families expressed in postnatal mouse cerebral cortex. *J Comp Neurol.* 2004;473 (4):496–510.
69. Wayman Ga, Impey S, Marks D, et al. Activity-dependent dendritic arborization mediated by CaM-kinase I activation and enhanced CREB-dependent transcription of Wnt-2. *Neuron.* 2006;50(6):897–909.
70. Davis EK, Zou Y, Ghosh A. Wnts acting through canonical and noncanonical signaling pathways exert opposite effects on hippocampal synapse formation. *Neural Dev.* 2008;3:32.
71. Gogolla N, Galimberti I, Deguchi Y, Caroni P. Wnt signaling mediates experience-related regulation of synapse numbers and mossy fiber connectivities in the adult hippocampus. *Neuron.* 2009;62:510–525.

72. Wisniewska MB. Physiological role of β-catenin/TCF signaling in neurons of the adult brain. *Neurochem Res.* 2013;38(6):1144–1155.
73. Wisniewska MB, Nagalski A, Dabrowski M, Misztal K, Kuznicki J. Novel β-catenin target genes identified in thalamic neurons encode modulators of neuronal excitability. *BMC Genom.* 2012;13(1):635.
74. L'Episcopo F, Tirolo C, Testa N, et al. Reactive astrocyte Wnt/beta-catenin signaling link nigrostriatal injury to repair in 1-methyl-4-phenyl-1,2,3,6-tetrahydropyridine model of Parkinson's disease. *Neurobiol Dis.* 2011;41(2):508–527.
75. L'Episcopo F, Tirolo C, Testa N, et al. Wnt/β-catenin signaling is required to rescue midbrain dopaminergic progenitors and promote neurorepair in ageing mouse model of Parkinson's disease. *Stem Cells.* 2014;32(8):2147–2163.
76. Chong ZZ, Shang YC, Hou J, Maiese K. Wnt1 neuroprotection translates into improved neurological function during oxidant stress and cerebral ischemia through AKT1 and mitochondrial apoptotic pathways. *Oxid Med Cell Longev.* 2010;3(2):153–165.
77. Zhang J, Götz S, Vogt Weisenhorn DM, Simeone A, Wurst W, Prakash N. A WNT1-regulated developmental gene cascade prevents dopaminergic neurodegeneration in adult En1 +/− mice. *Neurobiol Dis.* 2015;82:32–45.
78. Zhou W-J, Xu N, Kong L, et al. The antidepressant roles of Wnt2 and Wnt3 in stress-induced depression-like behaviors. *Transl Psychiatry.* 2016;6(9):e892.
79. Okamoto H, Voleti B, Banasr M, et al. Wnt2 expression and signaling is increased by different classes of antidepressant treatments. *Biol Psychiatry.* 2010;68(6):521–527.
80. Hiester BG, Galati DF, Salinas PC, Jones KR. Neurotrophin and Wnt signaling cooperatively regulate dendritic spine formation. *Mol Cell Neurosci.* 2013;56:115–127.
81. Lie D-C, Colamarino SA, Song H-J, et al. Wnt signalling regulates adult hippocampal neurogenesis. *Nature.* 2005;437(7063):1370–1375.
82. Okamoto M, Inoue K, Iwamura H, et al. Reduction in paracrine Wnt3 factors during aging causes impaired adult neurogenesis. *FASEB J.* 2011;25(10):3570–3582.
83. Varela-Nallar L, Grabowski CP, Alfaro IE, Alvarez AR, Inestrosa NC. Role of the Wnt receptor Frizzled-1 in presynaptic differentiation and function. *Neural Dev.* 2009;4(1):41.
84. Oliva Ca, Inestrosa NC. A novel function for Wnt signaling modulating neuronal firing activity and the temporal structure of spontaneous oscillation in the entorhinal-hippocampal circuit. *Exp Neurol.* 2015;269:43–55.
85. Hooper C, Sainz-Fuertes R, Lynham S, et al. Wnt3a induces exosome secretion from primary cultured rat microglia. *BMC Neurosci.* 2012;13(1):144.
86. Halleskog C, Schulte G. WNT-3A and WNT-5A counteract lipopolysaccharide-induced pro-inflammatory changes in mouse primary microglia. *J Neurochem.* 2013;125(6):803–808.
87. Arrázola MS, Ramos-Fernández E, Cisternas P, Ordenes D, Inestrosa NC. Wnt signaling prevents the Aβ oligomer-induced mitochondrial permeability transition pore opening preserving mitochondrial structure in hippocampal neurons. *PLoS One.* 2017;12(1):e0168840.
88. Farías GG, Alfaro IE, Cerpa W, et al. Wnt-5a/JNK signaling promotes the clustering of PSD-95 in hippocampal neurons. *J Biol Chem.* 2009;284(23):15857–15866.
89. Codocedo JF, Montecinos-Oliva C, Inestrosa NC. Wnt-related SynGAP1 is a neuroprotective factor of glutamatergic synapses against Aβ oligomers. *Front Cell Neurosci.* 2015;9:227.
90. Varela-Nallar L, Alfaro IE, Serrano FG, Parodi J, Inestrosa NC. Wingless-type family member 5A (Wnt-5a) stimulates synaptic differentiation and function of glutamatergic synapses. *Proc Natl Acad Sci USA.* 2010;107(49):21164–21169.

91. Muñoz FJ, Godoy JA, Cerpa W, Poblete IM, Huidobro-Toro JP, Inestrosa NC. Wnt-5a increases NO and modulates NMDA receptor in rat hippocampal neurons. *Biochem Biophys Res Commun.* 2014;444(2):189–194.
92. Cerpa W, Farías GG, Godoy Ja, Fuenzalida M, Bonansco C, Inestrosa NC. Wnt-5a occludes Abeta oligomer-induced depression of glutamatergic transmission in hippocampal neurons. *Mol Neurodegener.* 2010;5:3.
93. Parodi J, Montecinos-Oliva C, Varas R, et al. Wnt5a inhibits K+ currents in hippocampal synapses through nitric oxide production. *Mol Cell Neurosci.* 2015;68:314–322.
94. Ramírez VT, Ramos-Fernández E, Henríquez JP, Lorenzo A, Inestrosa NC. Wnt-5a/Frizzled9 receptor signaling through the Gαo-Gβγ complex regulates dendritic spine formation. *J Biol Chem.* 2016;291(36):19092–19107.
95. Chen C-M, Orefice LL, Chiu S-L, et al. Wnt5a is essential for hippocampal dendritic maintenance and spatial learning and memory in adult mice. *Proc Natl Acad Sci USA.* 2017;114(4):E619–E628.
96. Cuitino L, Godoy Ja, Farías GG, et al. Wnt-5a modulates recycling of functional GABAA receptors on hippocampal neurons. *J Neurosci.* 2010;30(25):8411–8420.
97. Tabatadze N, Tomas C, McGonigal R, Lin B, Schook A, Routtenberg A. Wnt transmembrane signaling and long-term spatial memory. *Hippocampus.* 2012;22(6):1228–1241.
98. Zhang G-L, Zhang J, Li S-F, et al. Wnt-5a prevents Aβ-induced deficits in long-term potentiation and spatial memory in rats. *Physiol Behav.* 2015;149:95–100.
99. Blakely BD, Bye CR, Fernando CV, et al. Ryk, a receptor regulating Wnt5a-mediated neurogenesis and axon morphogenesis of ventral midbrain dopaminergic neurons. *Stem Cells Dev.* 2013;22(15):2132–2144.
100. Ciani L, Boyle KA, Dickins E, et al. Wnt7a signaling promotes dendritic spine growth and synaptic strength through Ca2+/Calmodulin-dependent protein kinase II. *Proc Natl Acad Sci.* 2011;108(26):10732–10737.
101. Sahores M, Gibb A, Salinas PC. Frizzled-5, a receptor for the synaptic organizer Wnt7a, regulates activity-mediated synaptogenesis. *Development.* 2010;137(13):2215–2225.
102. Qu Q, Sun G, Murai K, et al. Wnt7a regulates multiple steps of neurogenesis. *Mol Cell Biol.* 2013;33(13):2551–2559.
103. Farías GG, Vallés AS, Colombres M, et al. Wnt-7a induces presynaptic colocalization of alpha 7-nicotinic acetylcholine receptors and adenomatous polyposis coli in hippocampal neurons. *J Neurosci.* 2007;27(20):5313–5325.
104. Ciani L, Marzo A, Boyle K, et al. Wnt signalling tunes neurotransmitter release by directly targeting Synaptotagmin-1. *Nat Commun.* 2015;6:8302.
105. Lucas FR, Salinas PC. WNT-7a induces axonal remodeling and increases synapsin i levels in cerebellar neurons. *Dev Biol.* 1997;192:31–44.
106. Fernando CV, Kele J, Bye CR, et al. Diverse roles for Wnt7a in ventral midbrain neurogenesis and dopaminergic axon morphogenesis. *Stem Cells Dev.* 2014;23(17):1991–2003.
107. Ahmad-Annuar A, Ciani L, Simeonidis I, et al. Signaling across the synapse: a role for Wnt and Dishevelled in presynaptic assembly and neurotransmitter release. *J Cell Biol.* 2006;174:127–139.
108. Slater PG, Ramirez VT, Gonzalez-Billault C, Varela-Nallar L, Inestrosa NC. Frizzled-5 receptor is involved in neuronal polarity and morphogenesis of hippocampal neurons. *PLoS One.* 2013;8(10):e78892.
109. Clark CEJ, Richards LJ, Stacker SA, Cooper HM. Wnt5a induces Ryk-dependent and -independent effects on callosal axon and dendrite growth. *Growth Factors.* 2014;32(1):11–17.
110. Park M, Shen K. WNTs in synapse formation and neuronal circuitry. *EMBO J.* 2012;31(12):2697–2704.

111. Paganoni S, Bernstein J, Ferreira A. Ror1-Ror2 complexes modulate synapse formation in hippocampal neurons. *Neuroscience.* 2010;165(4):1261–1274.
112. Cerpa W, Godoy Ja, Alfaro I, et al. Wnt-7a modulates the synaptic vesicle cycle and synaptic transmission in hippocampal neurons. *J Biol Chem.* 2008;283(9):5918–5927.
113. Hall AC, Lucas FR, Salinas PC. Axonal remodeling and synaptic differentiation in the cerebellum is regulated by WNT-7a signaling. *Cell.* 2000;100(5):525–535.
114. Lanoue V, Langford M, White A, Sempert K, Fogg L, Cooper HM. The Wnt receptor Ryk is a negative regulator of mammalian dendrite morphogenesis. *Sci Rep.* 2017;7:1–11.
115. Castelo-Branco G, Wagner J, Rodriguez FJ, et al. Differential regulation of midbrain dopaminergic neuron development by Wnt-1, Wnt-3a, and Wnt-5a. *Proc Natl Acad Sci USA.* 2003;100(22):12747–12752.
116. Jessberger S, Clark RE, Broadbent NJ, et al. Dentate gyrus-specific knockdown of adult neurogenesis impairs spatial and object recognition memory in adult rats. *Learn Mem.* 2009;16(2):147–154.
117. Kuwabara T, Hsieh J, Muotri A, et al. Wnt-mediated activation of NeuroD1 and retro-elements during adult neurogenesis. *Nat Neurosci.* 2009;12(9):1097–1105.
118. Luo Y, Shan G, Guo W, et al. Fragile X mental retardation protein regulates proliferation and differentiation of adult neural stem/progenitor cells. *PLoS Genet.* 2010;6(4).
119. Cerpa W, Gambrill A, Inestrosa NC, Barria A. Regulation of NMDA-receptor synaptic transmission by Wnt signaling. *J Neurosci.* 2011;31:9466–9471.
120. Yu X, Malenka RC. Beta-catenin is critical for dendritic morphogenesis. *Nat Neurosci.* 2003;6(11):1169–1177.
121. Alvarez VA, Sabatini BL. Anatomical and physiological plasticity of dendritic spines. *Annu Rev Neurosci.* 2007;30(1):79–97.
122. Holtmaat A, Svoboda K. Experience-dependent structural synaptic plasticity in the mammalian brain. *Nat Rev Neurosci.* 2009;10(9):647–658.
123. Bosch M, Hayashi Y. Structural plasticity of dendritic spines. *Curr Opin Neurobiol.* 2012;22(3):383–388.
124. Redmond L, Kashani AH, Ghosh A. Calcium regulation of dendritic growth via CaM kinase IV and CREB-mediated transcription. *Neuron.* 2002;34(6):999–1010.
125. Larson J, Wong D, Lynch G. Patterned stimulation at the theta frequency is optimal for the induction of hippocampal long-term potentiation. *Brain Res.* 1986;368(2):347–350.
126. Chen J, Park CS, Tang S-J. Activity-dependent synaptic Wnt release regulates hippocampal long term potentiation. *J Biol Chem.* 2006;281(17):11910–11916.
127. Jang M-H, Bonaguidi MA, Kitabatake Y, et al. Secreted frizzled-related protein 3 regulates activity-dependent adult hippocampal neurogenesis. *Cell Stem Cell.* 2013;12(2):215–223.
128. Budnik V, Salinas PC. Wnt signaling during synaptic development and plasticity. *Curr Opin Neurobiol.* 2011;21:151–159.
129. Maguschak KA, Ressler KJ. β-catenin is required for memory consolidation. *Nat Neurosci.* 2008;11(11):1319–1326.
130. Maguschak KA, Ressler KJ. Wnt Signaling in amygdala-dependent learning and memory. *J Neurosci.* 2011;31(37):13057–13067.
131. Fortress AM, Schram SL, Tuscher JJ, Frick KM. Canonical Wnt signaling is necessary for object recognition memory consolidation. *J Neurosci.* 2013;33(31):12619–12626.
132. Koshimizu H, Fukui Y, Takao K, et al. Adenomatous polyposis coli heterozygous knockout mice display hypoactivity and age-dependent working memory deficits. *Front Behav Neurosci.* 2011;5:85.
133. Ivaniutsin U, Chen Y, Mason JO, Price DJ, Pratt T. Adenomatous polyposis coli is required for early events in the normal growth and differentiation of the developing cerebral cortex. *Neural Dev.* 2009;4(1):3.

134. Tan Y, Yu D, Busto GU, Wilson C, Davis RL. Wnt signaling is required for long-term memory formation. *Cell Rep*. 2013;4(6):1082–1089.
135. Vargas JY, Fuenzalida M, Inestrosa NC. In vivo activation of Wnt signaling pathway enhances cognitive function of adult mice and reverses cognitive deficits in an Alzheimer's disease model. *J Neurosci*. 2014;34(6):2191–2202.
136. Hickey MA, Kosmalska A, Enayati J, et al. Extensive early motor and non-motor behavioral deficits are followed by striatal neuronal loss in knock-in Huntington's disease mice. *Neuroscience*. 2008;157(1):280–295.
137. Hermann D, Both M, Ebert U, et al. Synaptic transmission is impaired prior to plaque formation in amyloid precursor protein-overexpressing mice without altering behaviorally-correlated sharp wave-ripple complexes. *Neuroscience*. 2009;162(4):1081–1090.
138. Palop JJ, Mucke L. Amyloid-β–induced neuronal dysfunction in Alzheimer's disease: from synapses toward neural networks. *Nat Neurosci*. 2010;13(7):812–818.
139. Penzes P, Buonanno A, Passafaro M, Sala C, Sweet RA. Developmental vulnerability of synapses and circuits associated with neuropsychiatric disorders. *J Neurochem*. 2013;126 (2):165–182.
140. Benilova I, Karran E, De Strooper B. The toxic Aβ oligomer and Alzheimer's disease: an emperor in need of clothes. *Nat Neurosci*. 2012;15(3):349–357.
141. Pimplikar SW, Nixon Ra, Robakis NK, Shen J, Tsai L-H. Amyloid-independent mechanisms in Alzheimer's disease pathogenesis. *J Neurosci*. 2010;30(45):14946–14954.
142. Institute on Aging N. Alzheimer's Disease Medications Fact Sheet. 2017. Available from: https://order.nia.nih.gov/publication/alzheimers-disease-medications-fact-sheet
143. Inestrosa NC, Toledo EM. The role of Wnt signaling in neuronal dysfunction in Alzheimer's Disease. *Mol Neurodegener*. 2008;3(1):9.
144. De Ferrari GV, Papassotiropoulos A, Biechele T, et al. Common genetic variation within the low-density lipoprotein receptor-related protein 6 and late-onset Alzheimer's disease. *Proc Natl Acad Sci USA*. 2007;104(22):9434–9439.
145. Herz J. The LDL receptor gene family: (un)expected signal transducers in the brain. *Neuron*. 2001;29(3):571–581.
146. Caricasole A, Copani A, Caraci F, et al. Induction of Dickkopf-1, a negative modulator of the Wnt pathway, is associated with neuronal degeneration in Alzheimer's brain. *Neurobiol Dis*. 2004;24(26):6021–6027.
147. Rosi MC, Luccarini I, Grossi C, et al. Increased Dickkopf-1 expression in transgenic mouse models of neurodegenerative disease. *J Neurochem*. 2010;112(6):1539–1551.
148. Purro Sa, Dickins EM, Salinas PC. The secreted Wnt antagonist Dickkopf-1 is required for amyloid β-mediated synaptic loss. *J Neurosci*. 2012;32(10):3492–3498.
149. Marzo A, Galli S, Lopes D, et al. Reversal of synapse degeneration by restoring Wnt signaling in the adult hippocampus article reversal of synapse degeneration by restoring Wnt signaling in the adult hippocampus. *Curr Biol*. 2016;26(19):2551–2561.
150. Hurtado DE, Molina-Porcel L, Carroll JC, et al. Selectively silencing GSK-3 isoforms reduces plaques and tangles in mouse models of Alzheimer's disease. *J Neurosci*. 2012;32 (21):7392–7402.
151. Ly PTT, Wu Y, Zou H, et al. Inhibition of GSK3β-mediated BACE1 expression reduces Alzheimer-associated phenotypes. *224cehsp sp025/J Clin Investig*. 2013;123(1).
152. Long Z-MM. Zhao L, Jiang R, et al. Valproic acid modifies synaptic structure and accelerates neurite outgrowth via the glycogen synthase kinase-3?.? signaling pathway in an Alzheimer's disease model. *CNS Neurosci Ther*. 2015;21(11):887–897.
153. Avrahami L, Farfara D, Shaham-Kol M, Vassar R, Frenkel D, Eldar-Finkelman H. Inhibition of glycogen synthase kinase-3 ameliorates β-amyloid pathology and restores lysosomal acidification and mammalian target of rapamycin activity in the Alzheimer disease mouse model: in vivo and in vitro studies. *J Biol Chem*. 2013;288(2):1295–1306.

154. Tiwari SK, Seth B, Agarwal S, et al. Ethosuximide induces hippocampal neurogenesis and reverses cognitive deficits in an amyloid-β toxin-induced Alzheimer rat model via the phosphatidylinositol 3-kinase (PI3K)/Akt/Wnt/β-catenin pathway. *J Biol Chem.* 2015;290(47):28540–28558.
155. Takashima A. GSK-3 is essential in the pathogenesis of Alzheimer's disease. *J Alzheimers Dis.* 2006;9(3 Suppl):309–317.
156. Hooper C, Killick R, Lovestone S. The GSK3 hypothesis of Alzheimer's disease. *J Neurochem.* 2008;104(6):1433–1439.
157. Tapia-Rojas C, Burgos PV, Inestrosa NC. Inhibition of Wnt signaling induces amyloidogenic processing of amyloid precursor protein and the production and aggregation of Amyloid-β (Aβ)$_{42}$ peptides. *J Neurochem.* 2016;139(6):1175–1191.
158. Fiorentini A, Rosi MC, Grossi C, Luccarini I, Casamenti F. Lithium improves hippocampal neurogenesis, neuropathology and cognitive functions in APP mutant mice. *PLoS One.* 2010;5(12.).
159. Serrano FG, Tapia-Rojas C, Carvajal FJ, Hancke J, Cerpa W, Inestrosa NC. Andrographolide reduces cognitive impairment in young and mature AβPPswe/PS-1 mice. *Mol Neurodegener.* 2014;9(1):61.
160. Toledo EM, Inestrosa NC. Activation of Wnt signaling by lithium and rosiglitazone reduced spatial memory impairment and neurodegeneration in brains of an APPswe/PSEN1DeltaE9 mouse model of Alzheimer's disease. *Mol Psychiatry.* 2010;15(3):272–285.
161. Caracci MO, Avila ME, De Ferrari GV. Synaptic Wnt/GSK3beta signaling hub in Autism. *Neural Plast.* 2016;2016:9603751.
162. Lijam N, Paylor R, McDonald MP, et al. Social interaction and sensorimotor gating abnormalities in mice lacking Dvl1. *Cell.* 1997;90(5):895–905.
163. Vincent JB, Herbrick J-A, Gurling HMD, Bolton PF, Roberts W, Scherer SW. Identification of a novel gene on chromosome 7q31 that is interrupted by a translocation breakpoint in an autistic individual. *Am J Hum Genet.* 2000;67(2):510–514.
164. Wassink TH, Piven J, Vieland VJ, et al. Evidence supporting WNT2 as an autism susceptibility gene. *Am J Med Genet.* 2001;105(5):406–413.
165. McCoy PA, Shao Y, Wolpert CM, et al. No association between the WNT2 gene and autistic disorder. *Am J Med Genet.* 2002;114(1):106–109.
166. Zhang Y, Sun Y, Wang F, Wang Z, Peng Y, Li R. Downregulating the canonical Wnt/β-catenin signaling pathway attenuates the susceptibility to autism-like phenotypes by decreasing oxidative stress. *Neurochem Res.* 2012;37(7):1409–1419.
167. Wang Z, Xu L, Zhu X, et al. Demethylation of specific Wnt/β-catenin pathway genes and its upregulation in rat brain induced by prenatal valproate exposure. *Anat Rec.* 2010;293(11):1947–1953.
168. Chenn A, Walsh CA. Increased neuronal production, enlarged forebrains and cytoarchitectural distortions in beta-catenin overexpressing transgenic mice. *Cereb Cortex.* 2003;13(6):599–606.
169. Belinson H, Nakatani J, Babineau BA, et al. Prenatal β-catenin/Brn2/Tbr2 transcriptional cascade regulates adult social and stereotypic behaviors. *Mol Psychiatry.* 2016;21(10):1417–1433.
170. Todd PK, Malter JS. Fragile X mental retardation protein in plasticity and disease. *J Neurosci Res.* 2002;70(5):623–630.
171. Santoro MR, Bray SM, Warren ST. Molecular mechanisms of fragile X syndrome: a twenty-year perspective. *Annu Rev Pathol.* 2012;7:219–245.
172. Rudelli RD, Brown WT, Wisniewski K, et al. Adult fragile X syndrome. Clinico-neuropathologic findings. *Acta Neuropathol.* 1985;67(3–4):289–295.
173. Wisniewski KE, Segan SM, Miezejeski CM, Sersen EA, Rudelli RD. The Fra(X) syndrome: neurological, electrophysiological, and neuropathological abnormalities. *Am J Med Genet.* 1991;38(2–3):476–480.

174. Zhang A, Shen C-H, Ma SY, Ke Y, El Idrissi A. Altered expression of Autism-associated genes in the brain of fragile X mouse model. *Biochem Biophys Res Commun.* 2009;379(4):920–923.
175. Rosales-Reynoso MA, Ochoa-Hernández AB, Aguilar-Lemarroy A, Jave-Suárez LF, Troyo-Sanromán R, Barros-Núñez P. Gene expression profiling identifies WNT7A as a possible candidate gene for decreased cancer risk in fragile X syndrome patients. *Arch Med Res.* 2010;41(2):110–118.
176. Wang H, Fukushima H, Kida S, Zhuo M. Ca2+/Calmodulin-dependent protein kinase IV links group I metabotropic glutamate receptors to fragile X mental retardation protein in cingulate cortex. *J Biol Chem.* 2009;284(28):18953–18962.
177. Wang H, Wu L-J, Zhang F, Zhuo M. Roles of calcium-stimulated adenylyl cyclase and calmodulin-dependent protein kinase IV in the regulation of FMRP by group I metabotropic glutamate receptors. *J Neurosci.* 2008;28(17):4385–4397.
178. Waltes R, Duketis E, Knapp M, et al. Common variants in genes of the postsynaptic FMRP signalling pathway are risk factors for autism spectrum disorders. *Hum Genet.* 2014;133(6):781–792.
179. Zalfa F, Eleuteri B, Dickson KS, et al. A new function for the fragile X mental retardation protein in regulation of PSD-95 mRNA stability. *Nat Neurosci.* 2007;10(5):578–587.
180. Hering H, Sheng M. Direct interaction of Frizzled-1, -2, -4, and -7 with PDZ domains of PSD-95. *FEBS Lett.* 2002;521(1–3):185–189.
181. Todd PK, Mack KJ, Malter JS. The fragile X mental retardation protein is required for type-I metabotropic glutamate receptor-dependent translation of PSD-95. *Proc Natl Acad Sci USA.* 2003;100(24):14374–14378.
182. Gandhi RM, Kogan CS, Messier C, MacLeod LS. Visual–spatial learning impairments are associated with hippocampal PSD-95 protein dysregulation in a mouse model of fragile X syndrome. *Neuroreport.* 2014;25(4):255–261.
183. Eadie BD, Zhang WN, Boehme F, et al. Fmr1 knockout mice show reduced anxiety and alterations in neurogenesis that are specific to the ventral dentate gyrus. *Neurobiol Dis.* 2009;36(2):361–373.
184. Mak BC, Takemaru K, Kenerson HL, Moon RT, Yeung RS. The tuberin-hamartin complex negatively regulates beta-catenin signaling activity. *J Biol Chem.* 2003;278(8):5947–5951.
185. Boon EM, van der Neut R, van de Wetering M, Clevers H, Pals ST. Wnt signaling regulates expression of the receptor tyrosine kinase met in colorectal cancer. *Cancer Res.* 2002;62(18):5126–5128.
186. Campbell DB, Sutcliffe JS, Ebert PJ, et al. A genetic variant that disrupts MET transcription is associated with autism. *Proc Natl Acad Sci USA.* 2006;103(45):16834–16839.
187. O'Roak BJ, Vives L, Fu W, et al. Multiplex targeted sequencing identifies recurrently mutated genes in autism spectrum disorders. *Science.* 2012;338(6114):1619–1622.
188. Zhou J, Parada LF. PTEN signaling in autism spectrum disorders. *Curr Opin Neurobiol.* 2012;22(5):873–879.
189. Mohn JL, Alexander J, Pirone A, et al. Adenomatous polyposis coli protein deletion leads to cognitive and autism-like disabilities. *Mol Psychiatry.* 2014;19(10):1133–1142.
190. Lee SH, Peng IF, Ng YG, et al. Synapses are regulated by the cytoplasmic tyrosine kinase Fer in a pathway mediated by p120catenin, Fer, SHP-2, and beta-catenin. *J Cell Biol.* 2008;183(5):893–908.
191. Gershon S, Chengappa KR, Malhi GS. Lithium specificity in bipolar illness: a classic agent for the classic disorder. *Bipolar Disord.* 2009;11:34–44.
192. Müller-Oerlinghausen B, Lewitzka U. Lithium reduces pathological aggression and suicidality: a mini-review. *Neuropsychobiology.* 2010;62(1):43–49.

193. Timothy O'Brien W, Klein PS. Validating GSK3 as an in vivo target of lithium action: Table 1. *Biochem SocTrans.* 2009;37(5):1133–1138.
194. Timothy O, 'brien W, Harper AD, Jové F, et al. Glycogen synthase kinase-3β haploinsufficiency mimics the behavioral and molecular effects of lithium. *J Neurosci.* 2004;24 (30):6791–6798.
195. Sutton LP, Rushlow WJ. The effects of neuropsychiatric drugs on glycogen synthase kinase-3 signaling. *Neuroscience.* 2011;199:116–124.
196. Chen G, Rajkowska G, Du F, Seraji-Bozorgzad N, Manji HK. Enhancement of hippocampal neurogenesis by lithium. *J Neurochem.* 2000;75(4):1729–1734.
197. Wexler EM, Geschwind DH, Palmer TD. Lithium regulates adult hippocampal progenitor development through canonical Wnt pathway activation. *Mol Psychiatry.* 2008;13(3):285–292.
198. Gould TD, Chen G, Manji HK. In vivo evidence in the brain for lithium inhibition of glycogen synthase kinase-3. *Neuropsychopharmacology.* 2004;29(1):32–38.
199. Klein PS, Melton DA. A molecular mechanism for the effect of lithium on development. *Proc Natl Acad Sci USA.* 1996;93(16):8455–8459.
200. Stambolic V, Ruel L, Woodgett JR. Lithium inhibits glycogen synthase kinase-3 activity and mimics wingless signalling in intact cells. *Curr Biol.* 1996;6(12):1664–1668.
201. Gould TD, Einat H, O'Donnell KC, Picchini AM, Schloesser RJ, Manji HK. Beta-catenin overexpression in the mouse brain phenocopies lithium-sensitive behaviors. *Neuropsychopharmacology.* 2007;32(10):2173–2183.
202. Ahnaou A, Drinkenburg WHIM. Disruption of glycogen synthase kinase-3-beta activity leads to abnormalities in physiological measures in mice. *Behav Brain Res.* 2011;221 (1):246–252.
203. Prickaerts J. Transgenic mice overexpressing glycogen synthase kinase 3beta: a putative model of hyperactivity and mania. *J Neurosci.* 2006;26(35):9022–9029.
204. Luykx JJ, Boks MPM, Terwindt APR, Bakker S, Kahn RS, Ophoff RA. The involvement of GSK3β in bipolar disorder: Integrating evidence from multiple types of genetic studies. *Eur Neuropsychopharmacol.* 2010;20(6):357–368.
205. Van Erp TGM, Thompson PM, Kieseppä T, et al. Hippocampal morphology in lithium and non-lithium-treated bipolar i disorder patients, non-bipolar co-twins, and control twins. *Hum Brain Mapp.* 2012;33(3):501–510.
206. Foland LC, Altshuler LL, Sugar CA, et al. Increased volume of the amygdala and hippocampus in bipolar patients treated with lithium. *Neuroreport.* 2008;19 (2):221–224.
207. Yucel K, Taylor VH, McKinnon MC, et al. Bilateral hippocampal volume increase in patients with bipolar disorder and short-term lithium treatment. *Neuropsychopharmacology.* 2008;33(2):361–367.
208. Pandey GN, Rizavi HS, Tripathi M, Ren X. Region-specific dysregulation of GSK-3β and β-catenin in the postmortem brain of subjects with bipolar disorder and schizophrenia. *Bipolar Disord.* 2015;17(2):160–171.
209. Huang C, Fu X-H, Zhou D, Li J-M. The role of Wnt/β-catenin signaling pathway in disrupted hippocampal neurogenesis of temporal lobe epilepsy: A potential therapeutic target? *Neurochem Res.* 2015;40(7):1319–1332.
210. Scharfman HE, Goodman JH, Sollas AL. Granule-like neurons at the hilar/CA3 border after status epilepticus and their synchrony with area CA3 pyramidal cells: functional implications of seizure-induced neurogenesis. *J Neurosci.* 2000;20(16):6144–6158.
211. Kobayashi M, Buckmaster PS. Reduced inhibition of dentate granule cells in a model of temporal lobe epilepsy. *J Neurosci.* 2003;23(6):2440–2452.
212. Kobayashi M, Wen X, Buckmaster PS. Reduced inhibition and increased output of layer II neurons in the medial entorhinal cortex in a model of temporal lobe epilepsy. *J Neurosci.* 2003;23(24):8471–8479.

213. Liu S, Wang J, Zhu D, Fu Y, Lukowiak K, Lu YM. Generation of functional inhibitory neurons in the adult rat hippocampus. *J Neurosci.* 2003;23(3):732–736.
214. Busceti CL, Biagioni F, Aronica E, et al. Induction of the Wnt inhibitor, Dickkopf-1, is associated with neurodegeneration related to temporal lobe epilepsy. *Epilepsia.* 2007;48 (4):694–705.
215. Cappuccio I, Calderone A, Busceti CL, et al. Induction of Dickkopf-1, a negative modulator of the Wnt pathway, is required for the development of ischemic neuronal death. *J Neurosci.* 2005;25(10):2647–2657.
216. Lange C, Mix E, Frahm J, et al. Small molecule GSK-3 inhibitors increase neurogenesis of human neural progenitor cells. *Neurosci Lett.* 2011;488(1):36–40.
217. Pirone A, Alexander J, Lau LA, et al. APC conditional knock-out mouse is a model of infantile spasms with elevated neuronal β-catenin levels, neonatal spasms, and chronic seizures. *Neurobiol Dis.* 2017;98:149–157.
218. Poewe W, Seppi K, Tanner CM, et al. Parkinson disease. *Nat Rev Dis Prim.* 2017;3:17013.
219. Sousa KM, Villaescusa JC, Cajanek L, et al. Wnt2 regulates progenitor proliferation in the developing ventral midbrain. *J Biol Chem.* 2010;285(10):7246–7253.
220. Zhang L, Deng J, Pan Q, et al. Targeted methylation sequencing reveals dysregulated Wnt signaling in Parkinson disease. *J Genet Genomics.* 2016;43(10):587–592.
221. Rivetti di Val Cervo P, Romanov RA, Spigolon G, et al. Induction of functional dopamine neurons from human astrocytes in vitro and mouse astrocytes in a Parkinson's disease model. *Nat Biotechnol.* 2017;35(5):444–452.
222. Antonova E, Sharma T, Morris R, Kumari V. The relationship between brain structure and neurocognition in schizophrenia: a selective review. *Schizophr Res.* 2004;70 (2–3):117–145.
223. Zaidel DW, Esiri MM, Harrison PJ. Size, shape, and orientation of neurons in the left and right hippocampus: investigation of normal asymmetries and alterations in schizophrenia. *Am J Psychiatry.* 1997;154(6):812–818.
224. Zaidel DW, Esiri MM, Harrison PJ. The hippocampus in schizophrenia: lateralized increase in neuronal density and altered cytoarchitectural asymmetry. *Psychol Med.* 1997;27(3):703–713.
225. Ftouh S, Akbar MT, Hirsch SR, De Belleroche JS. Down-regulation of Dickkopf 3, a regulator of the Wnt signalling pathway, in elderly schizophrenic subjects. *J Neurochem.* 2005;94(2):520–530.
226. Cotter D, Kerwin R, al-Sarraji S, et al. Abnormalities of Wnt signalling in schizophrenia: evidence for neurodevelopmental abnormality. *Neuroreport.* 1998;9(7):1379–1383.
227. Kozlovsky N, Belmaker RH, Agam G. Low GSK-3beta immunoreactivity in postmortem frontal cortex of schizophrenic patients. *Am J Psychiatry.* 2000;157(5):831–833.
228. Kozlovsky N, Belmaker RH, Agam G. Low GSK-3 activity in frontal cortex of schizophrenic patients. *Schizophr Res.* 2001;52(1–2):101–105.
229. Kozlovsky N, Shanon-Weickert C, Tomaskovic-Crook E, Kleinman JE, Belmaker RH, Agam G. Reduced GSK-3?. mRNA levels in postmortem dorsolateral prefrontal cortex of schizophrenic patients. *J Neural Transm.* 2004;111(12):1583–1592.
230. Reif A, Fritzen S, Finger M, et al. Neural stem cell proliferation is decreased in schizophrenia, but not in depression. *Mol Psychiatry.* 2006;11(5):514–522.
231. Mao Y, Ge X, Frank CL, et al. Disrupted in schizophrenia 1 regulates neuronal progenitor proliferation via modulation of GSK3?./?.-catenin signaling. *Cell.* 2009;136(6):1017–1031.
232. Duan X, Chang JH, Ge S, et al. Disrupted-in-schizophrenia 1 regulates integration of newly generated neurons in the adult brain. *Cell.* 2007;130(6):1146–1158.
233. Levchenko A, Davtian S, Freylichman O, Zagrivnaya M, Kostareva A, Malashichev Y. Beta-catenin in schizophrenia: possibly deleterious novel mutation. *Psychiatry Res.* 2015;228(3):843–848.

234. Singh K. An emerging role for Wnt and GSK3 signaling pathways in schizophrenia. *Clin Genet.* 2013;83(6):511–517.
235. Min C, Cho D-I, Kwon K-J, Kim K-S, Shin CY, Kim K-M. Novel regulatory mechanism of canonical Wnt signaling by dopamine D2 receptor through direct interaction with β-catenin. *Mol Pharmacol.* 2011;80(1):68–78.
236. Sutton LP, Rushlow WJ. The dopamine D2 receptor regulates Akt and GSK-3 via Dvl-3. *Int J Neuropsychopharmacol.* 2012;15(7):965–979.
237. Sara P, Veronica A, Silvia B, et al. Impact of early and recent stress on white matter microstructure in major depressive disorder. *J Affect Disord.* 2018;225:289–297.
238. Malberg JE, Eisch AJ, Nestler EJ, Duman RS. Chronic antidepressant treatment increases neurogenesis in adult rat hippocampus. *J Neurosci.* 2000;20(24):9104–9110.
239. de Sousa RT, Zanetti MV, Talib LL, et al. Lithium increases platelet serine-9 phosphorylated GSK-3β levels in drug-free bipolar disorder during depressive episodes. *J Psychiatr Res.* 2015;62:78–83.
240. Hui J, Zhang J, Kim H, et al. Fluoxetine regulates neurogenesis in vitro through modulation of GSK-3/-catenin signaling. *Int J Neuropsychopharmacol.* 2015;18(5): pyu099.
241. Gassen NC, Hartmann J, Zannas AS, et al. FKBP51 inhibits GSK3β and augments the effects of distinct psychotropic medications. *Mol Psychiatry.* 2016;21(2):277–289.
242. Radu Enatescu V, Papava I, Enatescu I, et al. Circulating plasma miro RNAs in patients with major deppresive disorder treated with antidepressants: a pilot study. *Psychiatry Investig.* 2016;13(5):549–557.
243. Lopez JP, Fiori LM, Cruceanu C, et al. MicroRNAs 146a/b-5 and 425-3p and 24-3p are markers of antidepressant response and regulate MAPK/Wnt-system genes. *Nat Commun.* 2017;8:15497.

CHAPTER FOUR

Of Wnts and Ribosomes

Astrid S. Pfister[1], Michael Kühl
Institute of Biochemistry and Molecular Biology, Ulm University, Ulm, Germany
[1] Corresponding author. E-mail address: astrid.pfister@uni-ulm.de

Contents

Abstract

Wnt proteins are secreted glycoproteins that activate different intracellular signal transduction pathways. They regulate cell proliferation and are required for proper embryonic development. Misregulation of Wnt signaling can result in various diseases including cancer. In most circumstances, cell growth is essential for cell division and thus cell proliferation. Therefore, several reports have highlighted the key role of Wnt proteins for cell growth. Ribosomes represent the cellular protein synthesis machinery and cells need to be equipped with an appropriate number of ribosomes to allow cell growth. Recent findings suggest a role for Wnt proteins in regulating ribosome

Progress in Molecular Biology and Translational Science, Volume 153
ISSN 1877-1173
http://dx.doi.org/10.1016/bs.pmbts.2017.11.006

biogenesis and we here summarize these findings representing a previously unknown function of Wnt proteins. Understanding this role of Wnt signaling might open new avenues to slow down proliferation by drugs for instance in cancer therapy.

1. INTRODUCTION

1.1 Wnt Proteins, Wnt Pathways, and Wnt Targets

Wnt proteins are mitogenic growth factors constituting the Wnt family of extracellular glycoproteins. In vertebrates, Wnt ligands comprise a family of 19 members. The first member of the family, Wnt1, was originally identified as *int-1* in the mouse and was shown to function as a protooncogene.[1] Later, *int-1* was characterized as the homolog of the earlier identified *Drosophila* segment polarity gene *wingless*.[2,3] To avoid later confusion, the family of *int-1* and *wingless* genes was collectively denoted "Wnt" for *wingless*-related integration site, with int-1 being the founding member Wnt1.[4,5] Importantly, Wnt ligands drive cell proliferation and differentiation, and are therefore important for embryonic development and stem-cell biology. In contrast, dysregulation of Wnt signaling provokes the development of diverse pathologies and most prominently cancer.

Wnt proteins can activate different but interconnected intracellular signaling pathways through interaction with seven-pass transmembrane receptors of the Frizzled (Fz) family and different coreceptors, such as LRP5/6, Ror2, or Ryk.[6] These Wnt pathways were historically separated into canonical (β-catenin dependent) and noncanonical (β-catenin independent) branches. Of note, canonical and noncanonical Wnt ligands trigger recruitment of shared intracellular components, such as Dishevelled (Dvl), Axin and GSK3β to the respective receptor in order to activate distinct signaling cascades. Intriguingly, canonical and noncanonical Wnt ligands have been demonstrated to *per se* exert reciprocal Wnt pathway inhibition and thus the picture of a Wnt signaling network emerged.[6,7]

1.1.1 The Canonical Wnt/β-Catenin Signaling Pathway

The canonical Wnt/β-catenin-dependent signaling pathway regulates a variety of prosurvival processes, including cell proliferation, ribosome biogenesis, and inhibition of apoptosis. Here, we roughly summarize the key steps of pathway activation.[8]

In the absence of canonical Wnt ligands, cytosolic β-catenin, the central player of the pathway, is bound by components of the β-catenin destruction complex, composed of the tumor suppressor *Adenomatous polyposis coli* (APC), the scaffolding proteins Axin1/Axin2, the serine threonine kinase GSK3β and casein kinase 1α (CK1α). In this complex, β-catenin first gets earmarked by phosphorylation at specific serine and threonine residues and subsequently is polyubiquitinated by the E3-ligase β-TrCP to finally be degraded in the proteasome. Hence, cytosolic β-catenin levels are kept low and β-catenin dependent transcription of Wnt target genes is switched off.

Activation of the canonical Wnt pathway is characterized by canonical Wnt ligands interacting with their specific seven-pass transmembrane Fz receptors and LRP5/6 coreceptors of the LDL receptor protein family. Dvl is then recruited to the plasma membrane and also Axin and GSK3β are recruited to the receptor complex. LRP5/6 is phosphorylated at specific PPPSPxS motifs and flanking serine/threonine clusters leading to the formation of the so-called signalosome.[9] As a consequence, degradation of β-catenin is relieved and β-catenin is stabilized in the cytosol. β-catenin in turn translocates from the cytosol into the nucleus and interacts with transcription factors of the TCF/LEF1 family (T-cell factor/lymphoid enhancer factor), thereby driving the expression of Wnt target genes as a coactivator. Well-known canonical Wnt targets are for instance *c-Myc*, *CyclinD1*, *Axin2*, and *BIRC5*, encoding for survivin,[10–14] all of which regulate cell growth and survival. Note that β-catenin-mediated *Axin2* expression establishes a negative feedback mechanism.[15] Meanwhile, various regulators of β-catenin-dependent Wnt signaling have been uncovered. These include coreceptors, components of the secretory machinery, the destruction complex and nuclear cofactors, which have been thoroughly reviewed elsewhere.[8]

1.1.2 Noncanonical Wnt Signaling

In contrast to the described β-catenin-dependent, canonical Wnt signaling pathway, Wnt proteins are able to activate additional signaling pathways. Those are called noncanonical and are *per* definition independent of β-catenin. Instead, noncanonical Wnt signaling involves activation of kinases, such as jun-N-terminal kinase (JNK), protein kinase C (PKC), or calcium-calmodulin-dependent kinase II (CamKII).[16] In more detail, noncanonical Wnt signaling can be subdivided into the Wnt/calcium and Wnt/JNK signaling branches. The Wnt/calcium pathway is characterized by a G-protein dependent release of calcium resulting in an increase of intracellular calcium concentration. This leads to the activation of calcium sensitive enzymes including the protein kinases PKC, CamKII, or the protein phosphatase calcineurin. The Wnt/JNK pathway describes the activation of rho-GTPases that finally activate JNK. JNK itself not only modifies transcription factors but also together with rho-GTPases regulates the actin cytoskeleton. For a more detailed description of noncanonical Wnt signaling see.[6,16]

Some Wnt ligands preferentially activate either canonical or noncanonical Wnt signaling. For example, Wnt3A or Wnt1 are ligands activating rather canonical Wnt signaling, whereas Wnt5A or Wnt11 are considered to activate noncanonical signaling and the same holds true for receptors of the Fizzled family.[16,17] The underlying mechanism is not yet fully understood, but might reflect the presence of certain Frizzled receptors or defined coreceptors. Interestingly, some ligands like Wnt3A or Wnt11 are able to activate both major branches of Wnt signaling, likely in a concentration dependent manner.[18] Also, these signaling branches are nowadays considered to be interwoven to form a complex Wnt signaling network[6] with interesting properties, such as a switch-like behavior.[19,20]

1.2 Wnt as Regulator of Cell Cycle, Cell Size, and Cell Cycle Progression

Wnt ligands control various mitogenic processes, such as cell proliferation, G1 progression, cell differentiation, and cell growth.

As a prerequisite for proper cell division, cells need to replicate their DNA and grow in size to ensure equality of daughter and maternal cells. While DNA replication occurs during S phase of the cell cycle, cell growth mainly takes place during G1 phase.[21] Key characteristics of cell growth are a gain in overall protein levels and production of sufficient ribosomes.[22] Thus, a rapid cell cycle is efficiently achieved either in small cells, for instance embryonic stem cells, or in embryos with large stores of maternal ribosomes, such as *Danio rerio* or *Xenopus laevis*.

Canonical Wnt ligands promote cell proliferation at the transcriptional level, at the level of protein biosynthesis and protein degradation.[23] One key branch point is represented by the kinase GSK3β, which couples mitogenic Wnt stimulation to cell growth and cell-cycle progression. By inhibiting GSK3β, Wnt signaling is located upstream of mTOR (mammalian target of rapamycin) signaling and activates mTOR signaling, which drives protein synthesis and cell growth.[24]

Wnt/β-catenin signaling stimulates cell cycle progression on a transcriptional level by inducing *c-Myc* and *Cyclin D1* expression. Cyclin D1 is a central player for cell cycle progression. Findings of the past decades led to the accepted model that Cyclin D1 inhibits the tumor suppressor

Retinoblastoma (Rb), thereby releasing E2F transcription factors and allowing the expression of *CyclinE*, which constitutes the G1/S checkpoint.[25–30] By inducing the expression of *Cyclin D1* on a transcriptional level, Wnt signaling mediates accumulation of Cyclin D1 protein during G1 phase. As a consequence Cyclin E is produced, which needs to accumulate to allow cells to enter S phase in order to duplicate their DNA. c-Myc also positively regulates transcription by the three RNA polymerases I–III and thus promotes generation of new ribosomes during G1 phase of the cell cycle.[31]

Wnt/β-catenin signaling activity itself is linked to the progression of the cell cycle by several nontranscriptional mechanisms.[32,33] Mitogenic Wnt signaling peaks at the transcriptionally inactive G2/M phase, slows down protein degradation and increases cell size in order to prepare proliferating cells for cell division.[23,33,34] This Wnt/STOP (stabilization of proteins) pathway functions in a β-catenin independent manner to promote cell cycle progression by inhibiting GSK3β and GSK3β mediated protein degradation. In turn, mitotic effector proteins Polo-like kinase (Plk), Cyclin E, and c-Myc are stabilized and allow Wnt mediated cell cycle progression.[35] Stabilization of Cyclin E by Wnt/STOP signaling is a good example how two different Wnt signaling activities impinge on the same target. Wnt/β-catenin signaling induces transcription and translation of Cyclin E (through the intermediate step of activating Cyclin D1), whereas Wnt/STOP signaling prevents the degradation of Cyclin E.

1.3 Ribosomes: Function and Biogenesis

Ribosomes represent the cellular protein synthesis machinery and function as ribozymes, with rRNA taking over catalytic activity.[36] It is common textbook knowledge that ribosomes are composed of two subunits, a small and a large subunit, each consisting of rRNA and ribosomal proteins. The synthesis of ribosomes, so called ribosome biogenesis, represents a key event for proper cell function. Moreover, ribosomes are essential for cell growth and survival and therefore an appropriate number of ribosomes is a prerequisite for cell division and thus cell proliferation.

Ribosome biogenesis collectively denotes the steps of rRNA transcription, pre-rRNA processing and ribosome subunit assembly. Ribosome biogenesis is a highly complex and conserved process, which primarily occurs in a subnuclear compartment, the nucleolus. Nucleoli constitute the largest membrane free structures within the nucleus. They are closely connected to growth and proliferation, are highly dynamic and function as cellular stress sensors.[37]

Eukaryotic 80S ribosomes consist of the 40S and 60S subunits. Ribosome production requires the synthesis of four different rRNAs, the 5S rRNA, 5.8S rRNA, 18S rRNA, and 28S rRNA, as well as more than 70 ribosomal proteins (RPs).[38] RPs are synthesized from RNA polymerase II transcribed genes coding for these respective factors. With the exception of 5S rRNA, which is transcribed by RNA polymerase III in the nucleoplasm, the other rRNAs are processed from a single, polycistronic 47S pre-rRNA precursor transcribed by RNA polymerase I (Fig. 1). Processing requires several highly

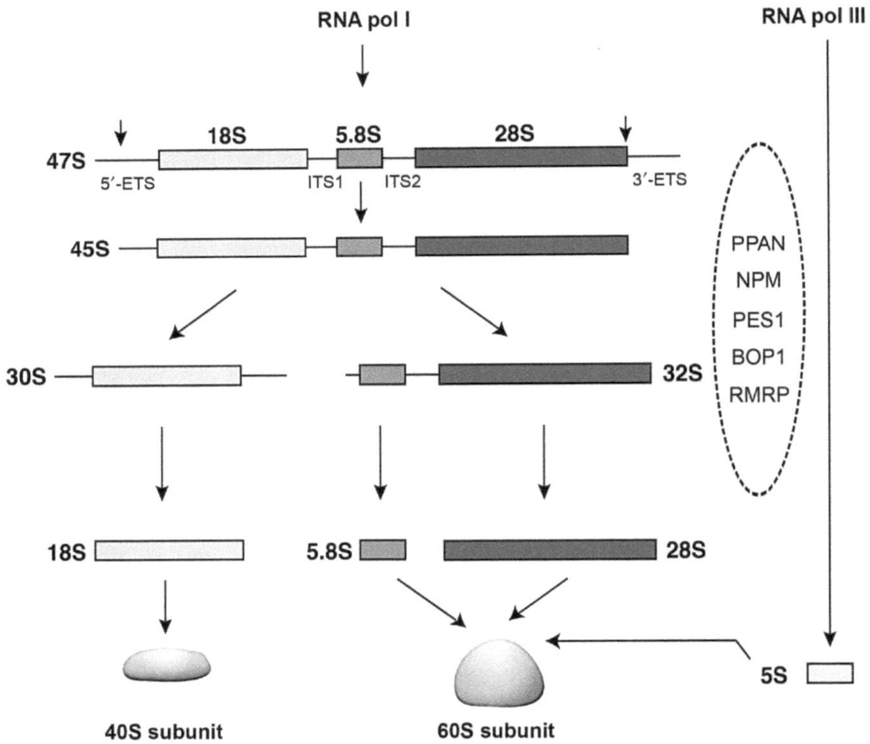

Fig. 1 A simplified model of mammalian ribosome biogenesis. The primary 47S pre-rRNA is transcribed by RNA polymerase I, and the 5S rRNA is transcribed by RNA polymerase III. Note, that processing of the 45S rRNA occurs via two routes, and the primary pathway is depicted here. 47S rRNA is processed via distinct intermediates into the mature 18S, 5.8S, and 28S rRNA species. Removal of the ITS1 (internal spacer) separates the small from large precursors, ETS denotes the external spacers. The Wnt targets PES1, NPM, PPAN, RMRP, and BOP1 function as ribosome biogenesis factors and are highlighted by the dashed circle. The mature 18S rRNA is incorporated into the 40S subunit and the 5.8S, 28S, and 5S rRNAs into the 60S subunit, respectively. The final maturation of the ribosome occurs in the cytoplasm.

complex and well-coordinated cleavages and modifications in nucleoli.[39–42] In yeast the corresponding long precursor pre-rRNA is the 35S pre-rRNA.[41,42] In humans, approximately 300 copies of the 47S rRNA gene (47s rDNA) are found in 5 clusters on 5 different chromosomes thereby generating 10 nucleoli organizing regions (NORs) in the diploid cell.[37] In particular during G1 phase nucleoli are detected,[43] as NORs fuse around the rDNA repeats to form a nucleolus.[37]

Processing of the large rRNA precursor has been well studied for over 4 decades in yeast as an attractive model organism and in lesser extent in higher eukaryotes.[42,44] This process requires the activity of more than 150 proteins assembled in different and dynamically changing larger complexes, so called preribosomal complexes, which have been well reviewed.[38,42,44,45] The mammalian 47S precursor is subsequently processed by multiple endo- and exonucleolytic cleavages, as well as by chemical modification steps. These cleavages include tightly coordinated removal of external (5′-ETS and 3′-ETS) and internal spacers (ITS1 and ITS2) thereby producing several rRNA intermediates.[40,42] Initially, the 45S precursor is generated, which is further processed and finally gives rise to the mature 18S, 28S and 5.8S rRNAs (compare Fig. 1). Note that the 18S rRNA transcript is incorporated into the small 40S ribosomal subunit, whereas the 5.8S, 28S, and 5S rRNAs are incorporated into the 60S subunit.[39,40]

Many components of the pre-60S particles involved in 45S rRNA precursor processing have been analyzed in more detail in yeast, including the yeast protein Nop7p/Yph1p, the vertebrate counterpart of Pescadillo (PES1).

PES1 (Pescadillo 1), NPM (Nucleophosmin), PPAN (Peter Pan), BOP1 (Block of proliferation 1), and RMRP (RNAse mitochondrial RNA processing) are induced by Wnt signaling and involved in processing of the large ribosomal subunit. The role of these Wnt target genes in ribosome biogenesis is depicted in Fig. 1 and their functions are introduced in more detail in the section later.

Other mammalian proteins analyzed include for instance WDR12 (WD repeat-containing protein 12) and TCOF1/treacle (see later). PES1, BOP1, and WDR12 form the nucleolar PeBoW complex, which is essential for maturation of the large 60S ribosomal subunit.[46] In line, dominant-negative mutants of the PeBoP complex show impaired synthesis of 28S rRNA.[46,47]

Despite the progress made by employing tandem affinity purification (TAP) in yeast,[38] our knowledge in the mammalian system still lags behind and many factors involved in the complex process of ribosome biogenesis remain uncharacterized.[42] However, advances in proteomic analyses already revealed several hundred proteins present in human nucleoli, some of which were implicated in ribosome biogenesis.[38,48–50]

The transcription factor c-Myc is an important regulator of RNA polymerase I, II, and III dependent gene expression and thereby essential for ribosome biogenesis.[51–53] It enhances the transcription of rRNA precursors and supports the synthesis of ribosomal proteins. In addition, c-Myc directly induces the expression of various genes involved in ribosome biogenesis, for instance PES1, NPM, and BOP1 and regulates the efficiency of rRNA processing.[54]

The oncogene c-Myc also represents one of the first identified direct canonical Wnt target genes characterized so far[10] and thereby establishes a strong link between Wnt signaling and the upregulation of ribosome biogenesis.

Overall, Wnt signaling and ribosome biogenesis are essential for guaranteeing proper cell proliferation and cell growth. The striking connection of Wnt ligands activating the expression of specific ribosome biogenesis factors and some of ribosome biogenesis factors signaling back on Wnt signaling is given in Fig. 2.

In the following sections we will discuss this relationship in more detail.

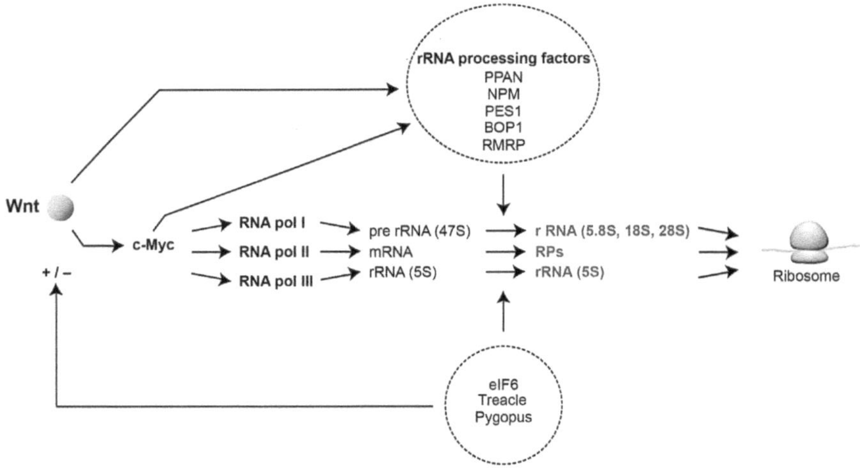

Fig. 2 The Wnt-ribosome-network at a glance. Wnt ligands function as central regulators of the ribosome network. Wnt ligands induce expression of specific Wnt targets, which mediate either directly or indirectly ribosome biogenesis. Wnt induces expression of c-Myc, thereby indirectly affecting ribosome biogenesis by inducing RNA polymerases I, II, and III required for ribosome biogenesis. More directly, Wnt mediates expression of several regulators of ribosome biogenesis, such as PPAN, NPM, PES1, BOP1, and RMRP. Some of these ribosome biogenesis factors, such as eIF6, treacle, and Pygopus signal back, and actively inhibit or activate Wnt signaling, thereby establishing a feedback mechanism.

1.4 Impaired Ribosome Biogenesis

Failure in the highly complex process of ribosome biogenesis and loss of nucleolar integrity triggers so-called nucleolar stress.[55,56] As a result, free ribosomal proteins (RPs) accumulate, some of which (e.g., RPL5/L5, RPL11/L11, RPL23/L23) can trigger p53-mediated apoptosis through interaction with Hdm2/mdm2 (Mouse double minute 2).[57] The ribosomal protein L11 also provides a negative feedback to the activity of c-Myc.[58,59] Thus, ribosome formation is tightly controlled and it has been revealed that proper ribosome formation sets a novel checkpoint for cell cycle progression.[60]

Virtually all factors orchestrating ribosome biogenesis, as well as ribosomal proteins are essential for survival. Defective ribosome biogenesis, either due to mutations in ribosomal proteins, or ribosome processing factors, causes a heterogeneous class of genetic diseases collectively termed ribosomopathies.[41] Despite the heterogeneity, many ribosomopathy syndromes share specific symptoms, such as skeletal abnormalities, anemia, and predisposition to cancer. Almost all syndromes are characterized by haploinsufficiency, suggesting that null mutations are embryonic lethal. However, why a haploinsufficient loss of broadly expressed ribosome biogenesis factors implicated in distinct processes has such specific defects still remains to be elucidated.[41] Given that ribosome biogenesis factors apparently have additional, nonribosomal functions, they might as well contribute to diverse pathologies.

Cancer cells are characterized by upregulated proliferation rates and a need for protein synthesis and ribosomes.[61] Thus, ribosomes have become novel targets for anticancer therapies, for instance by triggering nucleolar stress. Ribosomal proteins and ribosome biogenesis factors are overexpressed in cancer; however, it is still not understood whether this is cause or consequence of the transformed cells.[41]

2. TARGET GENES OF WNT SIGNALING INVOLVED IN RIBOSOME BIOGENESIS

Several components of the ribosome biogenesis machinery have recently been described to be target genes of Wnt signaling. These findings establish a strong link between Wnt signaling and cell growth, cell cycle progression, and cell proliferation. In this section, we want to highlight this novel regulatory link by shortly discussing these factors in more detail (Figs. 1 and 2).

2.1 Pescadillo as Downstream Factor of Wnt Signaling

Pescadillo (PES1) is a highly conserved nucleolar factor involved in embryonic development, ribosome biogenesis, DNA replication, and cell cycle progression. Pescadillo contains a central BRCT protein–protein interaction domain,[62] nuclear localization signals, and a potential SUMOylation site.[63,64] PES1 can directly bind to DNA and differentially regulates gene expression.[65,66] The BRCT domain is essential for the function of PES1 during ribosome formation.[67] The yeast counterpart, termed Nop7p/Yph1p (Yeast pescadillo homolog1) is highly expressed in the 60S preribosomal particle and plays a critical role in pre-rRNA processing, removal of the ITS1, 60S subunit assembly and transport.[68–70] An overview of the Nop7p interactome of is given by Granneman and coworkers.[70]

As mentioned earlier, human PES1 interacts with BOP1 and WDR12 and is a component of the PeBoW complex.[46] In addition, PES1 is required for the nuclear import and nucleolar localization of BOP1.[46] Depletion of Pescadillo impairs embryonic development in zebrafish[71] and Pescadillo knock out mice already die at the two-cell stage likely due to the failure of generating new ribosomes.[72] Loss of PES1 inhibits proliferation and reduces *Cyclin D1* in human breast cancer cells.[73] Moreover, loss of PES1 triggers chromosomal instability.[74]

Thus, human PES1 functions as protooncogene and is highly expressed in gastric cancer, colon tumor samples, breast, and brain cancer.[63,73,75,76]

It has been demonstrated that the PES1 promoter contains c-JUN binding sites and that *PES1* expression is positively regulated by JNK.[76] In *Xenopus*, Pes1 was shown to be a target gene of Wnt4[77] and its expression in neural tissue and in the developing pronephros depends on Wnt4 and its putative receptor Frizzled3.[78–80]

2.2 Peter Pan as Downstream Factor of Wnt Signaling

Peter Pan (PPAN) has been identified in *Drosophila* in a screen for larval growth defective mutants[81] and is highly conserved from yeast to human. Yeast expresses 2 orthologs termed Ssf1p and Ssf2p (Second step splicing factor), which are functionally redundant. Ssf1p cosediments with the 60S E1 preribosomal particle and prevents premature processing of the 60S ribosomal precursor.[82] Ssf1p is involved in early steps of rRNA precursor processing and copurifies with yeast Pescadillo, Nop7p.[82,83] In line, *Xenopus* PPan has been shown to directly bind to *Xenopus* Pes1.[79] Ssf1p/PPAN contains a N-terminal Brix domain (Biogenesis of ribosomes in *Xenopus*) including a σ70-motif required for rRNA interaction.[84] In humans one

PPAN homolog exists, which we have previously cloned.[85] Human PPAN interacts with the ribosome biogenesis factor NPM in HeLa cells and inhibits nucleolar stress, indicating a conserved function.[85] Interestingly, human PPAN, as well as *Xenopus* PPan not only localize to nucleoli but also to mitochondria.[79,85] Human PPAN is shuttling between the nucleus and cytoplasm by use of nuclear export (NES) and nuclear localization sequences (NLS), is phosphorylated and can be degraded in a Caspase-dependent manner.[85] We previously showed that PPAN inhibits mitochondrial apoptosis in *Xenopus* and human cells independently of p53.[79,85] Loss of human PPAN reduces NPM, stabilizes proapoptotic BAX and triggers nucleolar stress and apoptosis independently of p53.[85] In contrast, PPAN overexpression inhibits chemotherapy-induced apoptosis. PPAN knockdown also downregulates UBF1 (Upstream-Binding Factor 1),[85] the transcription factor for RNA-polymerase I,[86] suggesting a regulation of ribosome biogenesis by PPAN at multiple levels.

The expression of PPan in *Xenopus* neural tissue and in the developing pronephros is positively regulated by noncanonical Wnt4 and its putative receptor Frizzled3.[78–80] Also in a murine intestinal tumor model with overactive canonical Wnt signaling, microarray analysis revealed that PPAN expression is upregulated following APC loss and depends on c-Myc.[87] So far, PPAN has not been linked to the development of human diseases.

2.3 Nucleophosmin as Target Gene of Wnt Signaling

Nucleophosmin (NPM, B23, NO38, numartrin) is a multifunctional phoshoprotein mainly located to nucleoli. NPM can shuttle between the nucleus and cytoplasm and has a huge interactome involved in various prosurvival processes. To name a few, NPM regulates gene expression and repair, ribosome biogenesis, protein synthesis, DNA damage, cellular stress, cell cycle regulation, apoptosis, cytoskeletal regulation, and centrosome duplication.[88–90]

The best-studied cellular function of NPM is related to the production and export of mature ribosomes.[88] NPM functions as an endoribonuclease and cleaves pre-rRNA at a specific ITS2 site of the pre-5.8S rRNA.[91] NPM chaperones the export of ribosomal protein RPL5/L5 by use of a CRM1-dependent nuclear export sequence[92] and is required for export of 40S and 60S ribosomal subunits.[93]

NPM interacts with the nucleolar factor PES1 and controls its nucleolar localization[94] and also binds to BOP1 in HeLa cells.[93]

NPM has been shown to directly bind to nucleic acids,[95] to function as histone chaperone[96] and to drive p53-mediated transcription.[97] In addition, NPM directly interacts with c-Myc, is required for nucleolar c-Myc positioning and rDNA gene transcription and drives c-Myc-mediated proliferation.[98,99]

Importantly, NPM functions as a nucleolar stress sensor, as a loss of NPM impairs nucleolar structure and function and triggers p53-mediated apoptosis.[100] NPM interacts with PPAN and knockdown of PPAN downregulates NPM in human cells.[85] NPM inhibits mitochondrial apoptosis via counteracting BAX activation and mitochondrial translocation of p53.[101,102]

NPM has been described to exert both oncogenic and tumor suppressor functions depending on the cellular context. NPM is overexpressed in various tumors; it binds and stabilizes p53 and is frequently chromosomally translocated or mutated in hematological malignancies.[97] In acute myeloid leukemia (AML) the mutant NPMc+ serves as prognostic marker. NPMc+ was shown to be mislocalized in the cytosol and to sequester the residual nucleolar wildtype NPM into the cytosol.[103] When overexpressed in zebrafish, NPMc+ activates canonical Wnt signaling and results in expansion of hematopoietic progenitor cells.[104] In line, AML patient samples bearing the NPMc+ mutation show overactivation of Wnt signaling, as evidenced by increased *Axin2* expression.[104]

Interestingly, NPM was found to be a direct target gene of canonical Wnt signaling and shown to be upregulated by Wnt3A in colorectal cancer cells.[105]

This data demonstrates the necessity of a tight cellular regulation of the Wnt-ribosome network.

2.4 BOP1 as Target Gene of Wnt Signaling

BOP1 (Block of proliferation1) is a nucleolar WD40 repeat protein, which cosediments with the 32S precursor.[106] Expression of a N-terminal BOP1 deletion mutant triggers cell cycle arrest in G1 phase and impairs processing of 32S rRNA to form mature 5.8S and 28S rRNAs.[106] BOP1 mediates cleavage of the ITS1[107] and as a consequence, cells display a deficit in 60S subunits.[108]

Also the yeast homolog, Erb1p, was demonstrated to be essential for processing of the large ribosomal subunit.[109] BOP1 interacts with PES1 and requires PES1 for nucleolar localization.[46,110] BOP1 is overexpressed in colorectal cancers and triggers formation of multipolar spindles indicating an additional role in mitosis.[111]

Recently, BOP1 was identified as β-catenin/TCF4 Wnt target gene in colorectal cancer samples and was placed upstream of JNK.[112] Overexpression of BOP1 promotes epithelial mesenchymal transition (EMT), cell migration and experimental metastasis in mice.[112] The same holds true in hepatocellular carcinoma. Also here, BOP1 is overexpressed and promotes EMT and cell migration.[113]

2.5 RNAse MRP as Target Gene of Wnt Signaling

RNAse MRP (RNAse mitochondrial RNA processing, RMRP) is an abundant noncoding MRP RNA, which is related to RNAseP.[114] RMRP RNA is transcribed by RNA polymerase III and forms a stable ribonucleoprotein (RNP) together with other ribosomal proteins. RMRP functions as an endoribonuclease, it is encoded in the nucleus and localizes to the nucleolus and mitochondria. RMRP is required for pre-rRNA processing of 5.8S RNA, as well as for RNA processing of mitochondrial rRNA.[115–118] Loss of RMRP by CRISPR/Cas9 technology disrupts proliferation and causes an accumulation of rRNA in those cells.[119] Mutations are unviable in yeast, embryonic lethal in mice, and cause a spectrum of diseases in humans, including skeletal abnormalities, anemia, and increased predisposition to cancer.[41]

Upon Wnt stimulation, RMRP translocates to the cytoplasm and triggers mitochondrial rRNA processing. RMRP was shown to contain putative TCF binding elements in the upstream regions of the *RMRP* gene.[120] Interestingly, in cancer cells with high Wnt signaling, RMRP was increased by β-catenin. Thus, the authors suggested either an unusual induction of RNA polymerase III transcription by β-catenin, or enhancement of RNA polymerase II as usually the case.[120]

Aberrant induction of Wnt signaling—as in cancer—therefore leads to accumulation of a class of nucleolar factors, such as PES1, NPM, PPAN, and BOP1, all of which interact with each other and participate in 60S precursor processing. Wnt-on mode therefore guarantees rapid cell proliferation and growth by providing key players of the ribosome processing machinery. Whether a broader repertoire of ribosome processing factors, for instance WDR1 of the PeBoW, and specific RNAs, as observed for RMRP, are induced following Wnt stimulation remains to be determined.

Besides their traditional role as ribosome biogenesis factors, this nucleolar, Wnt-induced factors fulfill a serious of nontraditional prosurvival functions, such as direct inhibition of apoptosis and cell-cycle regulation.

3. REGULATORS OF RIBOSOME BIOGENESIS AND NUCLEOLAR FACTORS AS MODIFIERS OF WNT SIGNALING

Emerging evidence uncovers that nucleolar ribosome processing factors can also regulate Wnt signaling more directly by actively participating in Wnt signaling at different levels. Reciprocally, Wnt ligands and pathway components were demonstrated to localize to nucleoli, to affect nucleolar function and to control ribosome biogenesis. This dual role of several factors highlights the importance of connecting Wnt signaling and ribosomes for proper cell function.

3.1 Eukaryotic Initiation Factor 6

Eukaryotic initiation factor 6 (eIF6) localizes to nucleoli and in the cytoplasm where it regulates ribosome biogenesis and translation, respectively. The yeast homologue, Tif6, is a component of preribosomal particles and essential for biogenesis of the 60S ribosomal subunit. It binds to 60S particles and prevents the premature interaction with 40S particles.[121] However, yeast Tif6 does not regulate translation. Loss of eIF6 in mice is embryonic lethal, heterozygous mice are viable and are characterized by loss of cytosolic eIF6, whereas the nucleolar pool and 60S biogenesis is normal.[122] In contrast, loss of eIF6 in the liver induces accumulation of 80S RNA.[121,122] eIF6 is overexpressed in human tumors and was implicated in the autosomal-recessive ribosomopathy Schwachman Bodian Diamond syndrome (SBDS), given that eIF6 compensates a SBDS loss.[123]

Strikingly, eIF6 was shown to interact with the C-terminus of β-catenin and to inhibit canonical Wnt signaling.[124] eIF6 reduces levels of β-catenin protein, but also a negative regulation from β-catenin to eIF6 was observed.[124]

3.2 Treacle/TCOF1

The *TCOF1* gene encodes the nuclear phosphoprotein treacle. Treacle plays a critical role in transcription of rDNA and methylation of 18S rRNA.[41,125] Treacle is mutated in patients with the autosomal-dominant ribosomopathy Treacher Collins syndrome (TCS). Patients display severe craniofacial abnormalities, which are in 90% of cases caused by loss-of-function mutations in the *TCOF1* gene. Haploinsufficiency impairs ribosome biogenesis, triggers nucleolar stress, p53 stabilization, cell cycle arrest, and apoptosis.[126]

Treacle haploinsufficiency in mice reduces the number of ribosomes and triggers cell death of migratory neural crest cells, which give rise to craniofacial cartilage recapitulating the patients' phenotypes.[41]

Treacle interacts and activates UBF thereby regulating RNA polymerase I and in turn promotes rRNA transcription.[127] Treacle also binds to Nop56p, which is a component of the box C/D small nucleolar ribonucleoprotein complexes involved in early steps of 60S pre-rRNA processing.[128]

Interestingly, reduced expression of TCOF1 altered *Cyclin D1* expression and noncanonical Wnt5A, indicating a crosstalk between treacle and Wnt signaling pathways.[129] Thus, impaired ribosome function and dysregulated Wnt signaling might both be observed in patients with TCS.

3.3 Pygopus

Pygopus (Pygo) protein was originally identified in *Drosophila* as a nuclear component of the β-catenin transcription complex.[130,131] Pygo interacts with β-catenin and regulates chromatin remodeling and transcription. Pygo binds to histones trimethylated at lysine residues (H3K4me3) and can recruit histone acetyltransferases (HAT). As a consequence, Pygo promotes Wnt target gene expression and is upregulated in a variety of cancers.[132,133]

Human Pygopus (hPygo2) localizes to the nucleoli of several transformed cell lines and is associated with treacle and UBF1.[134] Pygo knockdown decreases 47S pre-rRNA and activates the Mdm2-p53 nucleolar stress response indicating a novel function in ribosome biogenesis.[134]

3.4 Thyroid Cancer1

TC1 (thyroid cancer1, C8orf4) localizes to nucleoli,[135] however, despite the nucleolar localization of TC1 no function in ribosome biogenesis has been demonstrated so far. TC1 is overexpressed in thyroid, gastric, and lung cancers and correlates with β-catenin expression.[136–139] TC1 activates canonical Wnt signaling by interacting with Chibby and releasing the inhibitory function of Chibby on β-catenin.[135] Chibby interacts with β-catenin and competes with LEF1 for β-catenin thereby inhibiting Wnt signaling.[140,141] Thus, TC1 promotes aggressiveness of cancers through canonical Wnt signaling.[138]

Whether, TC1 actively participates in ribosome biogenesis in the nucleoli remains to be determined.

Taken together, nucleolar factors are implicated in the development of the diverse class of ribosomopathies and are commonly deleted, upregulated, or mutated in cancer. Emerging evidence uncovers that nucleolar factors function as regulators of Wnt signaling for instance by counteracting or fostering β-catenin-dependent transcription. Interestingly, the NPMc+ mutant, which translocates to the cytosol due to ectopic NES sequences, was shown to increase levels of nuclear β-catenin and trigger *Axin2* expression.[104] However, the precise molecular mechanism underlying NPMc+ function remains to be determined.

4. NONCANONICAL WNT SIGNALING REGULATES NUCLEOLAR SIZE

The noncanonical Wnt5A was demonstrated to repress rDNA gene transcription by RNA polymerase I and to decrease nucleolar size.[142] These effects were shown to depend on Dishevelled1 (Dvl1), which binds to rRNA and accumulates in nucleolar organizer regions (NORs) upon Wnt5A stimulation. Thus, Wnt5A signals through Dvl1 to suppress rRNA transcription. In mammary tumors, loss of Wnt5A increases nucleolar size, cell proliferation, and impairs survival, whereas Wnt5A has tumor suppressive functions.[142]

Wnt ligands and pathway components seem to have a direct role on nucleolar function and ribosome biogenesis. Whereas canonical Wnt signaling induces expression of ribosome biogenesis factors, noncanonical Wnt5A inhibits rRNA transcription and reduces nucleolar size and/or function.

Importantly, nucleolar size matters tremendously with respect to aging.[143] Long-lived animals display smaller nucleoli and reduced ribosome biogenesis, and therefore small nucleoli are considered to represent a hallmark of increased longevity.[144] On the contrary, nucleolar expansion and increased ribosome biogenesis is connected to premature aging disorders.[145] Thus, Wnt5A seems to influence aging by limiting nucleolar size and global translation.

5. WNT REGULATES RIBOSOME COMPOSITION

Increasing evidence suggests that the composition of ribosomes is heterogeneous depending on the cell type and context. This diversity of specialized ribosomes, termed ribosome signature, drives differential translation.[146] It was recently demonstrated that canonical Wnt3 signaling regulates the ribosome signature of the neocortex in mice. Mass spectrometry of protein changes in 40S–60S–80S and polysomes most prominently revealed altered levels of ribosomal proteins associated with 60S and/or 40S subunits.[147] Wnt3 secretion from the thalamus spatiotemporally regulates the composition of a subset of ribosomal proteins and the specificity of mRNA translation. For instance, ribosomal protein RPL7/L7 was enriched at polysomes and in turn the translation of APC was inhibited, whereas FOXP2 was induced. The authors concluded that extracellular signaling might spatiotemporally control ribosome signature and differential translation to drive cellular specification.[147]

In the light of this finding, investigating the role of specific canonical and noncanonical Wnt ligands on ribosome signature and differential translation will be beneficial to understand Wnt-driven pathogenesis.

6. CLOSING REMARKS

Over the past decades a multitude of studies deepened our understanding of Wnt signaling and ribosome biogenesis. Several lines of evidence suggest a connection of both branches. It became obvious that Wnt proteins play a dual role in regulating ribosome formation. On the one hand Wnts promote ribosome formation indirectly through c-Myc mediated induction of RNA polymerase I–III expression. On the other hand Wnt ligands directly influence rRNA processing by inducing the expression of involved processing factors. It becomes clear that various signaling pathways might influence ribosome signature and function and vice versa.

The nucleolus functions as a multifunctional, cellular gatekeeper by controlling health and survival. Dysfunction of nucleolar activity triggers nucleolar stress, which causes a broad spectrum of pathophysiological malignancies. For instance, malfunction of nucleolar structure or function predisposes to certain types of cancer and is closely linked to skeletal disorders and anemia, as observed in ribosomopathies. Nucleolar stress is also coupled to neurodegenerative diseases, such as Alzheimer, Parkinson, and Huntington, as well as aging.[148] Given the emerging role of nucleolar factors in the regulation of Wnt signaling, novel strategies might be applied to antagonize these pathologies. For instance, several nucleolar factors are mutated in ribosomopathies and predispose to cancer. It was suggested that a malfunction of ribosome biogenesis leads to differential translation of subset of mRNAs by specialized ribosomes.[149] Given the more direct role in the regulation of cellular signaling by direct interaction with LEF/TCF and β-catenin in the nucleolus, these nucleolar and ribosome biogenesis factors might play a more direct role as anticipated. Modulating nucleolar activity has been proposed as proapoptotic anticancer therapy.[150] Henceforth, analyzing and manipulating nucleolar activity in these multifactorial pathologies will open new avenues for therapeutic approaches.

We are still far from a comprehensive understanding of the pathways and molecules involved. Thus, delineating the still poorly characterized cross-regulatory signaling networks, such as of Wnts and ribosomes will provide novel insights into the mechanisms of development and disease.

REFERENCES

1. Nusse R, Varmus HE. Many tumors induced by the mouse mammary tumor virus contain a provirus integrated in the same region of the host genome. *Cell.* 1982;31(1): 99–109.
2. Sharma RP, Chopra VL. Effect of the Wingless (wg1) mutation on wing and haltere development in *Drosophila* melanogaster. *Dev Biol.* 1976;48(2):461–465.
3. van Ooyen A, Nusse R. Structure and nucleotide sequence of the putative mammary oncogene int-1; proviral insertions leave the protein-encoding domain intact. *Cell.* 1984;39(1):233–240.
4. Nusse R, Brown A, Papkoff J, et al. A new nomenclature for int-1 and related genes: the Wnt gene family. *Cell.* 1991;64(2):231.
5. Nusse R, Varmus H. Three decades of Wnts: a personal perspective on how a scientific field developed. *EMBO J.* 2012;31(12):2670–2684.
6. Kestler HA, Kuhl M. From individual Wnt pathways towards a Wnt signalling network. *Philos Trans R Soc Lond B Biol Sci.* 2008;363(1495):1333–1347.
7. Grumolato L, Liu G, Mong P, et al. Canonical and noncanonical Wnts use a common mechanism to activate completely unrelated coreceptors. *Genes Dev.* 2010;24(22): 2517–2530.
8. MacDonald BT, Tamai K, He X. Wnt/beta-catenin signaling: components, mechanisms, and diseases. *Dev Cell.* 2009;17(1):9–26.
9. Bilic J, Huang YL, Davidson G, et al. Wnt induces LRP6 signalosomes and promotes dishevelled-dependent LRP6 phosphorylation. *Science.* 2007;316(5831): 1619–1622.
10. He TC, Sparks AB, Rago C, et al. Identification of c-MYC as a target of the APC pathway. *Science.* 1998;281(5382):1509–1512.
11. Shtutman M, Zhurinsky J, Simcha I, et al. The cyclin D1 gene is a target of the beta-catenin/LEF-1 pathway. *Proceedings of the National Academy of Sciences of the United States of America.* 1999;96(10):5522–5527
12. Tetsu O, McCormick F. Beta-catenin regulates expression of cyclin D1 in colon carcinoma cells. *Nature.* 1999;398(6726):422–426.
13. Zhang T, Otevrel T, Gao Z, et al. Evidence that APC regulates surviving expression: a possible mechanism contributing to the stem cell origin of colon cancer. *Cancer Res.* 2001;61(24):8664–8667.
14. Yan D, Wiesmann M, Rohan M, et al. Elevated expression of axin2 and hnkd mRNA provides evidence that Wnt/β-catenin signaling is activated in human colon tumors. *Proceedings of the National Academy of Sciences of the United States of America.* 2001;98(26): 14973–14978

15. Lustig B, Jerchow B, Sachs M, et al. Negative feedback loop of Wnt signaling through upregulation of conductin/axin2 in colorectal and liver tumors. *Mol Cell Biol.* 2002;22(4): 1184–1193.
16. Rao TP, Kuhl M. An updated overview on Wnt signaling pathways: a prelude for more. *Circul Res.* 2010;106(12):1798–1806.
17. Kuhl M, Sheldahl LC, Park M, Miller JR, Moon RT. The Wnt/Ca^{2+} pathway: a new vertebrate Wnt signaling pathway takes shape. *Trend Genet.* 2000;16(7):279–283.
18. Nalesso G, Sherwood J, Bertrand J, et al. WNT-3A modulates articular chondrocyte phenotype by activating both canonical and noncanonical pathways. *J Cell Biol.* 2011;193(3):551–564.
19. Strang JE, Schuler R, Kuhl M, Kestler HA. Switch-like behavior enables Wnt11 concentration specific response during dorso-ventral axis formation in *Xenopus laevis. J Theoretical Biol.* 2017;429:82–94 10.1016/j.jtbi.2017.06.027.
20. Kestler HA, Kuhl M. Generating a Wnt switch: it's all about the right dosage. *J Cell Biol.* 2011;193(3):431–433.
21. Grebien F, Dolznig H, Beug H, Mullner EW. Cell size control: new evidence for a general mechanism. *Cell Cycle.* 2005;4(3):418–421.
22. Baserga R. Is cell size important? *Cell Cycle.* 2007;6(7):814–816.
23. Niehrs C, Acebron SP. Mitotic and mitogenic Wnt signalling. *EMBO J.* 2012;31(12): 2705–2713.
24. Inoki K, Ouyang H, Zhu T, et al. TSC2 integrates Wnt and energy signals via a coordinated phosphorylation by AMPK and GSK3 to regulate cell growth. *Cell.* 2006;126(5):955–968.
25. Chellappan SP, Hiebert S, Mudryj M, Horowitz JM, Nevins JR. The E2F transcription factor is a cellular target for the RB protein. *Cell.* 1991;65(6):1053–1061.
26. Connell-Crowley L, Harper JW, Goodrich DW. Cyclin D1/Cdk4 regulates retinoblastoma protein-mediated cell cycle arrest by site-specific phosphorylation. *Mol Biol Cell.* 1997;8(2):287–301.
27. Duronio RJ, O'Farrell PH. Developmental control of the G1 to S transition in *Drosophila*: cyclin Eis a limiting downstream target of E2F. *Genes Dev.* 1995;9(12): 1456–1468.
28. Duronio RJ, O'Farrell PH, Xie JE, Brook A, Dyson N. The transcription factor E2F is required for S phase during *Drosophila* embryogenesis. *Genes Dev.* 1995;9(12): 1445–1455.
29. Nielsen NH, Emdin SO, Cajander J, Landberg G. Deregulation of cyclin E and D1 in breast cancer is associated with inactivation of the retinoblastoma protein. *Oncogene.* 1997;14(3):295–304.
30. Weinberg RA. The retinoblastoma protein and cell cycle control. *Cell.* 1995;81(3): 323–330.
31. Oskarsson T, Trumpp A. The Myc trilogy: lord of RNA polymerases. *Nat Cell Biol.* 2005;7(3):215–217.
32. Davidson G, Shen J, Huang YL, et al. Cell cycle control of Wnt receptor activation. *Dev Cell.* 2009;17(6):788–799.
33. Hadjihannas MV, Bernkopf DB, Bruckner M, Behrens J. Cell cycle control of Wnt/beta-catenin signalling by conductin/axin2 through CDC20. *EMBO Rep.* 2012;13(4): 347–354.
34. Acebron SP, Karaulanov E, Berger BS, Huang YL, Niehrs C. Mitotic Wnt signaling promotes protein stabilization and regulates cell size. *Mol Cell.* 2014;54(4):663–674.
35. Huang YL, Niehrs C. Polarized Wnt signaling regulates ectodermal cell fate in Xenopus. *Dev Cell.* 2014;29(2):250–257.
36. Nelson JW, Breaker RR. The lost language of the RNA world. *Sci Signal.* 2017;10 (483):1–10 10.1126/scisignal.aam8812.

37. Mangan H, Gailin MO, McStay B. Integrating the genomic architecture of human nucleolar organizer regions with the biophysical properties of nucleoli. *FEBS J.* 2017;1–9 10.1111/febs.14108.
38. Tschochner H, Hurt E. Pre-ribosomes on the road from the nucleolus to the cytoplasm. *Trends Cell Biol.* 2003;13(5):255–263.
39. Mullineux ST, Lafontaine DL. Mapping the cleavage sites on mammalian pre-rRNAs: where do we stand? *Biochimie.* 2012;94(7):1521–1532.
40. Lafontaine DL. Noncoding RNAs in eukaryotic ribosome biogenesis and function. *Nat Struct Mol Biol.* 2015;22(1):11–19.
41. Freed EF, Bleichert F, Dutca LM, Baserga SJ. When ribosomes go bad: diseases of ribosome biogenesis. *Mol BioSystems.* 2010;6(3):481–493.
42. Granneman S, Baserga SJ. Ribosome biogenesis: of knobs and RNA processing. *Exp Cell Res.* 2004;296(1):43–50.
43. Potmesil M, Goldfeder A. Identification and kinetics of G1 phase-confined cells in experimental mammary carcinomas. *Cancer Res.* 1977;37(3):857–864.
44. Fatica A, Tollervey D. Making ribosomes. *Curr Opin Cell Biol.* 2002;14(3):313–318.
45. Takahashi N, Yanagida M, Fujiyama S, Hayano T, Isobe T. Proteomic snapshot analyses of preribosomal ribonucleoprotein complexes formed at various stages of ribosome biogenesis in yeast and mammalian cells. *Mass Spectrom Rev.* 2003;22(5): 287–317.
46. Rohrmoser M, Holzel M, Grimm T, et al. Interdependence of Pes1, Bop1, and WDR12 controls nucleolar localization and assembly of the PeBoW complex required for maturation of the 60S ribosomal subunit. *Mol Cell Biol.* 2007;27(10):3682–3694.
47. Grimm T, Holzel M, Rohrmoser M, et al. Dominant-negative Pes1 mutants inhibit ribosomal RNA processing and cell proliferation via incorporation into the PeBoW-complex. *Nucleic Acids Res.* 2006;34(10):3030–3043.
48. Bensaddek D, Nicolas A, Lamond AI. Quantitative proteomic analysis of the human nucleolus. *Method Mol Biol.* 2016;1455:249–262.
49. Scherl A, Coute Y, Deon C, et al. Functional proteomic analysis of human nucleolus. *Mol Biol Cell.* 2002;13(11):4100–4109.
50. Andersen JS, Lyon CE, Fox AH, et al. Directed proteomic analysis of the human nucleolus. *Curr Biol.* 2002;12(1):1–11.
51. Arabi A, Wu S, Ridderstrale K, et al. c-Myc associates with ribosomal DNA and activates RNA polymerase I transcription. *Nat Cell Biol.* 2005;7(3):303–310.
52. Grandori C, Gomez-Roman N, Felton-Edkins ZA, et al. c-Myc binds to human ribosomal DNA and stimulates transcription of rRNA genes by RNA polymerase I. *Nat Cell Biol.* 2005;7(3):311–318.
53. Grewal SS, Li L, Orian A, Eisenman RN, Edgar BA. Myc-dependent regulation of ribosomal RNA synthesis during *Drosophila* development. *Nat Cell Biol.* 2005;7 (3):295–302.
54. Schlosser I, Holzel M, Murnseer M, Burtscher H, Weidle UH, Eick D. A role for c-Myc in the regulation of ribosomal RNA processing. *Nucleic Acids Res.* 2003;31(21): 6148–6156.
55. Boulon S, Westman BJ, Hutten S, Boisvert FM, Lamond AI. The nucleolus under stress. *Mol Cell.* 2010;40(2):216–227.
56. Suzuki A, Kogo R, Kawahara K, et al. A new PICTure of nucleolar stress. *Cancer Sci.* 2012;103(4):632–637.
57. Dai MS, Zeng SX, Jin Y, Sun XX, David L, Lu H. Ribosomal protein L23 activates p53 by inhibiting MDM2 function in response to ribosomal perturbation but not to translation inhibition. *Mol Cell Biol.* 2004;24(17):7654–7668.
58. Dai MS, Sears R, Lu H. Feedback regulation of c-Myc by ribosomal protein L11. *Cell Cycle.* 2007;6(22):2735–2741.

59. Dai MS, Arnold H, Sun XX, Sears R, Lu H. Inhibition of c-Myc activity by ribosomal protein L11. *EMBO J.* 2007;26(14):3332–3345.
60. Fumagalli S, Ivanenkov VV, Teng T, Thomas G. Suprainduction of p53 by disruption of 40S and 60S ribosome biogenesis leads to the activation of a novel G2/M checkpoint. *Genes Dev.* 2012;26(10):1028–1040.
61. Montanaro L, Trere D, Derenzini M. Nucleolus, ribosomes, and cancer. *Am J Pathol.* 2008;173(2):301–310.
62. Haque J, Boger S, Li J, Duncan SA. The murine Pes1 gene encodes a nuclear protein containing a BRCT domain. *Genomics.* 2000;70(2):201–210.
63. Kinoshita Y, Jarell AD, Flaman JM, et al. Pescadillo, a novel cell cycle regulatory protein abnormally expressed in malignant cells. *J Biol Chem.* 2001;276(9):6656–6665.
64. Li S, Wang M, Qu X, et al. SUMOylation of PES1 upregulates its stability and function via inhibiting its ubiquitination. *Oncotarget.* 2016;7(31):50522–50534.
65. Li J, Zhuang Q, Lan X, Zeng G, Jiang X, Huang Z. PES1 differentially regulates the expression of ERalpha and ERbeta in ovarian cancer. *IUBMB Life.* 2013;65(12): 1017–1025.
66. Sikorski EM, Uo T, Morrison RS, Agarwal A. Pescadillo interacts with the cadmium response element of the human heme oxygenase-1 promoter in renal epithelial cells. *J Biol Chem.* 2006;281(34):24423–24430.
67. Holzel M, Grimm T, Rohrmoser M, et al. The BRCT domain of mammalian Pes1 is crucial for nucleolar localization and rRNA processing. *Nucleic Acids Res.* 2007;35(3):789–800.
68. Oeffinger M, Leung A, Lamond A, Tollervey D. Yeast Pescadillo is required for multiple activities during 60S ribosomal subunit synthesis. *RNA.* 2002;8(5):626–636.
69. Du YC, Stillman B. Yph1p, an ORC-interacting protein: potential links between cell proliferation control, DNA replication, and ribosome biogenesis. *Cell.* 2002;109 (7):835–848.
70. Granneman S, Petfalski E, Tollervey D. A cluster of ribosome synthesis factors regulate pre-rRNA folding and 5.8S rRNA maturation by the Rat1 exonuclease. *EMBO J.* 2011;30(19):4006–4019.
71. Allende ML, Amsterdam A, Becker T, Kawakami K, Gaiano N, Hopkins N. Insertional mutagenesis in zebrafish identifies two novel genes, pescadillo and dead eye, essential for embryonic development. *Genes Dev.* 1996;10(24):3141–3155.
72. Lerch-Gaggl A, Haque J, Li J, Ning G, Traktman P, Duncan SA. Pescadillo is essential for nucleolar assembly, ribosome biogenesis, and mammalian cell proliferation. *J Biol Chem.* 2002;277(47):45347–45355.
73. Li J, Yu L, Zhang H, et al. Down-regulation of pescadillo inhibits proliferation and tumorigenicity of breast cancer cells. *Cancer Sci.* 2009;100(12):2255–2260.
74. Killian A, Le Meur N, Sesboue R, et al. Inactivation of the RRB1-Pescadillo pathway involved in ribosome biogenesis induces chromosomal instability. *Oncogene.* 2004; 23(53):8597–8602.
75. Li J, Zhou X, Lan X, Zeng G, Jiang X, Huang Z. Repression of PES1 expression inhibits growth of gastric cancer. *Tumour Biol.* 2016;37(3):3043–3049.
76. Xie W, Feng Q, Su Y, et al. Transcriptional regulation of PES1 expression by c-Jun in colon cancer. *PloS One.* 2012;7(7):e42253.
77. Maurus D, Heligon C, Burger-Schwarzler A, Brandli AW, Kuhl M. Noncanonical Wnt-4 signaling and EAF2 are required for eye development in *Xenopus laevis*. *EMBO J.* 2005;24(6):1181–1191.
78. Gessert S, Maurus D, Rossner A, Kuhl M. Pescadillo is required for *Xenopus laevis* eye development and neural crest migration. *Dev Biol.* 2007;310(1):99–112.
79. Bugner V, Tecza A, Gessert S, Kuhl M. Peter Pan functions independently of its role in ribosome biogenesis during early eye and craniofacial cartilage development in *Xenopus laevis*. *Development.* 2011;138(11):2369–2378.

80. Tecza A, Bugner V, Kuhl M, Kuhl SJ. Pescadillo homologue 1 and Peter Pan function during *Xenopus laevis* pronephros development. *Biol Cell.* 2011;103(10):483–498.
81. Migeon JC, Garfinkel MS, Edgar BA. Cloning and characterization of peter pan, a novel *Drosophila* gene required for larval growth. *Mol Biol Cell.* 1999;10(6):1733–1744.
82. Fatica A, Cronshaw AD, Dlakic M, Tollervey D. Ssf1p prevents premature processing of an early pre-60S ribosomal particle. *Mol Cell.* 2002;9(2):341–351.
83. Bogengruber E, Briza P, Doppler E, et al. Functional analysis in yeast of the Brix protein superfamily involved in the biogenesis of ribosomes. *FEMS Yeast Res.* 2003;3(1):35–43.
84. Wehner KA, Baserga SJ. The sigma(70)-like motif: a eukaryotic RNA binding domain unique to a superfamily of proteins required for ribosome biogenesis. *Mol Cell.* 2002; 9(2):329–339.
85. Pfister AS, Keil M, Kuhl M. The Wnt target protein peter pan defines a novel p53-independent nucleolar stress-response pathway. *J Biol Chem.* 2015;290(17):10905–10918.
86. Jantzen HM, Admon A, Bell SP, Tjian R. Nucleolar transcription factor hUBF contains a DNA-binding motif with homology to HMG proteins. *Nature.* 1990;344 (6269):830–836.
87. Sansom OJ, Meniel VS, Muncan V, et al. Myc deletion rescues Apc deficiency in the small intestine. *Nature.* 2007;446(7136):676–679.
88. Lindstrom MS. NPM1/B23: a multifunctional chaperone in ribosome biogenesis and chromatin remodeling. *Biochem Res Int.* 2011;2011:16[Article ID 195209].
89. Okuda M. The role of nucleophosmin in centrosome duplication. *Oncogene.* 2002; 21(40):6170–6174.
90. Colombo E, Alcalay M, Pelicci PG. Nucleophosmin and its complex network: a possible therapeutic target in hematological diseases. *Oncogene.* 2011;30(23): 2595–2609.
91. Savkur RS, Olson MO. Preferential cleavage in pre-ribosomal RNA byprotein B23 endoribonuclease. *Nucleic Acids Res.* 1998;26(19):4508–4515.
92. Yu Y, Maggi Jr LB, Brady SN, et al. Nucleophosmin is essential for ribosomal protein L5 nuclear export. *Mol Cell Biol.* 2006;26(10):3798–3809.
93. Maggi Jr LB, Kuchenruether M, Dadey DY, et al. Nucleophosmin serves as a rate-limiting nuclear export chaperone for the Mammalian ribosome. *Mol Cell Biol.* 2008;28 (23):7050–7065.
94. Zhang J, Yang Y, Wu J. B23 interacts with PES1 and is involved in nucleolar localization of PES1. *Acta Biochim Biophys Sinica.* 2009;41(12):991–997.
95. Wang D, Baumann A, Szebeni A, Olson MO. The nucleic acid binding activity of nucleolar protein B23.1 resides in its carboxyl-terminal end. *J Biol Chem.* 1994;269 (49):30994–30998.
96. Okuwaki M, Matsumoto K, Tsujimoto M, Nagata K. Function of nucleophosmin/B23, a nucleolar acidic protein, as a histone chaperone. *FEBS Lett.* 2001;506 (3):272–276.
97. Colombo E, Marine JC, Danovi D, Falini B, Pelicci PG. Nucleophosmin regulates the stability and transcriptional activity of p53. *Nature Cell Biol.* 2002;4(7):529–533.
98. Li Z, Hann SR. Nucleophosmin is essential for c-Myc nucleolar localization and c-Myc-mediated rDNA transcription. *Oncogene.* 2013;32(15):1988–1994.
99. Li Z, Boone D, Hann SR. Nucleophosmin interacts directly with c-Myc and controls c-Myc-induced hyperproliferation and transformation. *Proceedings of the National Academy of Sciences of the United States of America.* 2008;105(48):18794–18799.
100. Marquez-Lona EM, Tan Z, Schreiber SS. Nucleolar stress characterized by down-regulation of nucleophosmin: a novel cause of neuronal degeneration. *Biochem Biophys Res Commun.* 2012;417(1):514–520.
101. Lindenboim L, Borner C, Stein R. Nuclear proteins acting on mitochondria. *Biochim Biophys Acta.* 2011;1813(4):584–596.

102. Lo SJ, Fan LC, Tsai YF, et al. A novel interaction of nucleophosmin with BCL2-associated X protein regulating death evasion and drug sensitivity in human hepatoma cells. *Hepatology.* 2013;57(5):1893–1905.
103. Falini B, Bolli N, Shan J, et al. Both carboxy-terminus NES motif and mutated tryptophan(s) are crucial for aberrant nuclear export of nucleophosmin leukemic mutants in NPMc+AML. *Blood.* 2006;107(11):4514–4523.
104. Barbieri E, Deflorian G, Pezzimenti F, et al. Nucleophosmin leukemogenic mutant activates Wnt signaling during zebrafish development. *Oncotarget.* 2016;7(34): 55302–55312.
105. Willert J, Epping M, Pollack JR, Brown PO, Nusse R. A transcriptional response to Wnt protein in human embryonic carcinoma cells. *BMC Dev Biol.* 2002;2:8.
106. Strezoska Z, Pestov DG, Lau LF. Bop1 is a mouse WD40 repeat nucleolar protein involved in 28S and 5. 8S RRNA processing and 60S ribosome biogenesis. *Mol Cell Biol.* 2000;20(15):5516–5528.
107. Sloan KE, Mattijssen S, Lebaron S, Tollervey D, Pruijn GJ, Watkins NJ. Both endonucleolytic and exonucleolytic cleavage mediate ITS1 removal during human ribosomal RNA processing. *J Cell Biol.* 2013;200(5):577–588.
108. Pestov DG, Strezoska Z, Lau LF. Evidence of p53-dependent cross-talk between ribosome biogenesis and the cell cycle: effects of nucleolar protein Bop1 on G(1)/S transition. *Mol Cell Biol.* 2001;21(13):4246–4255.
109. Pestov DG, Stockelman MG, Strezoska Z, Lau LF. ERB1, the yeast homolog of mammalian Bop1, is an essential gene required for maturation of the 25S and 5.8S ribosomal RNAs. *Nucleic Acids Res.* 2001;29(17):3621–3630.
110. Lapik YR, Fernandes CJ, Lau LF, Pestov DG. Physical and functional interaction between Pes1 and Bop1 in mammalian ribosome biogenesis. *Mol Cell.* 2004;15(1): 17–29.
111. Killian A, Sarafan-Vasseur N, Sesboue R, et al. Contribution of the BOP1 gene, located on 8q24, to colorectal tumorigenesis. *Genes Chromosomes Cancer.* 2006;45(9):874–881.
112. Qi J, Yu Y, Akilli Ozturk O, et al. New Wnt/β-catenin target genes promote experimental metastasis and migration of colorectal cancer cells through different signals. *Gut.* 2016;65(10):1690–1701.
113. Chung KY, Cheng IK, Ching AK, Chu JH, Lai PB, Wong N. Block of proliferation 1 (BOP1) plays an oncogenic role in hepatocellular carcinoma by promoting epithelial-to-mesenchymal transition. *Hepatology.* 2011;54(1):307–318.
114. Gold HA, Topper JN, Clayton DA, Craft J. The RNA processing enzyme RNase MRP is identical to the Th RNP and related to RNase P. *Science.* 1989;245(4924): 1377–1380.
115. Clayton DA. A nuclear function for RNase MRP. *Proceedings of the National Academy of Sciences of the United States of America.* 1994;91(11):4615–4617.
116. Stohl LL, Clayton DA. *Saccharomyces cerevisiae* contains an RNase MRP that cleaves at a conserved mitochondrial RNA sequence implicated in replication priming. *Mol Cell Biol.* 1992;12(6):2561–2569.
117. Topper JN, Bennett JL, Clayton DA. A role for RNAase MRP in mitochondrial RNA processing. *Cell.* 1992;70(1):16–20.
118. Lygerou Z, Allmang C, Tollervey D, Seraphin B. Accurate processing of a eukaryotic precursor ribosomal RNA by ribonuclease MRP in vitro. *Science.* 1996;272(5259): 268–270.
119. Goldfarb KC, Cech TR. Targeted CRISPR disruption reveals a role for RNase MRP RNA in human preribosomal RNA processing. *Genes Dev.* 2017;31(1):59–71.
120. Park J, Jeong S. Wnt activated beta-catenin and YAP proteins enhance the expression of non-coding RNA component of RNase MRP in colon cancer cells. *Oncotarget.* 2015; 6(33):34658–34668.

121. Miluzio A, Beugnet A, Volta V, Biffo S. Eukaryotic initiation factor 6 mediates a continuum between 60S ribosome biogenesis and translation. *EMBO Rep.* 2009; 10(5):459–465.
122. Gandin V, Miluzio A, Barbieri AM, et al. Eukaryotic initiation factor 6 is rate-limiting in translation, growth and transformation. *Nature.* 2008;455(7213):684–688.
123. Brina D, Miluzio A, Ricciardi S, Biffo S. eIF6 anti-association activity is required for ribosome biogenesis, translational control and tumor progression. *Biochim Biophys Acta.* 2015;1849(7):830–835.
124. Ji Y, Shah S, Soanes K, et al. Eukaryotic initiation factor 6 selectively regulates Wnt signaling and beta-catenin protein synthesis. *Oncogene.* 2008;27(6):755–762.
125. Gonzales B, Henning D, So RB, Dixon J, Dixon MJ, Valdez BC. The Treacher Collins syndrome (TCOF1) gene product is involved in pre-rRNA methylation. *Hum Mol Genet.* 2005;14(14):2035–2043.
126. Sakai D, Trainor PA. Treacher Collins syndrome: unmasking the role of Tcof1/treacle. *Int J Biochem Cell Biol.* 2009;41(6):1229–1232.
127. Valdez BC, Henning D, So RB, Dixon J, Dixon MJ. The Treacher Collins syndrome (TCOF1) gene product is involved in ribosomal DNA gene transcription by interacting with upstream binding factor. Proceedings of the National Academy of Sciences of the United States of America. 2004;101(29):10709–10714.
128. Hayano T, Yanagida M, Yamauchi Y, Shinkawa T, Isobe T, Takahashi N. Proteomic analysis of human Nop56p-associated pre-ribosomal ribonucleoprotein complexes. Possible link between Nop56p and the nucleolar protein treacle responsible for Treacher Collins syndrome. *J Biol Chem.* 2003;278(36):34309–34319.
129. Dai J, Si J, Wang M, et al. Tcof1-related molecular networks in Treacher Collins syndrome. *J Craniofacial Surg.* 2016;27(6):1420–1426.
130. Kramps T, Peter O, Brunner E, et al. Wnt/wingless signaling requires BCL9/legless-mediated recruitment of pygopus to the nuclear beta-catenin-TCF complex. *Cell.* 2002;109(1):47–60.
131. Thompson BJ. A complex of Armadillo, Legless, and Pygopus coactivates dTCF to activate wingless target genes. *Curr Biol.* 2004;14(6):458–466.
132. Tzenov YR, Andrews P, Voisey K, et al. Selective estrogen receptor modulators and betulinic acid act synergistically to target ERalpha and SP1 transcription factor dependent Pygopus expression in breast cancer. *J Clin Pathol.* 2016;69(6):518–526.
133. Liu Y, Dong QZ, Wang S, et al. Abnormal expression of Pygopus 2 correlates with a malignant phenotype in human lung cancer. *BMC cancer.* 2013;13:346.
134. Andrews PG, He Z, Tzenov YR, Popadiuk C, Kao KR. Evidence of a novel role for Pygopus in rRNA transcription. *Biochem J.* 2013;453(1):61–70.
135. Jung Y, Bang S, Choi K, et al. TC1 (C8orf4) enhances the Wnt/beta-catenin pathway by relieving antagonistic activity of Chibby. *Cancer Res.* 2006;66(2):723–728.
136. Kim B, Koo H, Yang S, et al. TC1(C8orf4) correlates with Wnt/beta-catenin target genes and aggressive biological behavior in gastric cancer. *Clin Cancer Res.* 2006;12(11 Pt 1):3541–3548.
137. Chua EL, Young L, Wu WM, Turtle JR, Dong Q. Cloning of TC-1 (C8orf4), a novel gene found to be overexpressed in thyroid cancer. *Genomics.* 2000;69(3): 342–347.
138. Su K, Huang L, Li W, et al. TC-1 (c8orf4) enhances aggressive biologic behavior in lung cancer through the Wnt/beta-catenin pathway. *J Surg Res.* 2013;185(1): 255–263.
139. Zheng YW, Zhang L, Wang Y, et al. Thyroid cancer 1 (C8orf4) shows high expression, no mutation and reduced methylation level in lung cancers, and its expression correlates with beta-catenin and DNMT1 expression and poor prognosis. *Oncotarget.* 2017; 8(38):62880–62890.

140. Takemaru K, Yamaguchi S, Lee YS, Zhang Y, Carthew RW, Moon RT. Chibby, a nuclear beta-catenin-associated antagonist of the Wnt/Wingless pathway. *Nature.* 2003;422(6934):905–909.
141. Takemaru K, Fischer V, Li FQ. Fine-tuning of nuclear-catenin by Chibby and 14-3-3. *Cell Cycle.* 2009;8(2):210–213.
142. Dass RA, Sarshad AA, Carson BB, et al. Wnt5a signals through DVL1 to repress ribosomal DNA transcription by RNA polymerase I. *PLoS Genet.* 2016;12(8):e1006217.
143. Zlotorynski E. Ageing: live longer with small nucleoli. *Nat Rev Mol Cell Biol.* 2017;18(11):651.
144. Tiku V, Jain C, Raz Y, et al. Small nucleoli are a cellular hallmark of longevity. *Nat Commun.* 2016;8:16083.
145. Buchwalter A, Hetzer MW. Nucleolar expansion and elevated protein translation in premature aging. *Nat Commun.* 2017;8(1):328.
146. Kraushar ML, Popovitchenko T, Volk NL, Rasin MR. The frontier of RNA metamorphosis and ribosome signature in neocortical development. *Int J Dev Neurosci.* 2016;55:131–139.
147. Kraushar ML, Viljetic B, Wijeratne HR, et al. Thalamic WNT3 secretion spatiotemporally regulates the neocortical ribosome signature and mRNA translation to specify neocortical cell subtypes. *J Neurosci Off J Soc Neurosci.* 2015;35(31):10911–10926.
148. Parlato R, Kreiner G. Nucleolar activity in neurodegenerative diseases: a missing piece of the puzzle? *J Mol Med.* 2013;91(5):541–547.
149. Xue S, Barna M. Specialized ribosomes: a new frontier in gene regulation and organismal biology. *Nat Rev Mol Cell Biol.* 2012;13(6):355–369.
150. Woods SJ, Hannan KM, Pearson RB, Hannan RD. The nucleolus as a fundamental regulator of the p53 response and a new target for cancer therapy. *Biochim Biophys Acta.* 2015;1849(7):821–829.

CHAPTER FIVE

Wnt Signaling in Skeletal Muscle Development and Regeneration

Francesco Girardi, Fabien Le Grand[1]
Sorbonne Universitésy, UPMC Université Paris 06, INSERM UMRS974, Centre of Research in Myology, Paris, France
[1] Corresponding author. E-mail address: fabien.le-grand@inserm.fr

Contents

Abstract

Wnt is a family of signaling molecules involved in embryogenesis, adult tissue repair, and cancer. They activate canonical and noncanonical Wnt signaling cascades in target cells. Several studies, within the last decades, showed that several Wnt ligands are involved in myogenesis and both canonical and noncanonical Wnt pathways regulate muscle formation and the maintenance of adult tissue homeostasis. In this review, we provide a comprehensive overview of the roles of Wnt signaling during muscle development and an updated description of Wnt functions during muscle repair. Lastly, we discuss the crosstalk between Wnt and TGFβ signaling pathways in skeletal muscle.

Progress in Molecular Biology and Translational Science, Volume 153
ISSN 1877-1173
http://dx.doi.org/10.1016/bs.pmbts.2017.11.026

1. WNT SIGNALING PATHWAYS

Wnt proteins are secreted growth factors belonging to a conserved family of cysteine-rich glycoproteins. The signaling pathways mediated by these ligands are essential to a large variety of biological processes, ranging from organ formation during embryogenesis to adult stem cell homeostasis.[1,2] The human genome harbors 19 Wnt genes, each with a different expression pattern and playing a different role.[3] Depending on the involvement of β-catenin, Wnt signaling can be classified as either canonical (β-catenin-dependent) or noncanonical (β-catenin-independent).

The canonical Wnt cascade begins with the binding of Wnt proteins to specific 7-transmembrane domain receptors belonging to the Frizzled (Fzd) family and to the low-density lipoprotein receptor-related protein (LRP5/6) acting as a coreceptor.[4] In unstimulated conditions, β-catenin is phosphorylated by a cytoplasmic destruction complex formed by Axin, Adenomatous Polyposis Coli (APC), Casein Kinase I (CKI) and Glycogen Synthase Kinase 3β (GSK3β). Phosphorylated β-catenin is then ubiquinated and degraded by the proteasome. Upon Wnt-Fzd-LRP5/6 interaction, Axin is recruited to the phosphorylated tail of LRP5/6 by the scaffold protein Dishevelled (Dvl). This relocalization disassembles the destruction complex leading to the inhibition of β-catenin ubiquitination. Under these conditions, β-catenin accumulates within the cytoplasm and translocates to the nucleus, where it associates to the transcription factors, TCF and LEF, to activate the transcription of target genes.

As mentioned earlier, Wnt ligands can also activate the noncanonical pathways, which are called Planar Cell Polarity (Wnt/PCP) pathway and calcium-dependent (Wnt/Ca^{2+}) pathway. In the Wnt/PCP, Wnt ligands binding to Fzd receptors activate the small GTPase proteins, Rho and Rac, which in turn activate Rho-associated protein kinase (ROCK) and c-Jun N-terminal kinases (JNKs). These two kinases trigger a phosphorylation cascade that ultimately results in cytoskeletal reorganization and/or in activation of target genes. In turn, in the Wnt/Ca^{2+}, the Wnt-Fzd interaction induces the production of inositol triphosphate (IP3) and diacylglycerol (DAG) by the phospholipase C (PLC). Accumulation of IP3 triggers the release of intracellular calcium (Ca^{2+}), which in turn activates calcineurin (CaN), Ca^{2+} and calmodulin-dependent kinase (CAMKII), and protein kinase C (PKC). As a result, the NFAT, CREB, and NF-κB transcriptional regulators are activated and induce the expression of target genes.

2. SKELETAL MUSCLE DEVELOPMENT

Skeletal muscle is a highly plastic and complex tissue. The muscles are fixed to the bones by the tendons and have the function to move the body in space under the control of the central nervous system. They are constituted of individual components known as muscle fibers (myofibers), cylindrical multinucleated cells containing contracting myofibrils. Myogenesis, the process of muscle formation, is the result of a well-orchestrated coordination of transcriptional cascades and signaling pathways that spatio-temporally direct cell proliferation and differentiation, migration, and morphological changes.[5] In vertebrates, the skeletal muscles of the body and limbs derive from somites, which are segments of the paraxial mesoderm (Fig. 1A). More precisely, myogenesis initiates in the dermomyotome, a transient embryonic structure localized in the dorsal part of the somite. In the dermomyotome, two regions can be distinguished: the epaxial domain, which gives rise to the back muscle, and the hypaxial domain, from which originates the body wall muscles, the diaphragm, and the limb muscles. During embryonic and fetal development, myogenesis occurs in different sequential phases. In the first phase, postmitotic dermomyotome border cells loose their epithelial organization and delaminate underneath the remaining dorsal dermomyotome (Fig. 1A). Here, myogenic progenitors (myoblasts) differentiate to form the myotome, a structure composed of aligned myocytes and myotubes that serve as a scaffold for the following waves of myogenesis.[6,7] In the second phase, called primary myogenesis, proliferating muscle progenitors migrate from the central region of the dermomyotome and differentiate and fuse with the first myofibers thereby giving rise to skeletal muscles (Fig. 1A).[8] Myogenesis in the limbs will follow a similar process with a delay in developmental time as muscle progenitor cells first have to migrate from the ventral lip of the dermomyotome to the limb bud anlagen where they will form the primary muscle fibers. Secondary myogenesis occurs during fetal growth, when proliferating muscle progenitors continuously generate fetal myoblasts that fuse to form secondary myofibers around this scaffold of first generation of muscle fibers.

Dermomyotomal cells express Pax3, a paired-box transcription factor essential for the induction of skeletal myogenesis.[9] Before delamination, $Pax3^+$ cells initiate the expression of the myogenic regulatory factors (MRFs), which are a family of bHLH transcription factors that orchestrates the myogenic lineage. The first MRF to be activated is Myf5 that in turn promotes the expression of the MRF Myod1. Together, they activate the last

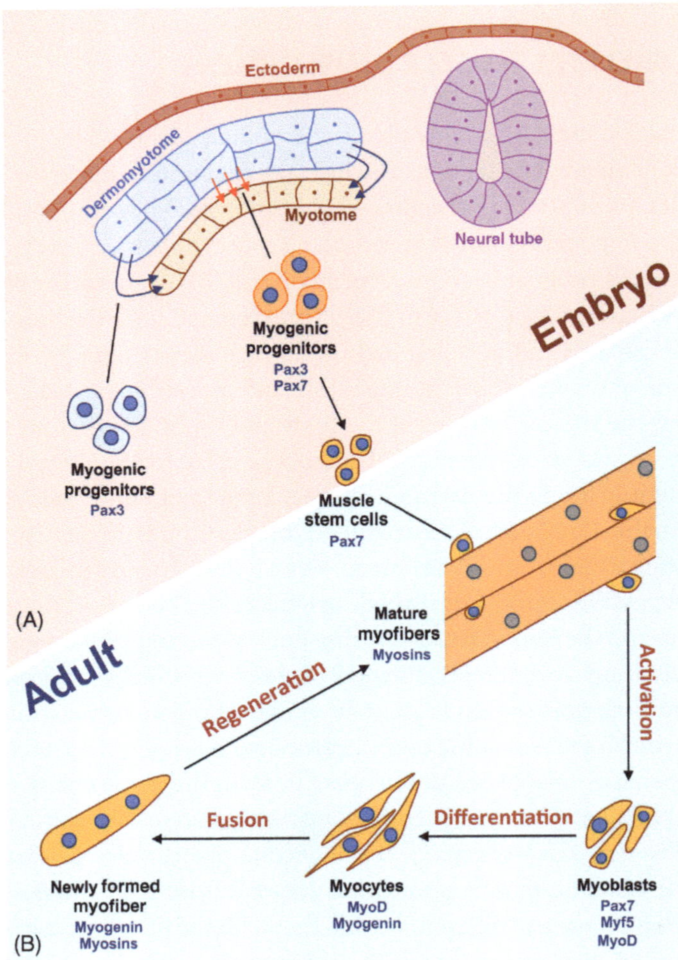

Fig. 1 Schematic representation of embryonic myogenesis and muscle regeneration. (A) Muscle formation is initiated in the dermomyotome and can be divided in sequential phases. During the first part, $Pax3^{+}$ cells delaminate from the extremities of the dermomyotome (blue arrows) and form the myotome. In the second phase (primary myogenesis), Pax3/Pax7 positive cells migrate from the central region of the dermomyotome to complete skeletal muscle formation (red arrows). At the end of fetal development, a group of Pax3/Pax7 positive cells give rise to adult muscle stem cells. (B) Satellite cells are located under the basal lamina that surrounds myofibers. After injury, satellite cells exit from quiescent and get activated giving rise to myoblasts. Myoblasts differentiate into myocytes, which in turn fuse to form new myofibers leading to muscle regeneration.

MRF Myogenin and this event is correlated with irreversible cell cycle exit and the activation of terminal differentiation. Another important factor during muscle formation is Pax7, a paralog of Pax3. Whereas Pax3 is activated in all myogenic progenitor cells, Pax7 expression is restricted to the central domain of the dermomyotome and its function is limited to primary, secondary, and adult myogenesis.[10] At the end of fetal development, $Pax3^{+}$/$Pax7^{+}$ myogenic progenitors locate underneath the basal lamina of myofibers, thereby giving rise to adult muscle stem cells (see Section 4).[11]

3. WNT SIGNALING IN SKELETAL MUSCLE DEVELOPMENT

As morphogens, Wnt proteins are involved in many embryonic processes, including somitogenesis, dermomyotome specification, and myogenesis (Fig. 2). Oscillations in canonical Wnt signaling in the presomitic mesoderm play a pivotal role in embryonic segmentation and somitogenesis ("Clock and Wavefront" mechanism) (Fig. 2A).[12] Consequently, correct activation of the Wnt signaling pathways is essential during the various steps of muscle formation. For instance, it has been shown that Wnt1 and Wnt3a are necessary for the induction of the dermomyotome (Fig. 2B).[13] By surgical ablation or by injection of protein-secreting cells in chick embryos, Marcelle and colleagues proved that the neural tube expresses Wnt1 and Wnt3a and that these two secreted factors mediate the formation of the dermomyotome.[13] Indeed, in Wnt1/Wnt3a mutant mouse embryos, the dermomyotome is reduced in size and lacks the median compartment, as shown by the reduction in the size of Pax3 and Myf5 expression domains.[14] In addition, implantation of Wnt1- and Wnt3a-secreting cells in chick embryos results in the expansion of the Pax3 expression domain and in the enlargement of the dermomyotome.[15] More recently, Hutcheson and colleagues demonstrated that canonical Wnt signaling is required for correct dermomyotome and myotome formation, as targeted β-catenin gene recombination in the Pax3 lineage abolishes the generation of embryonic myoblasts and results in a blockade of embryonic myogenesis.[16] Wnt proteins have also been shown to play a role in the maintenance of the epithelial organization of the dermomyotome.[17] Wnt6-mediated canonical cascade activation is required for the preservation of the epithelial structure of somites, while the absence of Wnt6 leads to the de-epithelialization (thus delamination) of the dermomyotome.[17]

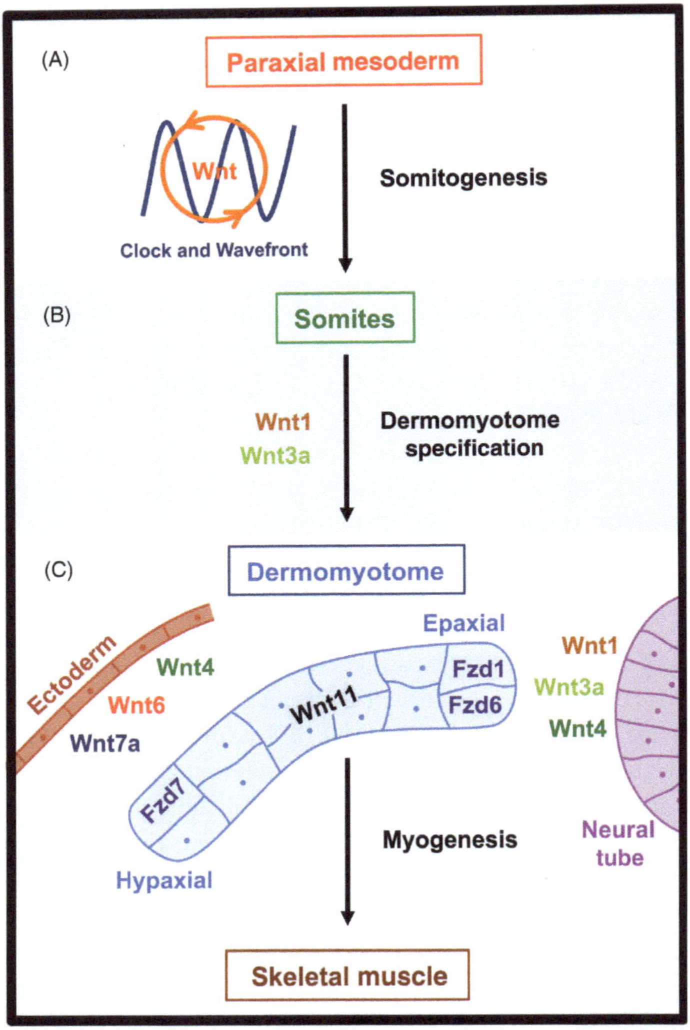

Fig. 2 Wnt functions from the paraxial mesoderm to skeletal muscle tissue. (A) Somites derive from the paraxial mesoderm and Wnt signaling plays a pivotal role in the "Clock and Wavefront" mechanism, a well-established model of somitogenesis. (B) After somite formation, Wnt1 and Wnt3a drive dermomyotome specification in the dorsal region of the somites. (C) Dermomyotome receives signals from the surrounding embryonic structures. Precisely, neural tube secretes Wnt1, Wnt3a, and Wnt4, while ectoderm expresses Wnt4, Wnt6, and Wnt7a. Dermomyotome epaxial region expresses Fzd1 and Fzd6, while in the hypaxial domain Fzd7 is present. Dermomyotome cells are characterized by Wnt11 expression. Together, all these Wnt ligands and receptors take part to embryonic myogenesis, leading to skeletal muscle formation.

The capacity of Wnt signaling to induce dermomyotome specification is strongly related to its ability to regulate the expression of Pax3, Pax7, and MRF genes. In explant cultures of chick segmental plate and somites, it has been shown that Wnt3a dramatically increases the levels of the two paired-box transcription factors and strongly enhances explant cell proliferation.[18] Subsequent in vivo experiments based on either Wnt electroporation or on engraftment of Wnt-secreting cells in chick embryos confirmed these results, demonstrating that Wnt1, Wnt3a, Wnt4, and Wnt6 regulate and maintain Pax3 and Pax7 expression in myogenic precursor cells, therefore controlling the number of dermomyotomal $Pax3^+/Pax7^+$ progenitors.[18–20]

As mentioned earlier, Wnt proteins are also able to modulate the expression of MRFs, thus playing an essential role during embryonic myogenesis (Fig. 2C).[21] Early studies showed that Wnt1 induces muscle formation through the preferential activation of Myf5, while Wnt6 or Wnt7a preferentially promote MyoD expression, which can occur independently from Myf5.[22] The identification of functional TCF/LEF binding sites in the Myf5 enhancers revealed that its Wnt1-dependent expression is mediated by the canonical Wnt cascade.[23] Conversely, Wnt7a promotes MyoD expression through the PKC-dependent noncanonical pathway, as proved by MyoD downregulation in the presence of PKC inhibitors in explant cultures.[24] Furthermore, it has been reported that Wnt1 and Wnt7a proteins can stimulate protein kinase A (PKA) signaling and CREB-mediated transcription of Pax3, Myf5, and MyoD in the somite, providing evidence for the involvement of multiple pathways.[25] Interestingly, a recent study demonstrated that migrating neural crest cells trigger β-catenin signaling and myogenesis in epithelial dermomyotome progenitor cells.[26] Mechanistically, Notch is activated in somitic cells by Delta-1 (Dll1)-expressing neural crest cells, and this leads to an epithelial to mesenchymal transition that allows the recruitment of β-catenin from the membrane independently of Wnt ligands. Once translocated into the nucleus, β-catenin acts as a transcriptional cofactor to activate Myf5 expression.[26] Most importantly, activation of the canonical Wnt cascade in dermomyotomal cells leads to the expression of Wnt11 in the medial border of the somites. Wnt11, then acts as a directional cue to orient myocyte elongation through the noncanonical Wnt/PCP pathway.[27] As such, the sequential action of the WNT/PCP and the WNT/β-catenin pathways is necessary for the formation of functional embryonic muscle fibers.

Myogenesis thus relies on a finely orchestrated spatial expression of Wnt proteins and Fzd receptors (Fig. 2C). Analysis of the expression pattern of

Wnt and Fzd isoforms in mouse or chick embryos revealed that Wnt1, Wnt3, and Wnt4 are specifically expressed in the dorsal neural tube,[28,29] while Wnt4, Wnt6, and Wnt7a are secreted from the dorsal ectoderm.[22,30] Fzd isoforms are also specifically distributed in the somites where they mediate different functions. Fzd1 and Fzd6 are expressed in the epaxial domain of the somite, where they promote Myf5 activation through the canonical Wnt cascade.[23] Whereas in the hypaxial region the receptor Fzd7 is present, which interacts with Wnt7a to activate MyoD expression.[31]

In summary, multiple Wnt ligands and receptors are expressed in different domains during embryonic myogenesis and all of them, together with many other signaling pathways, are part of a complex regulatory network that modulates and drives the formation of skeletal muscle.

4. SKELETAL MUSCLE REGENERATION

Adult skeletal muscle has a remarkable capacity to regenerate. Muscle regeneration (or adult myogenesis) relies on satellite cells (MuSCs). This adult muscle stem cell population is necessary for skeletal muscle growth and reconstruction following injury (Fig. 1B).[32] Satellite cells are engendered from fetal muscle progenitor cells at the end of embryonic development. They are characterized by the expression of Pax7 and no longer express Pax3.[33] Satellite cells are present in healthy muscle as mitotically quiescent cells[34] located in a niche between the basal lamina and the myofibers sarcolemma. In response to myofiber damage, satellite cells leave quiescence, become activated and proliferate to give rise to a transient amplifying population expressing Myf5 and MyoD (Fig. 1B). At this stage they are referred to as adult myoblasts. Akin to embryonic myogenesis, these myogenic precursors once they exit the cell cycle will express Myogenin and fuse with the damaged myofibers or with each other to form new myotubes. Then, additional myoblasts incorporate into the nascent myotubes to form new myofibers (Fig. 1B).[35] Generally, 3–4 weeks after injury, skeletal muscle returns to a normal morphological and histological condition and are characterized in mice by centrally located nuclei.[36]

In common with all adult stem cells, a hallmark of the muscle satellite cells is their ability to self-renew.[37] During regeneration, a correct balance between self-renewal and differentiation is essential. Symmetric divisions allow the maintenance of the satellite cell pool, while asymmetric expansion leads to the generation of myogenic progenitor cells.[38]

5. WNT SIGNALING IN SKELETAL MUSCLE REGENERATION

Unlike embryonic myogenesis, in which Wnt signaling is highly active, in mature skeletal muscle this pathway is poorly activated[39] (see Section 6). In contrast, early studies showed that some Wnts are expressed in freshly isolated myofibers and in uninjured muscles.[40,41] Even though Wnt expression in uninjured adult muscle is still unclear, its activation during muscle regeneration is well established (Fig. 3).[40–43] After injury, many Wnt

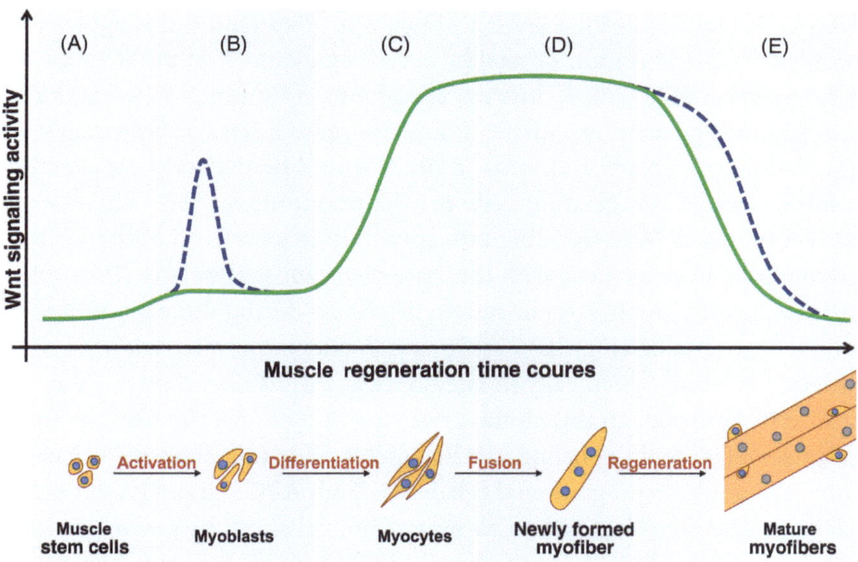

Fig. 3 Wnt signaling activity during muscle regeneration. (A) Wnt signaling (green line) is inactive in quiescent satellite cells and Adenomatous Polyposis Coli protein is essential to limit this signaling and to allow proper cell cycle entry.[43] (B) After satellite cell activation, Wnt7a (blue dashed line) stimulates symmetric division of satellite cells through the Planar Cell Polarity pathway,[44] while Axin1 inhibits Wnt/β-catenin cascade to maintain muscle progenitor proliferation.[46] (C) Canonical Wnt signaling inhibits Notch pathway causing the satellite cell transition from proliferation phase to differentiation phase.[42] (D) Wnt signaling promotes myoblast differentiation and fusion by inducing myogenic gene expression via Barx2 activation,[47] stimulating Follistatin,[48] inhibiting myostatin[49] and translocating the negative regulator Setdb1 into the cytoplasm.[50] (E) In mature myofibers Wnt7a regulates Akt/mTOR pathway increasing muscle size and strength.[51] Lastly, Wnt activity returns at low levels.[39] Overall, a timely regulation of Wnt/β-catenin signaling is essential for muscle regeneration.[45]

ligands (such as Wnt3a, Wnt5a, Wnt5b, Wnt7a, and Wnt7b) are expressed and secreted from multiple sources but in particular myofibers and activated satellite cells. Both canonical and noncanonical pathways have been implicated in muscle repair and accumulating evidences indicate that Wnt signaling is more particularly involved in the regulation of satellite cell differentiation and self-renewal.[42,44,45]

One of the first indications for a potential role of canonical Wnt signaling in adult muscle cell differentiation was suggested by the in vitro studies on immortalized cell lines, in which the Wnt1/β-catenin cascade and/or GSK3 inhibition were able to enhance myogenic differentiation.[52,53]

The involvement of the canonical Wnt cascade during satellite cell differentiation was further confirmed in mouse.[42] In particular, a switch from high Notch signaling levels to high Wnt/β-catenin signaling levels is essential for the transition of satellite cell descendants from proliferation to differentiation (Fig. 3C).[42] Indeed, exogenous activation of the canonical pathway in regenerating muscles stimulates muscle lineage progression in proliferating myoblasts and leads to premature differentiation (accelerated commitment of proliferating cells to postmitotic myocytes). Therefore a correct timing of Wnt signaling activation is fundamental to obtain correct tissue repair. In agreement with this notion, recent studies have shown that maintenance of low β-catenin activity is crucial during the early phases of muscle regeneration (Fig. 3B).[43,46] In vivo β-catenin overactivation via APC genetic depletion, specifically in satellite cells, results in complete failure of muscle tissue repair. In this context, we have shown that the massive upregulation of β-catenin signaling in APC-null satellite cells abolishes cell cycle entry and results in programmed cell death. Thus APC activity is essential to dampen Wnt signaling and to allow satellite cells to exit quiescence and proliferate (Fig. 3A).[43] Furthermore, siRNA-mediated knockdown of Axin1 in satellite cells leads to reduced proliferation and precocious myotube formation in vitro, indicating that Axin1, by limiting Wnt signaling, acts to maintain satellite cell expansion and prevent premature differentiation.[46] Lastly, our lab recently demonstrated that an adequate level of Wnt/β-catenin signaling is fundamental for controlling muscle progenitor function during muscle regeneration (Fig. 3).[45] In this study, satellite cell-specific β-catenin loss-of-function and gain-of-function mutations resulted in the alteration of the regeneration process in vivo. Satellite cell descendants deprived of canonical Wnt signaling differentiate less efficiently (reduced generation of Myogenin$^+$ cells), while constitutively active Wnt/β-catenin cascade leads to precocious growth arrest and premature differentiation. As a result, muscle regeneration is impaired in both conditional mutant mice. In conclusion, these studies suggest that a correct timing of activation together with a proper level of canonical Wnt pathway is crucial during muscle regeneration (Fig. 3).

At the molecular level, how Wnt signaling regulates muscle repair remains to be fully elucidated. Early studies on immortalized cell lines proposed that canonical Wnt signaling regulates MRF activity during myogenic differentiation,[54,55] but more recent papers have suggested novel and alternative mechanisms through which Wnt/β-catenin pathway controls muscle regeneration (Fig. 3D).[47,50] In differentiating myoblasts, Wnt signaling has been shown to be involved in the inhibition of Pax7 protein expression (negative regulator of differentiation) and in the activation and stabilization of Barx2, a homeobox protein that promotes myogenic differentiation gene expression via its association with MRFs. Barx2 also positively regulates Axin2 activity, suggesting the presence of a negative feedback loop modulating the Wnt cascade (Fig. 3D).[47] Finally, Beyer and colleagues showed that the histone lysine methyltransferase Setdb1 is a potent inhibitor of terminal differentiation and its activity is Wnt/β-catenin-dependent. Indeed, canonical Wnt signaling regulates satellite cell fate by driving the nuclear export of this chromatin modifier and this subcellular relocalization allows myoblast differentiation (Fig. 3D).[50] It should be noted that the promotion of myogenic differentiation driven by Wnt is also regulated by BCL9, an important mediator of β-catenin nuclear localization.[56] Loss of BCL9 (and its homologous BCL9-2) reduces canonical Wnt signaling activation and consequently impairs myogenic progenitor differentiation both in vitro and in vivo. Together, these results confirm not only the essential role of the Wnt/β-catenin cascade during muscle regeneration, but also the complexity of the regulatory network underlying tissue repair.

As mentioned earlier, noncanonical Wnt pathways are involved in adult myogenesis. In regenerating muscles, Wnt7a is expressed by newly formed regenerated myofibers and acts through the PCP pathway to regulate satellite cell self-renewal (Fig. 3B).[44] Specifically, Wnt7a interacts with its receptor Fzd7 and causes the redistribution of Vangl2 protein (an essential regulator of PCP pathway) on opposite poles of dividing satellite cells. This polarization induces satellite cell symmetric expansion and thus regulates self-renewed satellite cell numbers at the end of tissue regeneration. In addition, a recent work proposed the involvement of fibronectin (FN) and its receptor Syndecan-4 (Sdc4) in this process, expanding our knowledge about the molecular mechanism underlying satellite cell pool regulation.[57]

Sdc4 is a plasma membrane proteoglycan expressed by satellite cells and acts as Fzd7 coreceptor.[58] Following their activation, satellite cells express FN, which binds the Fzd7/Sdc4 receptor complex to potentiate Wnt7a signaling and thus promoting satellite cell symmetrical expansion. Wnt7a-mediated PCP pathway participates also to the modulation of satellite cell and myoblast motility.[59] By impacting cytoskeletal organization through activation of the small GTPase Rac1, Wnt7a enhances muscle stem cell migration and markedly improves cell engraftment in vivo. Moreover, Wnt7a has been implicated in the regulation of myofiber size and strength independently of the PCP pathway (Fig. 3E).[51] In postmitotic myotubes and myofibers, Wnt7a is able to activate the AKT/mTOR pathway leading to muscle hypertrophy and improved force generation. Taken together, these studies suggest that Wnt7a could have a potential therapeutic application. Further work is required to investigate the potential of Wnt7a as a biological protein that could be used to stimulate intrinsic repair in muscle-wasting diseases by coupling enhancement of satellite cell self-renewal and myofiber hypertrophy.

A family of canonical Wnt signaling activators called R-spondins (Rspo1, 2, 3, and 4) has recently been implicated in the regulation of adult myogenesis. In the myogenic cell line C2C12, R-spondin-dependent activation of the canonical Wnt cascade leads to the stimulation of myoblast differentiation and fusion.[60] Rspo2 induces the Wnt/β-catenin pathway, which in turn positively modulates the expression of the TGFβ antagonist Follistatin (Fst) (see Section 7).[61] In vivo, *Rspo1-null* mice exhibit a transient delay in the first phases of regeneration suggesting that Rspo1 is required during muscle repair.[62] Indeed, Rspo1 promotes myogenic progenitor cell differentiation through the Wnt/β-catenin cascade, but it also dampens the noncanonical PCP pathway specifically in differentiating myocytes. As a result of a perturbed equilibrium between these two Wnt signaling cascades, Rspo1-deficient myoblasts fuse with a better efficiency both in vitro and in vivo. Together, these data indicate that Rspo1 acts on both Wnt pathways and thus controls the balance between canonical and noncanonical signaling in differentiating muscle cells.

In conclusion, we are just beginning to understand the pivotal role of Wnt signaling during muscle regeneration. However, many questions, such as the spatio-temporal expression in uninjured myofibers, the precise molecular mechanism, and the molecular pathways that control the interplay between canonical and noncanonical Wnt signaling, remain unsolved.

6. CONTROL OF MYOFIBER TYPING BY WNT SIGNALING

Adult skeletal muscles in vertebrates are composed of different types of muscle fibers with distinct contraction speeds and metabolic profiles ranging from slow/oxidative to fast/glycolytic. There are four fiber types in mammalian muscle, one slow (type 1) and three fast (types 2A, 2X and 2B), each characterized by the expression of one specific myosin heavy chain isoform (MyHC). The Wnt signaling cascades have been shown to control the genesis of this fiber-type heterogeneity during development.[16,63–65]

The first indications of this role were reported in 2003, when Anakwe and colleagues showed that Wnt ligands control muscle fiber typing during avian wing development.[63] They showed that Wnt5 stimulation increases the number of slow and reduces the number of fast myofibers respectively, while Wnt11 promotes fast-type fiber formation. Another Wnt ligand involved in fiber-type specification is Wnt4, which has been shown to increase the number of fast MyHC fibers in the chick limb bud.[65] In contrast, Hutcheson and colleagues clearly demonstrated that β-catenin intrinsically regulates the formation of slow myofibers during fetal mouse development. Consequently, β-catenin gene recombination in Pax7-expressing muscle progenitor cells led to a decreased number of myofibers expressing slow MyHC, while most fetal myofibers expressed type 1 MyHC following targeted β-catenin gain-of-function.[16] Finally, a recent paper has shown that the downstream component of Wnt signaling pathway, TCF4, controls myofiber type by intrinsically regulating MyHC isoform expression in myogenic progenitors.[64] Moreover, ablation of TCF4-positive fibroblasts in fetal mouse myogenesis leads to a reduction of slow myofiber numbers. Together, these results indicate that Wnt signaling plays a role in fiber type determination, and also the presence of intrinsic and extrinsic mechanisms that drive myofiber diversification during development.

Intriguingly, Wnt signaling pathways have not been precisely linked to adult myofiber typing maintenance in adulthood. As such, distinct canonical Wnt activation domains have been proposed in healthy adult myofibers. By using a TCF-LacZ reporter mouse, Wnt signaling was detected exclusively in slow myofibers,[39] while the usage of Axin2-LacZ reporter mice showed that the Wnt cascade was active in fast myofibers only.[66] These conflicting results might be due to the different sensibilities of the two reporter alleles;

however, further studies are required to accurately define the Wnt signaling state in uninjured adult myofibers.

7. THE CROSSTALK BETWEEN WNT AND TGFβ SIGNALING IN SKELETAL MUSCLE

The transforming growth factor beta (TGFβ) superfamily is comprised of over 30 secreted factors, including TGFβ isoforms (TGFβ1, 2, and 3), bone morphogenetic proteins (BMPs), activins and myostatin. This cytokine family is ubiquitously expressed in different tissues and it plays a fundamental role in embryonic and adult myogenesis. The strongest regulator of skeletal muscle phenotype is myostatin, a potent inhibitor of muscle tissue growth and mass.[67] Other members of the TGFβ family are also essential in driving embryonic myogenesis. For instance, TGFβ1 has been shown to have an inhibitory effect on myoblast differentiation,[68] while BMP signaling is a positive regulator of muscle mass.[69] As such, myostatin negatively regulates myoblast proliferation and differentiation[70] and TGFβ1 inhibits the myogenic differentiation program at a gene expression level.[71] In contrast, modulation of TGFβ signaling using genetic approaches results in the alteration of satellite cell self-renewal and myoblast differentiation.[72,73] Taken together, these conflicting results indicate that the TGFβ signaling pathway has generally a negative role on muscle regeneration, but is still essential.

Even though the roles of Wnt and TGFβ signaling pathways during embryonic development have been intensively studied, the analysis of a potential interaction between these two pathways in embryonic myogenesis has not been fully addressed. Kuroda and colleagues showed that, during muscle development, the canonical Wnt cascade promotes slow myofiber formation by activating BMP4 expression. Nevertheless, further studies are required to properly dissect the interplay between Wnt and TGFβ cascades in the embryo.[39]

In contrast, multiple interactions between the two pathways have been described during adult myogenesis. In the C2C12 cell line, it has been reported that Wnt4 antagonizes myostatin-mediated phosphorylation of Smad2, thus promoting myoblast proliferation and differentiation.[65] Subsequent experiments confirmed this evidence and showed that Wnt4 overexpression in satellite cell-derived primary myoblasts enhances muscle cell fusion and myotube size. By acting through canonical signaling, Wnt4 limits myostatin function leading to Pax7-downregulation and MRF-upregulation (Fig. 3D).[49] An early study using C2C12 showed that myostatin modulates myoblast proliferation state by repressing Wnt4 expression.[74] Together, these experiments suggest a potential dynamic and reciprocal interplay between both pathways, but further studies are required to explore this hypothesis.[74] Wnt4 has also been described to interact with BMP signaling, but in this case in a β-catenin-independent manner. Indeed, noncanonical Wnt4 signaling can potentiate the BMP cascade leading to a negative regulation of myogenic differentiation.[75] Therefore additional analyses are needed to elucidate this dual role of canonical and noncanonical Wnt4 activation. Moreover, recent studies have shown that the Wnt/β-catenin cascade stimulates the expression of the TGFβ antagonist Fst, providing novel evidence of Wnt-TGFβ interaction in skeletal muscle cells.[48,61] Fst is known to neutralize several TGFβ members and it can positively regulate myogenic differentiation in a Wnt-dependent manner both in vitro and in vivo. In particular, Wnt signaling triggers the expression of Myogenin, which in turn binds to a regulatory E-Box element upstream of the Fst gene, leading to Fst expression and premature differentiation (Fig. 3D).[48] In addition, Rspo2 has also been reported to promote Fst transcription through the activation of the canonical Wnt pathway.[61] However, the mechanism by which Fst exerts its positive regulation on muscle stem cell differentiation remains unclear.

Finally, transcriptome analysis of primary myoblast cultures has revealed that TGFβ2 and 3 transcription are under the control of canonical Wnt signaling.[45] In fact, in regenerating muscles Wnt/β-catenin cascade activates TGFβ signaling and this step is crucial for proper muscle repair. As mentioned earlier, in vivo β-catenin conditional gene disruption in satellite cells results in perturbed myogenic differentiation, but a wild-type phenotype can be restored via exogenous TGFβ2 injection in injured mutant muscles, suggesting a positive role for TGFβ2 on myoblast differentiation. As the TGFβ pathway has generally been described as a negative regulator of adult myogenesis and is a well-established pro-fibrotic factor, this study calls into question the negative role of TGFβ signaling in muscle regeneration. At this point, an interesting hypothesis is that TGFβ, downstream of canonical Wnt signaling, may be required to instruct cell cycle arrest in proliferating muscle cells, as a prerequisite for terminal differentiation.

Together, these multiple interactions between Wnt and TGFβ signaling during adult myogenesis represent a clear indication that an elaborate interplay between these two pathways exists and is essential for the correct regeneration of muscle tissue.

8. WNT SIGNALING IMPLICATIONS IN MUSCULAR DISEASE AND TREATMENT

In the light of their pivotal role in myogenesis, Wnt proteins represent promising compounds for therapeutic purposes in muscle-related pathologies, such as cachexia or muscular dystrophies. These diseases are characterized by muscle weakening and wasting, and molecules such as Wnt7a that are able to improve muscle size and strength are potential candidates for the treatment of these specific conditions. Indeed, Wnt7a represents a current focus for therapeutic use, but the high hydrophobicity that characterizes all Wnt ligands hinders its production and application.[76] Recently, von Maltzahn and colleagues generated a truncated Wnt7a protein, which lacks the palmitoylation site and thus has reduced hydrophobicity.[77] This Wnt7a variant retained full biological functions in vivo and has an enhanced dispersion compared to the full-length protein, leading to muscle hypertrophy. This finding represents a significant advance for the development of Wnt7a-based therapeutic approaches.

Moreover, mis-regulation of canonical Wnt pathway has been described in both aged and dystrophic muscles. Modifications of the systemic environment that occurs during aging increases Wnt signaling activity in myogenic precursors, leading to impairment in muscle regeneration and increased fibrosis.[78] Similarly, the Duchenne Muscular Dystrophy mouse model (also known as the *mdx* mouse) shows a chronic upregulation of Wnt pathways in skeletal muscles.[79] In particular, the elevated Wnt activity in dystrophic muscles drives TGFβ2 activation, which in turn alters muscle progenitor cell fate promoting an alternate fibrotic identity. Although it has recently been shown that Wnt-dependent TGFβ2 activation in regenerating muscle is an essential step for proper muscle repair,[45] TGFβ2 upregulation in the context of the *mdx* mouse compromises the regenerative potential of the satellite cells and stimulates fibrogenic gene expression.[79] Taken together, these conflicting results confirm the crucial role of the Wnt/TGFβ crosstalk in skeletal muscle, but additional studies are needed to fully understand the precise function of the Wnt-TGFβ axis in adult myogenesis. Considering the negative effects of enhanced Wnt activity in aged and dystrophic muscles, downregulation of Wnt signaling could ameliorate these pathophysiological conditions. In support of this hypothesis, exogenous inhibition of the Wnt cascade through Dickkopf-1 (DKK1, Wnt antagonist) in aged regenerating muscles and in dystrophic *mdx* muscles results in reduced fibrosis.[78,80] For these reasons, in vivo modulation of Wnt signaling in pathological contexts represents a potential therapeutic strategy for the improvement of aged muscle and for the treatment of muscular dystrophies.

9. CONCLUDING REMARKS AND OUTSTANDING QUESTIONS

The multiple studies performed during the last decades clearly indicate the pivotal role of Wnt signaling in both embryonic and adult myogenesis. During muscle formation, many Wnt ligands and receptors have been described to have an impact on muscle progenitor cells and, although some mechanisms remain unclear, their roles and functions are currently well established. However, as embryonic myogenesis relies on a complex network of signaling pathways, it would be interesting to analyze potential crosstalk between these different pathways during embryogenesis (e.g. TGFβ signaling pathway), to further expand our knowledge on muscle development.

The role of Wnt signaling in adult myogenesis starts to be elucidated; however, many questions remain unresolved. For instance, conflicting Wnt expression domains have been proposed in mature healthy muscle fibers[39,66] and a clarification of these contrasting results could represent an important step to understand the Wnt signaling in uninjured myofibers and its role in the maintenance of adult tissue homeostasis in physiological conditions. Furthermore, the novel regulatory mechanisms proposed by recent papers revealed a complicated system underlying the myogenic modulation driven by Wnt signaling.[47,50] Hence, further studies are required to completely dissect the extensive molecular mechanism through which Wnt ligands control adult myogenesis. Lastly, the exquisite balance between canonical and noncanonical Wnt signaling represents an important point of interest. Based on the current knowledge, canonical and noncanonical pathways are complementary and exert different functions. Wnt/β-catenin cascade appears to be mainly involved in the regulation of satellite cell commitment and terminal differentiation,[42,43,45] while the noncanonical Wnt7a pathway controls satellite cell self-renewal and migration, and myofiber size and strength.[44,51,59] However, our understanding concerning the precise mechanism that regulates canonical and noncanonical cascades during muscle regeneration remains poor.

Satellite cell self-renewal is an intriguing and barely understood mechanism. Our observations and those of others concur to the fact that after muscle regeneration, the satellite cell niche reforms to the exact extent as that prior to regeneration.[81] We previously identified the noncanonical Wnt7a-PCP signaling as a pathway controlling satellite cell self-renewal, via control of stem cell division orientation.[44] On the other hand, the impact of canonical Wnt signaling on the return of satellites to the niche has not been clearly investigated. It is an ongoing question in multiple research fields whether both canonical and noncanonical Wnt pathways act sequentially, in parallel, or compete to regulate stem cell self-renewal. Consequently, we propose two hypotheses: first, it is possible that both Wnt signaling pathways compete, and that Wnt/β-catenin inhibits Wnt/PCP; second, Wnt/β-catenin signaling may feed into the Wnt/PCP cascade by regulating the transcription of PCP components. In vivo epistasis experiments and in vitro transcriptomic analyses may validate one of these hypotheses in the future.

One of the key properties that distinguish adult stem cells from their differentiated progenies is their ability to remain in the quiescent state for an extensive period of time. In contrast to various stem cell populations in fast-turnover tissues, the satellite cells are a population of primarily quiescent adult stem cells in adult skeletal muscle. Interestingly, canonical Wnt signaling pathways have been previously shown to control hematopoietic stem cell quiescence[82] and to be negatively regulated in quiescent hair follicle stem cells.[83] Our studies demonstrate that upregulation of canonical Wnt signaling in satellite cells following APC gene disruption prevented the mutant cells to enter the cell cycle in activating conditions.[43] Thus suggesting that β-catenin signaling may promote quiescence. In contrast, analysis of mice with targeted β-catenin gene disruption in satellite cells only showed that the number of Pax7 cells remained constant one month following the induction of Cre/Lox activity in adult mice.[45] However, more precise analyses, such as evaluation of the cell cycle status of mutant satellite cells, should be performed to clearly demonstrate an implication of canonical Wnt signaling in satellite cell quiescence. Moreover, intramuscular injection of recombinant Wnt7a protein to uninjured skeletal muscle promoted satellite cell amplification, suggesting that noncanonical Wnt ligand may be a satellite cell "activating" factor. Unfortunately, it is also possible that the needle injury was sufficient to send a stress signal throughout the muscle tissue that would trigger the satellite cell to leave quiescence. Hence, the development of new experimental setups, such as transgenic over-expression of Wnt7a, or targeted Fzd7 gene disruption is now needed to evaluate if noncanonical Wnt-PCP signaling plays a role in breaking satellite cell quiescence.

Another important question that will have to be resolved in the next years is what is/are the source(s) of Wnt proteins in skeletal muscle, either in the healthy adult skeletal muscle or during regeneration? Earlier studies have mainly used RT-PCR analysis of cultured muscle cells or bulk muscle tissue. While providing the blueprint for the expression pattern of the Wnt family members, positional information (a.k.a which cell type) was not gathered and temporal information was only patchy. Furthermore, it is now clear that satellite cells need to receive clues from neighboring cells, such as macrophages[84] or fibro-adipogenic progenitors[85] in order to function correctly. In this new era of single-cell analyses, we hope to soon be able to access single-cell RNA-sequencing and single-cell proteomic datasets of multiple muscle-resident cell types that will allow us to construct a precise mapping of the sources of Wnt signaling in skeletal muscle.

ACKNOWLEDGEMENTS

We thank P. Maire and G. Butler-Browne for critical reading of the manuscript. The work in the Le Grand Lab is supported by funding from INSERM, CNRS, the Agence National pour la Recherche (ANR, project ANR-14-CE11-0026) and the Association Française contre les Myopathies/AFM Telethon. F. Girardi is supported by a fellowship from the Fondation pour la Recherche Médicale (FRM, project ECO20160736081).

REFERENCES

1. Croce JC, McClay DR. Evolution of the Wnt pathways. *Methods Mol Biol.* 2008;469:3–18.
2. Holland JD, Klaus A, Garratt AN, Birchmeier W. Wnt signaling in stem and cancer stem cells. *Curr Opin Cell Biol.* 2013;25(2):254–264.
3. Nusse R, Varmus H. Three decades of Wnts: a personal perspective on how a scientific field developed. *EMBO J.* 2012;31(12):2670–2684.
4. MacDonald BT, Tamai K, He X. Wnt/β-catenin signaling: components, mechanisms, and diseases. *Dev Cell.* 2009;17(1):9–26.
5. Bryson-Richardson RJ, Currie PD. The genetics of vertebrate myogenesis. *Nat Rev Genet.* 2008;9(8):632–646.
6. Gros J, Scaal M, Marcelle C. A two-step mechanism for myotome formation in chick. *Dev Cell.* 2004;6(6):875–882.
7. Kahane N, Cinnamon Y, Kalcheim C. The origin and fate of pioneer myotomal cells in the avian embryo. *Mech Dev.* 1998;74(1–2):59–73.
8. Relaix F, Rocancourt D, Mansouri A, Buckingham M. A Pax3/Pax7-dependent population of skeletal muscle progenitor cells. *Nature.* 2005;435(7044):948–953.
9. Blake JA, Ziman MR. Pax genes: regulators of lineage specification and progenitor cell maintenance. *Development.* 2014;141(4):737–751.
10. Buckingham M, Rigby PWJ. Gene regulatory networks and transcriptional mechanisms that control myogenesis. *Dev Cell.* 2014;28(3):225–238.
11. Gros J, Manceau M, Thomé V, Marcelle C. A common somitic origin for embryonic muscle progenitors and satellite cells. *Nature.* 2005;435(7044):954–958.

12. Jensen PB, Pedersen L, Krishna S, Jensen MH. A Wnt oscillator model for somitogenesis. *Biophys J.* 2010;98(6):943–950.
13. Marcelle C, Stark MR, Bronner-Fraser M. Coordinate actions of BMPs, Wnts, Shh and noggin mediate patterning of the dorsal somite. *Development.* 1997;124 (20):3955–3963.
14. Ikeya M, Takada S. Wnt signaling from the dorsal neural tube is required for the formation of the medial dermomyotome. *Development.* 1998;125(24):4969–4976.
15. Wagner J, Schmidt C, Nikowits W, Christ B. Compartmentalization of the somite and myogenesis in chick embryos are influenced by wnt expression. *Dev Biol.* 2000;228 (1):86–94.
16. Hutcheson DA, Zhao J, Merrell A, Haldar M, Kardon G. Embryonic and fetal limb myogenic cells are derived from developmentally distinct progenitors and have different requirements for β-catenin. *Genes Dev.* 2009;23(8):997–1013.
17. Linker C, Lesbros C, Gros J, Burrus LW, Rawls A, Marcelle C. β-Catenin-dependent Wnt signalling controls the epithelial organisation of somites through the activation of paraxis. *Development.* 2005;132(17):3895–3905.
18. Galli LM, Willert K, Nusse R, et al. A proliferative role for Wnt-3a in chick somites. *Dev Biol.* 2004;269(2):489–504.
19. Otto A, Schmidt C, Patel K. Pax3 and Pax7 expression and regulation in the avian embryo. *Anat Embryol.* 2006;211(4):293–310.
20. Schmidt C, Stoeckelhuber M, McKinnell I, Putz R, Christ B, Patel K. Wnt 6 regulates the epithelialisation process of the segmental plate mesoderm leading to somite formation. *Dev Biol.* 2004;271(1):198–209.
21. Cossu G, Borello U. Wnt signaling and the activation of myogenesis in mammals. *EMBO J.* 1999;18(24):6867–6872.
22. Tajbakhsh S, Borello U, Vivarelli E, et al. Differential activation of Myf5 and MyoD by different Wnts in explants of mouse paraxial mesoderm and the later activation of myogenesis in the absence of Myf5. *Development.* 1998;125(21):4155–4162.
23. Borello U, Berarducci B, Murphy P, et al. The Wnt/β-catenin pathway regulates Gli-mediated Myf5 expression during somitogenesis. *Development.* 2006;133(18): 3723–3732.
24. Brunelli S, Relaix F, Baesso S, Buckingham M, Cossu G. Beta catenin-independent activation of MyoD in presomitic mesoderm requires PKC and depends on Pax3 transcriptional activity. *Dev Biol.* 2007;304(2):604–614.
25. Chen AE, Ginty DD, Fan C-M. Protein kinase A signalling via CREB controls myogenesis induced by Wnt proteins. *Nature.* 2005;433(7023):317–322.
26. Sieiro D, Rios AC, Hirst CE, Marcelle C. Cytoplasmic NOTCH and membrane-derived β-catenin link cell fate choice to epithelial–mesenchymal transition during myogenesis. *eLife Sci.* 2016;5:e14847.
27. Gros J, Serralbo O, Marcelle C. WNT11 acts as a directional cue to organize the elongation of early muscle fibres. *Nature.* 2009;457(7229):589–593.
28. Cauthen CA, Berdougo E, Sandler J, Burrus LW. Comparative analysis of the expression patterns of Wnts and Frizzleds during early myogenesis in chick embryos. *Mech Dev.* 2001;104(1–2):133–138.
29. Münsterberg AE, Kitajewski J, Bumcrot DA, McMahon AP, Lassar AB. Combinatorial signaling by Sonic hedgehog and Wnt family members induces myogenic bHLH gene expression in the somite. *Genes Dev.* 1995;9(23):2911–2922.
30. Schubert FR, Mootoosamy RC, Walters EH, et al. Wnt6 marks sites of epithelial transformations in the chick embryo. *Mech Dev.* 2002;114(1–2):143–148.
31. Borello U, Buffa V, Sonnino C, Melchionna R, Vivarelli E, Cossu G. Differential expression of the Wnt putative receptors Frizzled during mouse somitogenesis. *Mech Dev.* 1999;89(1–2):173–177.

32. Lepper C, Partridge TA, Fan C-M. An absolute requirement for Pax7-positive satellite cells in acute injury-induced skeletal muscle regeneration. *Development.* 2011;138 (17):3639–3646.
33. Seale P, Sabourin LA, Girgis-Gabardo A, Mansouri A, Gruss P, Rudnicki MA. Pax7 is required for the specification of myogenic satellite cells. *Cell.* 2000;102(6):777–786.
34. Tedesco FS, Dellavalle A, Diaz-Manera J, Messina G, Cossu G. Repairing skeletal muscle: regenerative potential of skeletal muscle stem cells. *J Clin Invest.* 2010;120(1):11–19.
35. Yin H, Price F, Rudnicki MA. Satellite cells and the muscle stem cell niche. *Physiol Rev.* 2013;93(1):23–67.
36. Chargé SBP, Rudnicki MA. Cellular and molecular regulation of muscle regeneration. *Physiol Rev.* 2004;84(1):209–238.
37. Collins CA, Olsen I, Zammit PS, et al. Stem cell function, self-renewal, and behavioral heterogeneity of cells from the adult muscle satellite cell niche. *Cell.* 2005;122 (2):289–301.
38. Kuang S, Kuroda K, Le Grand F, Rudnicki MA. Asymmetric self-renewal and commitment of satellite stem cells in muscle. *Cell.* 2007;129(5):999–1010.
39. Kuroda K, Kuang S, Taketo MM, Rudnicki MA. Canonical Wnt signaling induces BMP-4 to specify slow myofibrogenesis of fetal myoblasts. *Skelet Muscle.* 2013;3:5.
40. Otto A, Schmidt C, Luke G, et al. Canonical Wnt signalling induces satellite-cell proliferation during adult skeletal muscle regeneration. *J Cell Sci.* 2008;121 (17):2939–2950.
41. Polesskaya A, Seale P, Rudnicki MA. Wnt signaling induces the myogenic specification of resident CD45+ adult stem cells during muscle regeneration. *Cell.* 2003;113 (7):841–852.
42. Brack AS, Conboy IM, Conboy MJ, Shen J, Rando TA. A temporal switch from Notch to Wnt signaling in muscle stem cells is necessary for normal adult myogenesis. *Cell Stem Cell.* 2008;2(1):50–59.
43. Parisi A, Lacour F, Giordani L, Colnot S, Maire P, Le Grand F. APC is required for muscle stem cell proliferation and skeletal muscle tissue repair. *J Cell Biol.* 2015;210(5):717–726.
44. Le Grand F, Jones AE, Seale V, Scimè A, Rudnicki MA. Wnt7a activates the planar cell polarity pathway to drive the symmetric expansion of satellite stem cells. *Cell Stem Cell.* 2009;4(6):535–547.
45. Rudolf A, Schirwis E, Giordani L, et al. β-Catenin activation in muscle progenitor cells regulates tissue repair. *Cell Rep.* 2016;15(6):1277–1290.
46. Figeac N, Zammit PS. Coordinated action of Axin1 and Axin2 suppresses β-catenin to regulate muscle stem cell function. *Cell Signal.* 2015;27(8):1652–1665.
47. Hulin J-A, Nguyen TDT, Cui S, et al. Barx2 and Pax7 regulate Axin2 expression in myoblasts by interaction with β-catenin and chromatin remodelling. *Stem Cells.* 2016;34 (8):2169–2182.
48. Jones AE, Price FD, Le Grand F, et al. Wnt/β-catenin controls follistatin signalling to regulate satellite cell myogenic potential. *Skelet Muscle.* 2015;5:14.
49. Bernardi H, Gay S, Fedon Y, Vernus B, Bonnieu A, Bacou F. Wnt4 activates the canonical β-catenin pathway and regulates negatively myostatin: functional implication in myogenesis. *Am J Physiol Cell Physiol.* 2011;300(5):C1122–C1138.
50. Beyer S, Pontis J, Schirwis E, et al. Canonical Wnt signalling regulates nuclear export of Setdb1 during skeletal muscle terminal differentiation. *Cell Discov.* 2016;2:16037.
51. von Maltzahn J, Bentzinger CF, Rudnicki MA. Wnt7a-Fzd7 signalling directly activates the Akt/mTOR anabolic growth pathway in skeletal muscle. *Nat Cell Biol.* 2012;14 (2):186–191.
52. Rochat A, Fernandez A, Vandromme M, et al. Insulin and Wnt1 pathways cooperate to induce reserve cell activation in differentiation and myotube hypertrophy. *Mol Biol Cell.* 2004;15(10):4544–4555.

53. van der Velden JLJ, Langen RCJ, Kelders MCJM, Wouters EFM, Janssen-Heininger YMW, Schols AMWJ. Inhibition of glycogen synthase kinase-3β activity is sufficient to stimulate myogenic differentiation. *AmJ Physiol Cell Physiol.* 2006;290(2):C453–C462.
54. Kim C-H, Neiswender H, Baik EJ, Xiong WC, Mei L. β-Catenin interacts with MyoD and regulates its transcription activity. *Mol Cell Biol.* 2008;28(9):2941–2951.
55. Ridgeway AG, Petropoulos H, Wilton S, Skerjanc IS. Wnt signaling regulates the function of MyoD and myogenin. *J Biol Chem.* 2000;275(42):32398–32405.
56. Brack AS, Murphy-Seiler F, Hanifi J, et al. BCL9 is an essential component of canonical Wnt signaling that mediates the differentiation of myogenic progenitors during muscle regeneration. *Dev Biol.* 2009;335(1):93–105.
57. Bentzinger CF, Wang YX, von Maltzahn J, Soleimani VD, Yin H, Rudnicki MA. Fibronectin regulates Wnt7a signaling and satellite cell expansion. *Cell Stem Cell.* 2013;12(1):75–87.
58. Muñoz R, Moreno M, Oliva C, Orbenes C, Larraín J. Syndecan-4 regulates non-canonical Wnt signalling and is essential for convergent and extension movements in *Xenopus* embryos. *Nat Cell Biol.* 2006;8(5):492–500.
59. Bentzinger CF, von Maltzahn J, Dumont NA, et al. Wnt7a stimulates myogenic stem cell motility and engraftment resulting in improved muscle strength. *J Cell Biol.* 2014;205 (1):97–111.
60. Han XH, Jin Y-R, Seto M, Yoon JK. A WNT/β-catenin signaling activator, R-spondin, plays positive regulatory roles during skeletal myogenesis. *J Biol Chem.* 2011;286 (12):10649–10659.
61. Han XH, Jin Y-R, Tan L, Kosciuk T, Lee J-S, Yoon JK. Regulation of the follistatin gene by RSPO-LGR4 signaling via activation of the WNT/β-catenin pathway in skeletal myogenesis. *Mol Cell Biol.* 2014;34(4):752–764.
62. Lacour F, Vezin E, Bentzinger CF, et al. R-spondin1 controls muscle cell fusion through dual regulation of antagonistic Wnt signaling pathways. *Cell Rep.* 2017;18 (10):2320–2330.
63. Anakwe K, Robson L, Hadley J, et al. Wnt signalling regulates myogenic differentiation in the developing avian wing. *Development.* 2003;130(15):3503–3514.
64. Mathew SJ, Hansen JM, Merrell AJ, et al. Connective tissue fibroblasts and Tcf4 regulate myogenesis. *Development.* 2011;138(2):371–384.
65. Takata H, Terada K, Oka H, Sunada Y, Moriguchi T, Nohno T. Involvement of Wnt4 signaling during myogenic proliferation and differentiation of skeletal muscle. *Dev Dyn.* 2007;236(10):2800–2807.
66. Huraskin D, Eiber N, Reichel M, et al. Wnt/β-catenin signaling via Axin2 is required for myogenesis and, together with YAP/Taz and Tead1, active in IIa/IIx muscle fibers. *Development.* 2016;143(17):3128–3142.
67. McPherron AC, Lawler AM, Lee S-J. Regulation of skeletal muscle mass in mice by a new TGF-p superfamily member. *Nature.* 1997;387(6628):83–90.
68. Angelis MG., Molinari S, Donne AL, et al. Differential response of embryonic and fetal myoblasts to TGF beta: a possible regulatory mechanism of skeletal muscle histogenesis. *Development.* 1994;120(4):925–933.
69. Sartori R, Schirwis E, Blaauw B, et al. BMP signaling controls muscle mass. *Nat Genet.* 2013;45(11):1309–1318.
70. Langley B, Thomas M, Bishop A, Sharma M, Gilmour S, Kambadur R. Myostatin inhibits myoblast differentiation by down-regulating MyoD expression. *J Biol Chem.* 2002;277(51):49831–49840.
71. Kollias HD, McDermott JC. Transforming growth factor-β and myostatin signaling in skeletal muscle. *J Appl Physiol.* 2008;104(3):579–587.
72. Filvaroff EH, Ebner R, Derynck R. Inhibition of myogenic differentiation in myoblasts expressing a truncated type II TGF-beta receptor. *Development.* 1994;120(5):1085–1095.

73. Ge X, Vajjala A, McFarlane C, Wahli W, Sharma M, Kambadur R. Lack of Smad3 signaling leads to impaired skeletal muscle regeneration. *Am J Physiol Endocrinol Metab.* 2012;303(1):E90–E102.
74. Steelman CA, Recknor JC, Nettleton D, Reecy JM. Transcriptional profiling of myostatin-knockout mice implicates Wnt signaling in postnatal skeletal muscle growth and hypertrophy. *FASEB J.* 2006;20(3):580–582.
75. Terada K, Misao S, Katase N, Nishimatsu S, Nohno T. Interaction of Wnt signaling with BMP/Smad signaling during the transition from cell proliferation to myogenic differentiation in mouse myoblast-derived cells. *Int J Cell Biol.* 2013;2013:e616294.
76. Reichsman F, Smith L, Cumberledge S. Glycosaminoglycans can modulate extracellular localization of the wingless protein and promote signal transduction. *J Cell Biol.* 1996;135 (3):819–827.
77. von Maltzahn J, Zinoviev R, Chang NC, Bentzinger CF, Rudnicki MA. A truncated Wnt7a retains full biological activity in skeletal muscle. *Nat Commun.* 2013;4:2869.
78. Brack AS, Conboy MJ, Roy S, et al. Increased Wnt signaling during aging alters muscle stem cell fate and increases fibrosis. *Science.* 2007;317(5839):807–810.
79. Biressi S, Miyabara EH, Gopinath SD, Carlig PMM, Rando TA. A Wnt-TGFβ2 axis induces a fibrogenic program in muscle stem cells from dystrophic mice. *Sci Transl Med.* 2014;6(267):267ra176.
80. Trensz F, Haroun S, Cloutier A, Richter MV, Grenier G. A muscle resident cell population promotes fibrosis in hindlimb skeletal muscles of mdx mice through the Wnt canonical pathway. *Am J Physiol Cell Physiol.* 2010;299(5):C939–C947.
81. Abou-Khalil R, Le Grand F, Pallafacchina G, et al. Autocrine and paracrine angiopoietin 1/Tie-2 signaling promotes muscle satellite cell self-renewal. *Cell Stem Cell.* 2009;5 (3):298–309.
82. Fleming HE, Janzen V, Lo Celso C, et al. Wnt signaling in the niche enforces hematopoietic stem cell quiescence and is necessary to preserve self-renewal in vivo. *Cell Stem Cell.* 2008;2(3):274–283.
83. Lien W-H, Polak L, Lin M, Lay K, Zheng D, Fuchs E. In vivo transcriptional governance of hair follicle stem cells by canonical Wnt regulators. *Nat Cell Biol.* 2014;16(2):179–190.
84. Saclier M, Cuvellier S, Magnan M, Mounier R, Chazaud B. Monocyte/macrophage interactions with myogenic precursor cells during skeletal muscle regeneration. *FEBS J.* 2013;280(17):4118–4130.
85. Farup J, Madaro L, Puri PL, Mikkelsen UR. Interactions between muscle stem cells, mesenchymal-derived cells and immune cells in muscle homeostasis, regeneration and disease. *Cell Death Dis.* 2015;6:e1830.

CHAPTER SIX

Wnt Signaling in Kidney Development and Disease

Yongping Wang*, Chengji J. Zhou[†], Youhua Liu*[,‡,1]

* National Clinical Research Center of Kidney Disease, Nanfang Hospital, Southern Medical University, Guangzhou, People's Republic of China
[†] University of California Davis, Sacramento, CA, United States
[‡] University of Pittsburgh School of Medicine, Pittsburgh, PA, United States
[1] Corresponding author. E-mail address: liuy@upmc.edu

Contents

Abstract

Wnt signal cascade is an evolutionarily conserved, developmental pathway that regulates embryogenesis, injury repair, and pathogenesis of human diseases. It is well established that Wnt ligands transmit their signal via canonical, β-catenin-dependent and noncanonical, β-catenin-independent mechanisms. Mounting evidence has revealed that Wnt signaling plays a key role in controlling early nephrogenesis and is implicated in the development of various kidney disorders. Dysregulations of Wnt expression cause a variety of developmental abnormalities and human diseases, such as congenital anomalies of the kidney and urinary tract, cystic kidney, and renal

Progress in Molecular Biology and Translational Science, Volume 153
ISSN 1877-1173
http://dx.doi.org/10.1016/bs.pmbts.2017.11.019

carcinoma. Multiple Wnt ligands, their receptors, and transcriptional targets are upregulated during nephron formation, which is crucial for mediating the reciprocal interaction between primordial tissues of ureteric bud and metanephric mesenchyme. Renal cysts are also associated with disrupted Wnt signaling. In addition, Wnt components are important players in renal tumorigenesis. Activation of Wnt/β-catenin is instrumental for tubular repair and regeneration after acute kidney injury. However, sustained activation of this signal cascade is linked to chronic kidney diseases and renal fibrosis in patients and experimental animal models. Mechanistically, Wnt signaling controls a diverse array of biologic processes, such as cell cycle progression, cell polarity and migration, cilia biology, and activation of renin–angiotensin system. In this chapter, we have reviewed recent findings that implicate Wnt signaling in kidney development and diseases. Targeting this signaling may hold promise for future treatment of kidney disorders in patients.

1. INTRODUCTION

Wnt signaling is one of the most highly conserved pathways throughout evolution, and it plays important roles in a diverse array of biologic processes, such as embryonic development, metabolism, tumorigenesis, and stem cell renewal.[1–3] The first *Wnt-1* gene, also known as *Wingless* in *Drosophila* and *Int-1* in mammalians, was discovered based on the activation by integration of virus DNA in mouse breast tumors in 1982.[4] Sequence analyses revealed that *Wnt-1* was actually the homolog of the *Wingless* (*wg*) gene in *Drosophila* that controls segment polarity in the formation of the body axis during embryonic development. Since then, totally 19 Wnt ligands in mammalian genomes are identified, which possess unique structural features and distinct expression patterns. The signal is initiated through the engagement of Wnt ligands to the extracellular domain of frizzled (Fzd) receptor and coreceptors, the low-density lipoprotein receptor-related protein 5 and 6 (LRP5 and LRP6).[5] Depending on the involvement of key intracellular molecule β-catenin, Wnt signaling is typically classified as either β-catenin-dependent, canonical or β-catenin-independent, noncanonical pathways. The latter includes Wnt/planar cell polarity (PCP) and Wnt/Ca^{2+} signaling routes.[5–7]

Wnt ligands and their receptors are expressed in developing kidney during nephron formation.[8,9] Although this signal is relatively silent in adult, it becomes reactivated rapidly after a wide variety of injuries.[10–12] In the past decades, tremendous efforts have been made to delineate the expression, function, and targets of Wnt/β-catenin signaling in the developing kidneys, as well as in damaged kidneys after injury.[6,13] In this chapter, we provide a comprehensive review on the expression, regulation, and function of Wnt/β-catenin signaling in the context of kidney development and disorders.

2. WNT SIGNALING: COMPONENTS AND ROUTES

The canonical pathway of Wnt signaling is mediated by intracellular protein β-catenin (Fig. 1). In the absence of Wnt ligands-receptors engagement, cellular β-catenin levels are limited due to the phosphorylation-triggered proteasome degradation (Fig. 1A). Upon activation, Wnt ligands bind to the Fzd and LRP5/6 receptors, thereby activating the cytoplasmic protein dishevelled (Dvl). Dvl then inhibits the β-catenin destruction complex, consisting of the scaffold proteins Axin, adenomatous polyposis coli (APC), serine/threonine kinases of glycogen synthase kinase-3β (GSK-3β), and casein kinase 1 (CK1). Unphosphorylated β-catenin accumulates in the cytoplasm and enters the nucleus, where it interacts with T cell factor (TCF)/lymphoid enhancer-binding factor (LEF) family of transcription factors to regulate Wnt target genes (Fig. 1A).[14]

Recent studies indicate that pro(renin) receptor (PRR) may function as an amplifier of the canonical Wnt/β-catenin signaling (Fig. 1A).[15] PRR is a transmembrane protein with multiple distinct functions.[16] We reported that PRR is an essential component of the Wnt receptor complex and is obligatory for its signal transduction. Meanwhile, Wnt/β-catenin controls PRR gene and induces its expression in kidney tubular epithelial cells. Therefore, PRR induction and Wnt/β-catenin activation constitute a self-perpetuating cycle, which leads to the amplification of Wnt/β-catenin signaling.[15,17]

Wnts are also able to transmit their signals via β-catenin-independent, noncanonical pathways (Fig. 1B–C).[3,6] The Wnt/PCP pathway is evolutionarily conserved and plays a major role in embryonic tissue patterning, cell polarization, migration, and morphogenesis (Fig. 1B).[18] The Wnt/PCP ligands, such as Wnt5, Wnt7, and Wnt11, bind to the 7-transmembrane Fzd receptor and then recruit the cytoplasmic scaffold protein Dvl to the plasma membrane. Additional downstream components of the Wnt/PCP pathway include Van Gogh homolog 2 (Vangl2), Celsr1, Prickle, protein tyrosine kinase 7 (Ptk7), and Scrib, which work together to establish planar polarity. The Wnt/PCP effectors then regulate cytoskeleton dynamics directing asymmetric distribution of cellular components and migration. Many targets, such as Daam1, Rho, Rac, Rho kinase, C-Jun N-terminal kinase (JNK) are involved in these final steps (Fig. 1B).[19]

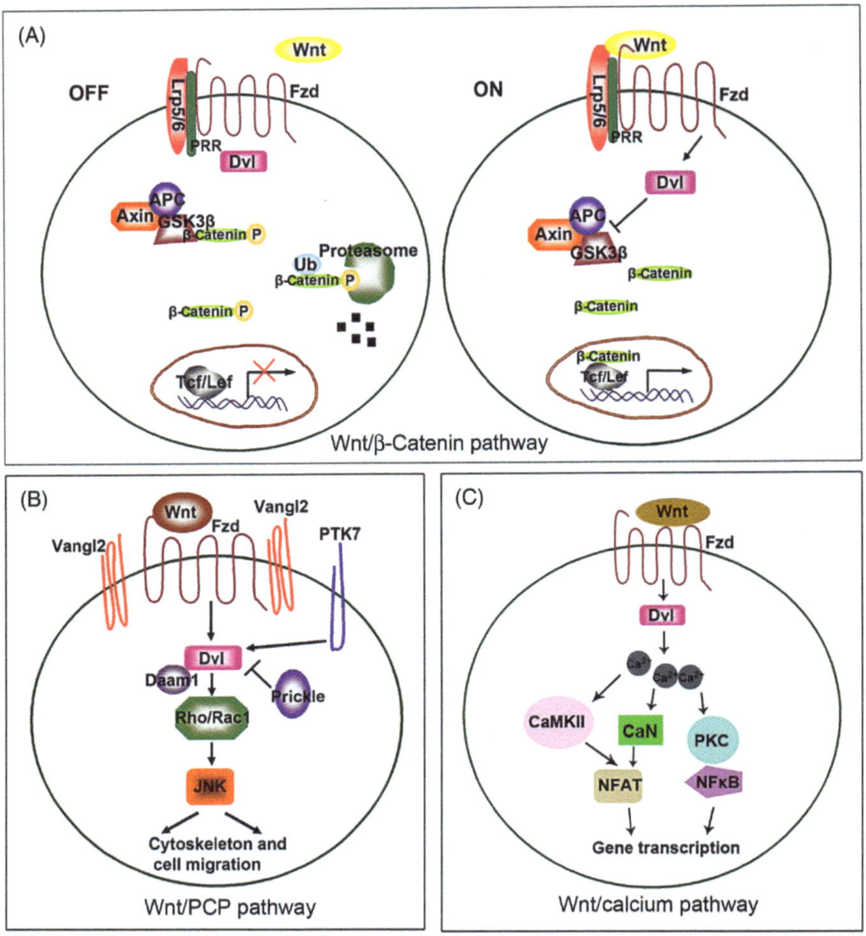

Fig. 1 Different Wnt signaling pathways. (A) Canonical Wnt/β-catenin signal pathway. When the signal is in "ON" state, Wnt ligands bind to Frizzled (Fzd) receptors and LRP5/6 coreceptors, which causes Dishevelled (Dvl) to inhibit the β-catenin destruction complex consisting of adenomatous polyposis coli (APC), Axin, and GSK-3β. Pro(renin) receptor (PRR) is also an obligatory component of the Wnt receptor complex and required for canonical Wnt/β-catenin signaling. Stabilized β-catenin is able to enter the nucleus and acts with T cell factor (TCF)/lymphoid enhancer-binding factor (LEF) transcription factor for specific gene expression. When the signal is "OFF," Dvl cannot act on the destruction complex and β-catenin is then phosphorylated by GSK-3β, leading phosphorylated β-catenin for degradation by ubiquitin (Ub)-proteasome system. (B) Noncanonical Wnt/PCP signal pathway. In Wnt/PCP pathway, Wnt ligands can bind Fzd, Vangl2, and PTK7, then acts with Rho/Rac1 small GTPases and JNK kinases through Dvl. Subsequently, cytoskeleton organization and cell migration are regulated in response to the signal. Daam1 acts with Dvl for downstream effectors, while Prickle inhibits this

Another noncanonical Wnt signaling is the Wnt/Ca^{2+} pathway, which leads to the release of intracellular Ca^{2+}, possibly via G-proteins (Fig. 1C).[3,6] This pathway involves activation of phospholipase C (PLC) and protein kinase C (PKC). Elevated Ca^{2+} can activate the phosphatase calcineurin, which leads to dephosphorylation of the nuclear factor of activated T-cells (NFAT) and its accumulation in the nucleus (Fig. 1C). The Ca^{2+}-mediated pathway has critical roles in dorsal/ventral patterning, gastrulation, and cardiac development. While the involvement of Wnt/β-catenin and Wnt/PCP signaling in regulating kidney development and disease is well studied, little is known about the role of Wnt/Ca^{2+} pathway in these processes.

3. WNT SIGNALING AND KIDNEY DEVELOPMENT

3.1 Major Events in Nephrogenesis

Significant advances have been made in deciphering molecular events in kidney development, making it possible to recognize the temporal and spatial pattern of embryonic kidney and its progenitors (Fig. 2).[20] The mammalian kidney originates from intermediate mesoderm (IM) after gastrulation, and then lies along the anteroposterior axis between paraxial mesoderm and IM.[21] There are three stages of kidney development in a temporal and craniocaudal sequence, and each is marked by the successive development of a more advanced kidney: pronephros, mesonephros, and metanephros (Fig. 2). The pronephros, also named primitive kidneys, develops at embryonic day 8.5 (E8.5) in a mouse embryo and at approximately the sixth somite in humans, and consists of 6–10 pairs of tubules.[22] This organ is considered as transient structures in mammalians and completely disappears, but it is essential for the next coming cells and tissues, including the mesonephric kidney. The mesonephros, also named middle kidney, develops by the formation of mesonephric tubules from the IM and is physiologically functional during early embryonic life with the structure of mesonephric tubules and ducts [Wolffian ducts (WD)] at E10 in mouse embryo. The mesonephric

◀ function. (C) Noncanonical Wnt/Ca^{2+} pathway. Wnt ligand binding activates Fzd/Dvl and induces a rise in intracellular Ca^{2+} levels. Elevated intracellular Ca^{2+} activates calcium calmodulin-dependent protein kinase II (CaMKII), calcinurin (CaN), and protein kinase C (PKC), which activate nuclear transcription factors (NFAT and NF-κB) and promote the expression of downstream target genes.

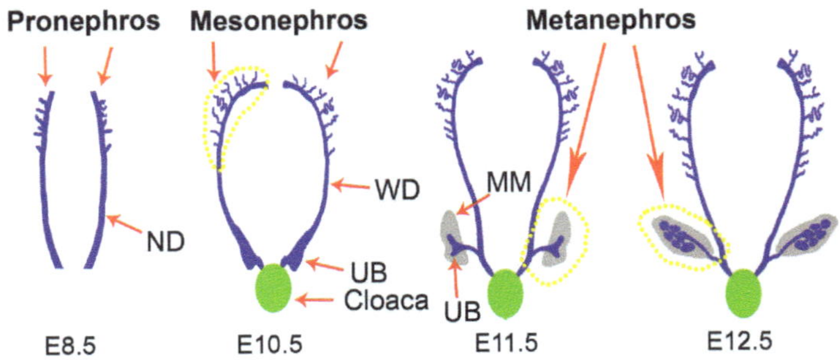

Fig. 2 Diagram depicts the major events in mammalian nephrogenesis. There are three stages of kidney development in a temporal sequence: pronephros, mesonephros, and metanephros. The pronephros develops at embryonic day 8.5 (E8.5) in mice. The mesonephros develops by the formation of mesonephric tubules from the intermediate mesoderm. The metanephros, begins at E10.5 in mice, finally becomes the permanent and functional kidney in mammalians. *MM*, Metanephric mesenchyme; *ND*, nephric duct; *UB*, ureteric bud; *WD*, Wolffian ducts.

duct reaches the cloaca with the proximal end and convoluted tubules blending into the duct. An important aggregate adjacent to the duct is called the metanephric mesenchyme (MM). The mesonephros gradually degenerates, but it also contributes to other organs, such as reproductive tracts of the testes, epididymal ducts, and vas deferens in all vertebrates. The metanephros, also named permanent kidney, begins in week 5 of gestation in humans and at E10.5 in mice. Metanephros is also derived from the IM, and finally becomes the functional adult kidney in mammalians (Fig. 2).[23]

The process of metanephric nephrogenesis is initiated by the invasion and reciprocal interaction between the ureteric bud (UB) and MM.[24] The UB arises from an outgrowth of the Wolffian duct and invades the MM, and induces the surrounding mesenchyme to undergo a mesenchymal to epithelial transition (MET), further forming the renal vesicles (RVs). The RVs subsequently differentiate through comma- and S-shaped body stages, and give rise to glomerular podocytes, parietal epithelium cells, proximal, and distal tubule cells[25]; whereas the UB undergoes branching morphogenesis to form collect duct. As such, the metanephros finally gives rise to ureteric tree and mature kidney.[26]

3.2 Wnt Ligands and Nephron Formation

Wnt signaling plays critical roles in several processes during kidney development, such as UB induction and nephrongenesis (Table 1).[27] Using

Table 1 Wnt Components Involved in Kidney Development.

Wnt Components	Developmental Roles	KO Renal Defects	References
Wnt1	Inducing tubulogenesis	?	34
Wnt2b	Inducing ureter branching	?	37
Wnt4	Initiating renal development	Müller duct regression and renal dysgenesis	38–44
Wnt5a	Regulating metanephric mesenchyme	CAKUT	45–48
Wnt6	Inducing ureter bud branching	?	49
Wnt7b	Maintaining cortico-medullary axis	Fails to form medullary zone and concentrate urine	50
Wnt9b	Regulating progenitors in metanephric mesenchyme	Disrupts tubular cell division and increases diameter	51–55
Wnt11	Autocrine factor in nephric duct and ureteric epithelium	Disrupts UB branching and causes renal hypoplasia	56–58
Lrp6	Acting through Ret signal pathway for renal defects	Hypoplasia cystic kidney	59
Frizzle4	Coordinating with Frizzle8	Frizzle4/8 double KO exhibits reduced ureteric bud growth	61
Frizzle8	Coordinating with Frizzle4 to maintain normal ureteric epithelial function	Frizzle4/8 double KO exhibits reduced ureteric bud growth	60
GSK-3β	Inducing early nephrogenesis	?	62,63
β-Catenin	Critical for nephron progenitor maintenance and nephrogenesis	Conditional deletion at different lineages causes various renal defects, including aplasia, hypoplasia and cysts	64–68
Vangl2	Indispensable for normal morphogenesis of both UB and MM-derived structures	Impairs branching morphogenesis and causes cystic kidney	73
Fat4	Regulating cell division and tubular elongation via acting with vangl2	Cystic kidney	75
Daam1	Promoting pronephric tubulogenesis	Disrupts tubule elaboration and branching in pronephic proximal segments	76

CAKUT, Congenital anomalies of the kidney and urinary tract; MM, metanephric mesenchyme; UB, ureteric bud; ?, undefined.

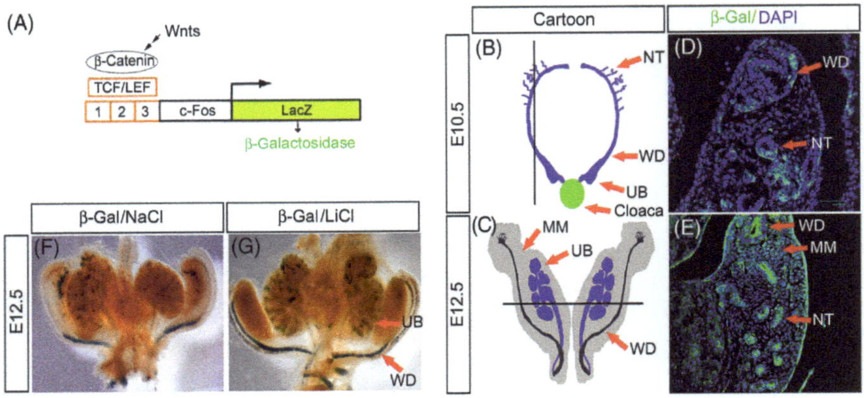

Fig. 3 Wnt/β-catenin signaling is activated during kidney development. (A) Diagram depicts the construction of the Wnt/β-catenin responsive reporter TOPgal mice. (B–E) Wnt/β-catenin is activated in different time points during kidney development. The *black lines* in Panels B and C indicate dissection planes for corresponding embryonic kidneys of Panels D and E, respectively. At E10.5 (D) and E12.5 (E), Wnt/β-catenin signaling reporter TOPgal is activated in the nephric tube (NT), Wolffian ducts (WD), and metanephric mesenchyme (MM). (F–G) GSK-3β inhibition by lithium chloride leads to β-catenin-mediated gene expression in TOPgal reporter mice. Maternal injection of the Wnt agonist LiCl significantly enhanced the activity of the reporter β-galactosidase (G), compared with the NaCl-treated mouse embryonic kidneys (F). *LEF*, Lymphoid enhancer-binding factor; *TCF*, T cell factor; *UB*, ureteric bud.

β-catenin responsive TCF/β-gal reporter mice (Fig. 3A),[28] canonical Wnt signaling activation was detected in the epithelia of branching UB and MM.[29] This signal is quickly decreased in mature nephrons and disappeared in postnatal kidney. The Wnt inhibitor, Dickkopf-1 (Dkk1), could disrupt UB growing in cultured fetal kidney explants, which confirms the intense canonical Wnt signaling in branching nephrogenesis.[29] Active β-catenin signal is detectable within the mesenchymal progenitor pool of RVs in mice, which is both necessary and sufficient to regulate RVs and induces the expected molecular responses.[30] Both microarray and genetic analysis revealed that there is Wnt-dependent pathway regulated by protein kinase A (PKA) via Wnt induction during progression of the MM to tubular differentiation.[31] Wnt signaling also patterns the proximal-distal nephron axis through enriching in the distal and decreasing in the proximal region of the forming nephron in chick models.[32] It is concluded that high Wnt levels induce tubular components, whereas areas with low level of Wnt ligands give rise to glomerular elements.[32]

Deficiencies in Wnt ligands have been linked to serious renal developmental defects.[33] The first significant observation indicates that Wnt1 could replace UB for the induction of tubulogenesis.[34] Thus far, a total of 7 Wnt ligands have been found during kidney ontogeny, including Wnt2b, Wnt4, Wnt5a, Wnt6, Wnt7b, Wnt9b, and Wnt11 (Table 1).[35,36] Wnt2b and Wnt4 are expressed in the kidney mesenchymal cells.[37,38] Studies suggest that Wnt4 plays central roles in the initial stages of renal development,[39] and it does so by both canonical and noncanonical pathways.[40,41] Furthermore, Wnt4 is able to coordinate with bone morphogenetic protein 4 (BMP4) to regulate the smooth muscle cells in medullary stroma for renal vascular development and maturation.[42] Patients with a *Wnt4* mutation are associated with Müllerian-duct regression and renal defects.[43,44]

Wnt5a is also critical for early kidney development, as suggested by recent studies.[45] Wnt5a is shown to act through Ror2 for the induction of metanephric mesenchyme and renal morphogenesis via a noncanonical pathway.[46,47] Mutation of *Wnt5a* could lead to urogenital defects of the congenital anomalies of the kidney and urinary tract (CAKUT), which disrupts multiple tissue differentiation and development.[48] Wnt6, Wnt7b, Wnt9b, and Wnt11 are all expressed in the UB during the early stages of development. Aside from its role in UB branching morphogenesis, Wnt6 could rescue the Wnt4 mutant embryos and upregulate Wnt4 transcription for ureter epithelialization.[49] Wnt7b acts in cortico-medullary axis, and embryonic kidneys fail to form medullary zone and are unable to concentrate urine normally in Wnt7b$^{-/-}$ null mice.[50]

Wnt9b is one of the major ligands in the organization of mammalian urogenital system.[51] The Wnt9b-expressing cells functionally act for the inductive response in metanephric mesenchyme, which is critical for the development of mesonephric/metanephric tubules and extension of the Müllerian duct.[51] In order to keep the balance between maintenance of renal mesenchymal progenitor and mesenchymal to epithelial transition (MET) differentiation, Wnt9b controls subsets of progenitor cell's state for differentiation or undifferentiation.[52] Further evidences demonstrate that active Wnt9b/β-catenin signaling in Six2 positive progenitors is required for their renewal/proliferation.[52] The transgenic mice with overexpression of Wnt9b in Six2 lineage display a disrupted cell fate decisions and severe renal defects, such as CAKUT.[53] After deletion of β-catenin specifically in the kidney stroma, the fibroblasts in kidney capsule, cortex, and medulla, the expression of Wnt9b was markedly reduced in adjacent ureteric epithelial cells, suggesting the role of Wnt9b as effector downstream of β-catenin during induction of nephron progenitors.[54] It is interesting to note that Wnt9b participates in the planar cell polarity of epithelium and contributes to the size of tubular diameters. The mice with attenuation of *Wnt9b* exhibit randomly oriented cell divisions and significantly increase tubular diameter, which act through the noncanonical Rho/JNK pathway during kidney morphogenesis.[55]

Wnt11 was first identified to be expressed in the nephric duct prior to the outgrowth of UB in control of its branching.[56] This expression is independent of Wnt4 and MM specific factors, suggesting that it acts as an autocrine factor inside the ureteric epithelium.[56] The *Wnt11* gene in human embryos displays its restricted expression in the tips of the UB. There is an important feedback loop among Wnt11, Ret tyrosine kinase receptor, and glial cell-derived neurotrophic factor (GDNF).[57] Deficiency of the *Wnt11* gene disrupted the UB branching morphogenesis and led to renal hypoplasia in embryonic mice, which was rescued by maintaining GDNF expression.[57] Furthermore, Wnt11 expression is reduced in the absence of Ret/GDNF signaling.[57] These results illustrate the synergistic interaction of Wnt11/GDNF/Ret and their positive autoregulatory mechanism during normal ureteric branching morphogenesis. The role of Wnt11 in later stage of mammalian kidney organogenesis is established and supported by the anomalies in kidney tubular system and secondary glomerular cysts in *Wnt11*$^{-/-}$ null mice, which is consistent with its early involvement in UB branching.[58] These effects presumably act through several downstream genes implicated in kidney development, such as *Wnt9b*, *Six2*, *Foxd1*, *Hox10*, and *Dvl2*.[58]

3.3 Wnt Signaling Effectors and Kidney Development

Our previous studies exhibit that the key Wnt coreceptor LRP6 plays critical roles during early kidney development.[59] The *LRP6* knockout mice exhibit severe urogenital defects with hypoplasia cystic kidney (Table 1). Further studies reveal that the UB inductive factor of Ret acts downstream of LRP6-mediated Wnt signaling, which is most likely responsible for renal defects in mutant mice.[59] After inhibition of *Xenopus frizzled-8* (*Xfz8*), one of the Fzd receptor of Wnts, defects in pronephric tubule branching are evident, suggesting a role for *Xfz8* in maintaining normal epithelium structure of pronephric duct and tubules.[60] The phenotypes of targeted mutations in *Fzd4* and *Fzd8* result in disrupted UB growth and reduction of kidney size, but there might be homeostatic network regulations in the mutants of *Fzd4*$^{-/-}$ or *Fzd4*$^{-/-}$; *Fzd8*$^{-/-}$ cells compared with wild-type controls, which can account for discrepancies in renal phenotypes.[61] Through inhibition of GSK-3β, the Wnt agonist lithium is able to induce early stages of epithelial differentiation in isolated nephrogenic mesenchyme, indicating the possible molecular mechanisms in the early events of nephrogenesis.[62] These results are further confirmed by independent group using GSK-3β inhibitor lithium or 6-bromoindirubin-3′-oxime (BIO), which result in abundant epithelial differentiation and full segregation of nephrons.[63] Furthermore, stabilized β-catenin and upregulated LEF1 and TCF1 are capable of inducing nephron differentiation in isolated kidney mesenchymes as well.[63]

As β-catenin is the principal intracellular mediator in canonical Wnt signaling, more attentions have been paid to elucidate its roles during kidney development. Active form of β-catenin is detectable as early as E6.5 in mouse embryo, and contributes to anterior–posterior axis, primitive streak, and mesoderm formation.[64] In vitro and in vivo studies indicate that Six2-dependent self-renewal and β-catenin-directed differentiation institutes a regulatory complex, which keeps the balance between nephron progenitor maintenance and nephrogenesis.[65] Conditional deletion of β-catenin in mouse WD epithelium with Hoxb7-Cre causes defect in branching morphogenesis and results in renal aplasia or hypoplasia.[66] Multiple putative targets of β-catenin/TCF transcripts, such as cyclin D1 and Emx2, are detected along renal epithelial differentiation in organ culture of rat embryonic kidney.[67] In order to delineate the role of β-catenin at the late S-shaped body stage, mice with conditional deletion of β-catenin was created in the developing kidney by using Pax8-Cre. The phenotypes of these mice include abnormal kidneys with reduced renal function, hypoplastic renal parenchyma with a thin cortex, missing of superficial layer of renal tubules, and absence of parietal epithelial cells in Bowman's capsule.[68] These results suggest an indispensable role of Wnt/β-catenin signaling in the late stages of nephrogenesis, although one cannot completely exclude the possibility that some effects of β-catenin ablation may result from its role in regulating adherent junctions.

3.4 Wnt/PCP Pathway and Nephron Maturation

Increasing evidence suggests a critical role for Wnt/PCP signaling in kidney development and function. The PCP is defined for the cells in tissue perpendicular to the apical–basal axis, which was first described in *Drosophila.*[69] Deficiency of core PCP genes, such as *Vangl2*, *Celsr1*, *Fzd3/6*, and *Dvl1/2*, have been shown to disrupt asymmetric localization, convergent extension, hair cell organization, neural tube closure, and to cause cystic kidney in mammals.[70] Some PCP components are expressed in the renal developing epithelia, such as UB, RVs, and S-shaped body, suggesting their functions in cell division orientation, movements, adhesion, and contribution to morphogenesis of the mature nephron.[71] Several systematic analyses have implicated the function of PCP in UB branching and elongation to determine tubular diameter, length, and shape.[72,73]

Transgenic mouse models with mutations in Wnt ligands of the PCP pathway (*Wnt5a*, *Wnt7b*, *Wnt9b*, and *Wnt11*) exhibit UB branching defects. The noncanonical Wnt5a/Ror2 signaling in IM-derived nephric duct controls UB expansion, and *Wnt5a* knockout mice exhibit duplicated ureters and kidneys because of abnormal GDNF signaling and spatiotemporally aberrant interaction between MM and WD.[46,47] The absence of Wnt7b ligand displayed normal development of cortical epithelium but was incapable of forming medullary zone and concentrated urine.[50] This phenotype is explained by the disorganization of cell division planes in the collecting duct epithelium of the emerging medullary zone. Likewise, Wnt9b is suggested to be essential in establishing the length and diameter of the kidney tubules, while deletion of *Wnt9b* during kidney morphogenesis disrupts PCP in the epithelium and leads to polycystic kidney disease (PKD).[55] Besides its role in convergent extension in fish, Wnt11 regulates Ret/GDNF loop for UB branching and metanephric development.[57]

The key PCP molecule *Vangl2* is expressed in UB, collecting duct, MM, and nephron, whose mutation impairs branching morphogenesis and causes cystic kidney.[74] After podocyte-specific ablation of *Vangl2* gene, the mutants show significantly smaller glomeruli with maturation defects.[75] These results indicate the important role of *Vangl2* in podocytes morphology and function.[75] Another *Drosophila* PCP protein, fat, is essential in vertebrate PCP pathway. Deletion of *fat4* caused cystic kidney owing to disrupted cell division and tubule elongation upon interaction with Vangl2.[76] The PCP component ofDaam1 is expressed at developing pronephric anlagen and interacts with the novel weak-similarity Rho-GEF (WGEF). Knockdown of *Daam1* led to downregulation oflate pronephric epithelial markers, possible through Daam1/Rho-GEF axis to inhibit pronephric tubulogenesis.[77] Therefore, Wnt/PCP is fundamental for kidney development and function and requires to be the focus of more research.

3.5 Wnt Signaling and Cystogenesis

Wnt/β-catenin signaling is involved in renal polycystic lesions, which was first illustrated by overexpression of β-catenin in epithelial cells of the kidney. The transgenic mice exhibited severe polycystic kidney with abnormal renal function, similar to human autosomal dominant polycystic kidney disease (ADPKD).[78] After conditional deletion of APC in renal epithelium, there is multiple cysts formation with increasing levels of β-catenin protein.[79] Aquaporin-1 (AQP1), the molecular marker of renal proximal tubules and thin descending limb of Henle, could interact with β-catenin in models of ADPKD. The *AQP1* knockout mice increased cysts development with upregulation of Wnt/β-catenin signal.[80] These results show that the activation of Wnt/β-catenin signaling is required for renal cystogenesis. Consistently, conditional stabilization of β-catenin with Hoxb7-Cre from mouse tubular epithelium led to cystic kidney and hydronephrosis.[66] On the other hand, the cystic kidney and abnormalities are also induced by loss of the Jouberin (Jbn) protein, owing to a downregulation of endogenous Wnt activity.[81] Jbn was found to interact with and facilitate β-catenin nuclear accumulation, resulting in positive modulation of its downstream transcription.[81] Our previous studies also demonstrate that global deletion of *LRP6* causes renal hypoplasia and cysts formation, further confirming the role of LRP6-mediated Wnt/β-catenin signaling in the process.[59] However, there are controversies in the literature regarding the role of canonical Wnt/β-catenin signaling in cystogenesis. Mice with *Inversin* gene mutation had developmental abnormalities and renal cysts formation, although canonical Wnt/β-catenin signaling was not altered as defined in the Wnt reporter *BATlacZ* mice.[82] Additional studies also indicate normal TCF/β-catenin activity in two different models of PKD.[83] Therefore, the unique and precise regulation of Wnt/β-catenin signaling is required during kidney development and disruption of which leads to a cystic abnormality in a content-dependent manner.

Polycystic kidneys are also caused by the genetic mutations, which are related to primary cilia and Wnt.[84] The impaired cilia led to a disturbed fluid flow and calcium influx that initiate cyst formation. Because the tubular epithelial cells undergo oriented cell division during tubular elongation, the Wnt/PCP signal might be critical for the cyst formation.[85] There is evidence that maturation of nephrons is associated with mitotic orientation and intrinsic PCP signal, and abnormality of this pathway causes polycystic kidney in animal models.[86] It is interesting to point out that gene mutation of cilia-associated protein *Inversin* led to nephronophthisis type II, which is an autosomal recessive cystic kidney disease with extensive renal cysts formation.[87] It becomes apparent that Inversin acts between canonical and noncanonical Wnt signaling for normal tubular differentiation.[87] Based on these observations, it is conceivable that Wnt/β-catenin signaling is sufficient to induce cysts formation, but compromised Wnt/PCP pathway is also involved.[88] Therefore, both canonical and noncanonical Wnt signaling are important in kidney development, and disruption of either pathway can cause cystic kidney disease.[89] As many proteins, including Vangl, Fz, and Dvl, have been found to localize in the ciliary basal body,[90] changes in their expression and structure would eventually cause defective primary cilia in cystic kidney disease.

4. WNT SIGNALING AND KIDNEY DISEASES

4.1 Wnt Signaling in Acute Kidney Injury

Progresses have been made in recent years to demonstrate a key role of Wnt/β-catenin signaling in kidney injury and fibrosis, which is summarized in several recent reviews (Table 2).[6,13] In both acute kidney injury (AKI) and chronic kidney disease (CKD) animal models, Wnt/β-catenin is activated, which could function as either renal protective or detrimental mechanisms.[91,92] The AKI is described as a rapid decrease in kidney function, which involves hemodynamic alterations, inflammation, and endothelial and epithelial cell injury.[93] The outcome of AKI is either adaptive leading to the restoration of epithelial integrity or maladaptive progressing to CKD.[94] At different time points after AKI, the Wnt4-mediated β-catenin is activated, which contributes to recovery from renal injury. The activated β-catenin is able to promote cell cycle progression through its transcriptional targets, such as cyclin D1 in LLC-PK1 cells.[95] Using the *BATgal* and *Axin2-lacz* Wnt signaling reporter mice, it has been shown that activation of Wnt7b signaling is renoprotective after ischemia reperfusion injury (IRI), and macrophages are the sources of Wnt7b during ischemic injury and repair process.[96] In vitro, β-catenin signaling reduces Bax-mediated apoptosis and improves cell survival after induction of metabolic stress in proximal tubular epithelial cells. Constitutively activated β-catenin also regulated Bax activation and translocation to mitochondria upon acute stress injury.[97] Consistent with these in vitro findings, mice with tubule-specific *β-catenin* knockout are more

Table 2 Wnt Components and Kidney Diseases.

Diseases	Wnt Components	Roles	References
AKI	Wnt4	Promotes tubular cell cycle progression	94
	Wnt7b	Protective effects after IRI	95
	Wnt agonist	Improves renal function after IRI	97
	Tubular β-catenin	Protects against ischemic and toxic AKI	96–98
CKD	Wnt ligands	Increased renal expressions after UUO	10,102,103
	Wnt1	Tubular induction of Wnt1 causes renal fibrosis	103
	Dkk1	Inhibition of renal fibrosis after UUO	10
	sFRP4	Inhibition of renal fibrosis	12
	Wntless	Tubular KO reduces renal fibrosis after UUO	102
	Klotho	Sequesters Wnts and inhibits kidney fibrosis	113,114
	Fibroblast β-Catenin	Activation of β-catenin in interstitial pericytes/fibroblasts causes renal fibrosis	101
	Tubular β-catenin	Tubular deletion of β-catenin fails to affect renal fibrosis after UUO	135
	PRR	Amplifies Wnt/β-catenin signaling and aggravates renal fibrosis	15
Podocytopathy	Wnt1	Wnt1 overexpression induces podocyte injury and albuminuria	11
	β-Catenin	Podocyte-specific deletion of β-catenin reduces proteinuria	11,13,115,118
	β-Catenin	Podocyte-specific stabilization of β-catenin causes podocytopathy	116
Cancer	Wnt1	Positively correlates with tumor progression in RCC	127
	Wnt7a	Tumor suppressor properties in RCC	129
	Wnt10a	Independent risk factors for RCC carcinogenesis	128
	Fzd5, Fzd8	Biomarkers in RCC	130
	Fzd7	High expression in RCC	131
	β-Catenin	Abnormal accumulation in RCC carcinogenesis, mutation involved in WTs	127

AKI, acute kidney injury; CKD, chronic kidney disease; IRI, ischemia reperfusion injury; PRR, pro(renin) receptor; RCC, renal cell carcinoma; UUO, unilateral ureteral obstruction.

susceptible to AKI following either ischemic or toxic insults.[92] These conditional knockout mice exhibit higher mortality rate and increased serum creatinine, and worsen the morphological injury. There is more tubular cell apoptosis in the β-catenin null kidneys exhibiting high levels of p53 and Bax.[92] Another study using the Wnt agonist, a synthetic pyrimidine, also showed renal protective effects of Wnt signaling after IRI in rats through attenuating inflammation and oxidative stress, suggesting the potential pharmacological application of manipulating Wnt activity on preventing kidney injury.[98]

Severe or repeated AKI will lead to incomplete renal recovery and often progresses to CKD. We have shown that mice subjected to 20 min IRI exhibit transient Wnt/β-catenin activation, moderate AKI with complete recovery of renal function; whereas 30 min IRI causes sustained activation of this signaling and severe AKI, and eventually progresses to CKD characterized by renal fibrosis.[99] A sustained activation of Wnt/β-catenin accelerates AKI to CKD progression characterized by interstitial myofibroblast activation and excessive extracellular matrix deposition, while blockade of Wnt/β-catenin prevents AKI to CKD progression.[99] Therefore, the magnitude and duration of Wnt/β-catenin activation plays a decisive role in determining the outcomes of AKI.[6,100]

4.2 Wnt Signaling in Chronic Kidney Disease

Wnt/β-catenin signaling is also activated in fibrotic kidney initiated by unilateral ureteral obstruction (UUO), as well as in many other models of CKD (Table 2).[10,12,101,102] The regulation and actions of Wnt/β-catenin signaling in the pathogenesis of fibrotic CKD has been reviewed previously.[91,101] Secreted frizzled-related protein 4 (sFRP4), an endogenous extracellular Wnt antagonist, inhibits β-catenin activation and reduces renal fibrosis after UUO presumably via preventing Wnts binding to their receptors.[12] A systematic analysis for the Wnt ligands in mouse model of UUO revealed that 16 out of 19 different Wnts are induced in the kidney in different levels at some points during the injury.[10] Notably, the cellular source of Wnt ligands in the fibrotic kidneys is largely from the tubular epithelium.[103,104] The activation of Wnt signaling leads to dramatic accumulation of β-catenin and upregulation of its target genes, such as *c-Myc*, *Twist*, *TCF1*, and *fibronectin* in renal epithelial cells.[10] Both in vitro and in vivo studies illustrate that Wnt/β-catenin controls several key fibrosis-related downstream genes, such as *Snail1*, *fibronectin*, *plasminogen activator inhibitor-1*

(*PAI-1*), *matrix metalloproteinase 7* (*MMP-7*), and multiple components of the renin–angiotensin system (RAS), including *angiotensinogen, renin, angiotensin converting enzyme* (*ACE*), *and angiotensin receptor type 1* (*AT1*).[11,105–108] Targeted inhibition of Wnt/β-catenin signaling by ICG-001, a small molecule peptidomimetic that selectively inhibits β-catenin signaling in a CBP-dependent fashion,[109,110] suppresses matrix expression and ameliorates renal interstitial fibrosis.[111] Therefore, inhibition of Wnt/β-catenin signaling may be an effective strategy to ameliorate kidney injury and fibrotic lesions in various models of CKD.[91]

The role of Wnt/β-catenin in renal fibrogenesis is supported by using the soluble form of Klotho as a secreted Wnt antagonist in renal fibrotic models. Klotho is antiaging protein that is highly expressed in renal tubular epithelium of normal kidney, which is suppressed in the elderly and in patients with CKD.[112,113] Klotho is present as full-length, transmembrane protein or secreted, soluble form. Both membranous and soluble forms of Klotho can bind with multiple Wnt ligands and represses their target genes transcription.[114] Overexpression of Klotho clearly hampers the activation of Wnt signal in mice with UUO, accompanied by reducing extracellular matrix deposition, bypassing the G2/M arrest and diminishing fibrotic cytokine production.[114,115] TGF-β1, a master regulator of tissue fibrosis, could suppress Klotho expression and concomitantly activates β-catenin to promote myofibroblast activation and renal fibrosis.[114] Klotho also attenuates Wnt1-triggered activation of RAS in a dose-dependent manner in renal epithelial cells.[116] It is concluded that Klotho protects against renal fibrosis through sequestering and antagonizing Wnt/β-catenin signaling.

4.3 Wnt Signaling in Podocytopathy and Proteinuria

Activation of Wnt/β-catenin signaling also play a crucial role in mediating podocyte dysfunction, proteinuria, and glomerulosclerosis, leading to end stage renal disease (ESRD).[13] Glomerular β-catenin is activated in adriamycin-induced kidney injury characterized by podocyte damage and albuminuria. In contrast, podocyte-specific β-catenin knockout mice are protected against albuminuria after adriamycin treatment.[11,117] These results confirm a critical role for Wnt/β-catenin signaling in the development of podocytopathy and proteinuria. Further studies using the same conditional knockout mice show that β-catenin is instrumental in disrupting podocyte slit diaphragm (SD), leading to an impaired glomerular filtration barrier.[117,118] Studies also demonstrated that Wnt/β-catenin plays a role in mediating the transforming growth factor-β1 (TGF-β1)-induced podocyte injury and proteinuria in vitro and in vivo.[119] Overexpression of Wnt antagonist *Dkk-1* gene alleviates TGF-β1-triggered podocyte damage and albuminuria. The action of β-catenin on podocytes includes its role in triggering ubiquitin-mediated degradation of Wilms' tumor 1 (WT1), a key podocyte-specific transcription factor that is critical to maintain podocyte integrity and function.[120] In that regard, β-catenin specifically targets WT1 for protein degradation, finally leading to podocyte dedifferentiation and mesenchymal transformation.[120]

Given the importance of β-catenin activation in mediating podocyte injury, it is conceivable that strategies modulating its activity may be effective in ameliorating podocytopathy and proteinuria. Indeed, vitamin D analogs, such as paricalcitol, preserves podocyte integrity and function, inhibits proinflammatory cytokines, suppresses expression of the fibrogenic genes, and also ameliorates established proteinuria in mice after adriamycin injection.[121] Studies also indicate that angiotensin II (Ang II), the principal effector of RAS activation, acts at upstream of Wnt/β-catenin signaling, impairs podocyte integrity and causes albuminuria in mice.[122] The Wnt antagonist Dkk1 attenuates Ang II-induced podocyte injury in vitro and in vivo.[122] In addition, high glucose is shown to impair podocyte integrity by suppressing *podocin* and *nephrin* and inducing *(pro)renin receptor* (*PRR*), *Wnt3a, β-catenin, and Snail1*. These results suggest a potential role of PRR-Wnt/β-catenin-Snail1 pathway in high glucose-triggered podocyte injury.[123] More interestingly, all RAS genes, including *angiotensinogen, renin, PRR, ACE, and AT1* are found to be novel downstream targets of Wnt/β-catenin, as overexpressions of either β-catenin or different Wnt ligands induces, but β-catenin inhibitor ICG-001 inhibits, their expression.[15,108,124] Thus, there might be a vicious cycle between Wnt/β-catenin and RAS in the process of podocyte injury, and disruption of this cycle formation will preserve podocyte function and prevent proteinuria.[13]

4.4 Wnt Signaling and Human Kidney Disease

Aside from its roles in cultured cells and animal models, Wnt/β-catenin is also activated in clinical human samples of CKD.[103] Active β-catenin is detectable in the podocytes of human kidney biopsy of CKD, such as diabetic nephropathy (DN) and focal segmental glomerulosclerosis (FSGS).[11] Increased nuclear β-catenin accumulation is discovered in the peripheral blood leukocytes from immunoglobulin A nephropathy (IgAN) patients, suggesting that a hyperactivation of Wnt signal might contribute to the pathogenesis of IgAN.[125] Upregulation of Wnt/β-catenin components is documented in the glomeruli and podocytes of diabetic kidney disease (DKD) patients, who exhibit glomerular basement membrane (GBM) abnormalities and albuminuria.[118] There is increased activation of Wnt signaling in human collapsing glomerulopathies, indicating a potential role of Wnt pathway in the regulation of mature podocytes cell cycle progression and proliferation in renal disease state.[126] In lupus nephritis patients, active Wnt/β-catenin signal is accompanied by increasing levels of plasmatic Dkk-1,[127] probably reflecting that Dkk1 is a target of the β-catenin signaling.[128] These alterations suggest a close relationship between hyperactive Wnt signaling and the pathogenesis of lupus nephritis.[127] Collectively, the activation of Wnt signaling in AKI and CKD after injury triggers adaptive or maladaptive responses depending on the injury severity and duration. This activation could be indispensable for kidney repair and regeneration, or promotes disease progression and fibrosis in humans as well.[6,100]

4.5 Wnt Signaling and Kidney Cancer

As there are similarities between embryonic growth and abnormal cell proliferation in cancer, the Wnt signaling molecules are considered as candidates for the development of kidney cancers. The renal cell carcinoma (RCC) is the major form of kidney tumor with high frequency. Studies showed that Wnt1 is upregulated and associated with increased tumor size, more advanced stage and invasiveness in clear cell RCC (ccRCC).[129] After screening normal kidney and RCC cell lines and tissues, investigators found that Wnt10a is significantly induced and acts as an independent risk factor for RCC carcinogenesis and progression.[130] However, *Wnt7a* gene is inactivated in ccRCC and exhibits tumor suppressor properties, and there is positive correlation between tumor stage and *Wnt7a* hypermethylation.[131] These observations demonstrate unique differences for canonical Wnts (Wnt1, Wnt10a) and noncanonical Wnts (Wnt7a) during tumorigenesis. The mRNA levels of Wnt receptors *Fzd5* and *Fzd8* are increased in RCC accompanied by active β-catenin signal and high levels of Wnt target cyclin D1, so Fzd5 and Fzd8 are considered as biomarkers in RCC.[132] Another receptor, Fzd7, also shows higher levels of expression compared with surrounding normal tissues in RCC, which might be activated by a

Wnt3a-mediated pathway.[133] Altered expression of β-catenin is also detected in RCC, and abnormal accumulation of β-catenin, at least partially, contributes to renal carcinogenesis.[129] Overall, Wnt signaling is constitutively activated in RCC and functional loss of Wnt antagonists may be one of the reasons for that process, and could be candidate for targeted therapies.[134] Additional evidences showed that Wnt signal not only participates in RCC, but also in the pathogenesis of embryonic kidney-derived tumor, Wilms' tumor (WT) or nephroplastoma, one of the most common pediatric malignancies.[135] Because Wnt4 serves as critical renal developmental factor, the mutations of its downstream *β-catenin* gene might cause WTs. A protein named WTX, which is mutant and encoded by Wilms' tumor gene in the X chromosome, could form a complex with β-catenin to promote its ubiquitination and degradation.[136] Thus, both RCC and WTs require β-catenin signal participation for the process of tumorigenesis.

5. CONCLUSIONS

Normal kidney development requires complex and precise cell-cell communications, in which Wnt signaling is one of the main mediators. Extensive studies in last decades have revealed that Wnt signaling is indispensable not only in normal nephrogenesis but also in kidney repair and regeneration after AKI, as well as in the evolution of various CKD characterized by tissue fibrosis and renal insufficiency. Both canonical and noncanonical Wnt signaling are instrumental for UB induction, nephron formation, and maturation. Dysregulation of Wnt signaling is implicated in a wide variety of kidney disorders ranging from fibrosis, cystic formation, proteinuria to tumorigenesis. There is a tremendous progress in the field during last several years through the discovery of novel targets of Wnt/β-catenin signaling and delineating new mechanisms of its action. As there are many Wnt antagonists, such as small molecule inhibitor ICG-001 and soluble Klotho are currently available, it is hopeful that manipulation of this signaling pathway by diverse strategies will eventually translate into effective therapies for patients with various kidney disorders.

ACKNOWLEDGMENTS

This work was supported by the National Science Foundation of China Grants 81521003 and 81770715, and the National Institutes of Health (NIH) Grants DK064005 and DK106049.

REFERENCES

1. Schuijers J, Clevers H. Adult mammalian stem cells: the role of Wnt, Lgr5 and R-spondins. *EMBO J.* 2012;31:2685–2696.
2. Nusse R, Varmus H. Three decades of Wnts: a personal perspective on how a scientific field developed. *EMBO J.* 2012;31:2670–2684.
3. Nusse R, Clevers H. Wnt/beta-catenin signaling, disease, and emerging therapeutic modalities. *Cell.* 2017;169:985–999.
4. Nusse R, Varmus HE. Many tumors induced by the mouse mammary tumor virus contain a provirus integrated in the same region of the host genome. *Cell.* 1982;31:99–109.
5. Clevers H, Nusse R. Wnt/beta-catenin signaling and disease. *Cell.* 2012;149: 1192–1205.
6. Zhou D, Tan RJ, Fu H, Liu Y. Wnt/beta-catenin signaling in kidney injury and repair: a double-edged sword. *Lab Invest.* 2016;96:156–167.
7. Grumolato L, Liu G, Mong P, Mudbhary R, Biswas R, Arroyave R, Vijayakumar S, Economides AN, Aaronson SA. Canonical and noncanonical Wnts use a common mechanism to activate completely unrelated coreceptors. *Genes Dev.* 2010;24: 2517–2530.
8. Clevers H. Wnt/beta-catenin signaling in development and disease. *Cell.* 2006;127: 469–480.
9. Angers S, Moon RT. Proximal events in Wnt signal transduction. *Nat Rev Mol Cell Biol.* 2009;10:468–477.
10. He W, Dai C, Li Y, Zeng G, Monga SP, Liu Y. Wnt/beta-catenin signaling promotes renal interstitial fibrosis. *J Am Soc Nephrol.* 2009;20:765–776.
11. Dai C, Stolz DB, Kiss LP, Monga SP, Holzman LB, Liu Y. Wnt/beta-catenin signaling promotes podocyte dysfunction and albuminuria. *J Am Soc Nephrol.* 2009;20: 1997–2008.
12. Surendran K, Schiavi S, Hruska KA. Wnt-dependent beta-catenin signaling is activated after unilateral ureteral obstruction, and recombinant secreted frizzled-related protein 4 alters the progression of renal fibrosis. *J Am Soc Nephrol.* 2005;16:2373–2384.
13. Zhou L, Liu Y. Wnt/beta-catenin signalling and podocyte dysfunction in proteinuric kidney disease. *Nat Rev Nephrol.* 2015;11:535–545.
14. MacDonald BT, Tamai K, He X. Wnt/beta-catenin signaling: components, mechanisms, and diseases. *Dev Cell.* 2009;17:9–26.
15. Li Z, Zhou L, Wang Y, Miao J, Hong X, Hou FF, Liu Y. (Pro)renin receptor Is an amplifier of Wnt/beta-catenin signaling in kidney injury and fibrosis. *J Am Soc Nephrol.* 2017;28:2391–2408.
16. Nguyen G, Muller DN. The biology of the (pro)renin receptor. *J Am Soc Nephrol.* 2010;21:18–23.
17. Cruciat CM, Ohkawara B, Acebron SP, Karaulanov E, Reinhard C, Ingelfinger D, Boutros M, Niehrs C. Requirement of prorenin receptor and vacuolar H+-ATPase-mediated acidification for Wnt signaling. *Science.* 2010;327:459–463.
18. Gray RS, Roszko I, Solnica-Krezel L. Planar cell polarity: coordinating morphogenetic cell behaviors with embryonic polarity. *Dev Cell.* 2011;21:120–133.
19. Sokol SY. Spatial and temporal aspects of Wnt signaling and planar cell polarity during vertebrate embryonic development. *Semin Cell Dev Biol.* 2015;42:78–85.
20. Costantini F, Kopan R. Patterning a complex organ: branching morphogenesis and nephron segmentation in kidney development. *Dev Cell.* 2010;18:698–712.
21. Brown AC, Muthukrishnan SD, Guay JA, Adams DC, Schafer DA, Fetting JL, Oxburgh L. Role for compartmentalization in nephron progenitor differentiation. *Proc Natl Acad Sci USA.* 2013;110:4640–4645.
22. Pietila I, Vainio SJ. Kidney development: an overview. *Nephron Exp Nephrol.* 2014;126:40.

23. Dressler GR. Advances in early kidney specification, development and patterning. *Development.* 2009;136:3863–3874.
24. Walker KA, Bertram JF. Kidney development: core curriculum 2011. *Am J Kidney Dis.* 2011;57:948–958.
25. Yu J, McMahon AP, Valerius MT. Recent genetic studies of mouse kidney development. *Curr Opin Genet Dev.* 2004;14:550–557.
26. Lechner MS, Dressler GR. The molecular basis of embryonic kidney development. *Mech Dev.* 1997;62:105–120.
27. Bridgewater D, Cox B, Cain J, Lau A, Athaide V, Gill PS, Kuure S, Sainio K, Rosenblum ND. Canonical WNT/beta-catenin signaling is required for ureteric branching. *Dev Biol.* 2008;317:83–94.
28. DasGupta R, Fuchs E. Multiple roles for activated LEF/TCF transcription complexes during hair follicle development and differentiation. *Development.* 1999;126: 4557–4568.
29. Iglesias DM, Hueber PA, Chu L, Campbell R, Patenaude AM, Dziarmaga AJ, Quinlan J, Mohamed O, Dufort D, Goodyer PR. Canonical WNT signaling during kidney development. *Am J Physiol Renal Physiol.* 2007;293:F494–F500.
30. Park JS, Valerius MT, McMahon AP. Wnt/beta-catenin signaling regulates nephron induction during mouse kidney development. *Development.* 2007;134:2533–2539.
31. Gallegos TF, Kouznetsova V, Kudlicka K, Sweeney DE, Bush KT, Willert K, Farquhar MG, Nigam SK. A protein kinase A and Wnt-dependent network regulating an intermediate stage in epithelial tubulogenesis during kidney development. *Dev Biol.* 2012;364:11–21.
32. Schneider J, Arraf AA, Grinstein M, Yelin R, Schultheiss TM. Wnt signaling orients the proximal-distal axis of chick kidney nephrons. *Development.* 2015;142:2686–2695.
33. Halt K, Vainio S. Coordination of kidney organogenesis by Wnt signaling. *Pediatr Nephrol.* 2014;29:737–744.
34. Herzlinger D, Qiao J, Cohen D, Ramakrishna N, Brown AM. Induction of kidney epithelial morphogenesis by cells expressing Wnt-1. *Dev Biol.* 1994;166:815–818.
35. Merkel CE, Karner CM, Carroll TJ. Molecular regulation of kidney development: is the answer blowing in the Wnt? *Pediatr Nephrol.* 2007;22:1825–1838.
36. Pulkkinen K, Murugan S, Vainio S. Wnt signaling in kidney development and disease. *Organogenesis.* 2008;4:55–59.
37. Lin Y, Liu A, Zhang S, Ruusunen T, Kreidberg JA, Peltoketo H, Drummond I, Vainio S. Induction of ureter branching as a response to Wnt-2b signaling during early kidney organogenesis. *Dev Dyn.* 2001;222:26–39.
38. Kispert A, Vainio S, McMahon AP. Wnt-4 is a mesenchymal signal for epithelial transformation of metanephric mesenchyme in the developing kidney. *Development.* 1998;125:4225–4234.
39. Shan J, Jokela T, Skovorodkin I, Vainio S. Mapping of the fate of cell lineages generated from cells that express the Wnt4 gene by time-lapse during kidney development. *Differentiation.* 2010;79:57–64.
40. Saulnier DM, Ghanbari H, Brandli AW. Essential function of Wnt-4 for tubulogenesis in the *Xenopus* pronephric kidney. *Dev Biol.* 2002;248:13–28.
41. Tanigawa S, Wang H, Yang Y, Sharma N, Tarasova N, Ajima R, Yamaguchi TP, Rodriguez LG, Perantoni AO. Wnt4 induces nephronic tubules in metanephric mesenchyme by a non-canonical mechanism. *Dev Biol.* 2011;352:58–69.
42. Itaranta P, Chi L, Seppanen T, Niku M, Tuukkanen J, Peltoketo H, Vainio S. Wnt-4 signaling is involved in the control of smooth muscle cell fate via Bmp-4 in the medullary stroma of the developing kidney. *Dev Biol.* 2006;293:473–483.
43. Mandel H, Shemer R, Borochowitz ZU, Okopnik M, Knopf C, Indelman M, Drugan A, Tiosano D, Gershoni-Baruch R, Choder M, Sprecher E. SERKAL syndrome: an

autosomal-recessive disorder caused by a loss-of-function mutation in WNT4. *Am J Hum Genet.* 2008;82:39–47.
44. Biason-Lauber A, Konrad D, Navratil F, Schoenle EJ. A WNT4 mutation associated with Mullerian-duct regression and virilization in a 46,XX woman. *N Engl J Med.* 2004;351:792–798.
45. Huang L, Xiao A, Choi SY, Kan Q, Zhou W, Chacon-Heszele MF, Ryu YK, McKenna S, Zuo X, Kuruvilla R, Lipschutz JH. Wnt5a is necessary for normal kidney development in zebrafish and mice. *Nephron Exp Nephrol.* 2014;128:80–88.
46. Yun K, Ajima R, Sharma N, Costantini F, Mackem S, Lewandoski M, Yamaguchi TP, Perantoni AO. Non-canonical Wnt5a/Ror2 signaling regulates kidney morphogenesis by controlling intermediate mesoderm extension. *Hum Mol Genet.* 2014;23:6807–6814.
47. Nishita M, Qiao S, Miyamoto M, Okinaka Y, Yamada M, Hashimoto R, Iijima K, Otani H, Hartmann C, Nishinakamura R, Minami Y. Role of Wnt5a-Ror2 signaling in morphogenesis of the metanephric mesenchyme during ureteric budding. *Mol Cell Biol.* 2014;34:3096–3105.
48. Pietila I, Prunskaite-Hyyrylainen R, Kaisto S, Tika E, van Eerde AM, Salo AM, Garma L, Miinalainen I, Feitz WF, Bongers EM, Juffer A, Knoers NV, Renkema KY, Myllyharju J, Vainio SJ. Wnt5a deficiency leads to anomalies in ureteric tree development, tubular epithelial cell organization and basement membrane integrity pointing to a role in kidney collecting duct patterning. *PLoS One.* 2016;11:e0147171.
49. Itaranta P, Lin Y, Perasaari J, Roel G, Destree O, Vainio S. Wnt-6 is expressed in the ureter bud and induces kidney tubule development in vitro. *Genesis.* 2002;32:259–268.
50. Yu J, Carroll TJ, Rajagopal J, Kobayashi A, Ren Q, McMahon AP. A Wnt7b-dependent pathway regulates the orientation of epithelial cell division and establishes the cortico-medullary axis of the mammalian kidney. *Development.* 2009;136:161–171.
51. Carroll TJ, Park JS, Hayashi S, Majumdar A, McMahon AP. Wnt9b plays a central role in the regulation of mesenchymal to epithelial transitions underlying organogenesis of the mammalian urogenital system. *Dev Cell.* 2005;9:283–292.
52. Karner CM, Das A, Ma Z, Self M, Chen C, Lum L, Oliver G, Carroll TJ. Canonical Wnt9b signaling balances progenitor cell expansion and differentiation during kidney development. *Development.* 2011;138:1247–1257.
53. Kiefer SM, Robbins L, Rauchman M. Conditional expression of Wnt9b in Six2-positive cells disrupts stomach and kidney function. *PLoS One.* 2012;7:e43098.
54. Boivin FJ, Sarin S, Lim J, Javidan A, Svajger B, Khalili H, Bridgewater D. Stromally expressed beta-catenin modulates Wnt9b signaling in the ureteric epithelium. *PLoS One.* 2015;10:e0120347.
55. Karner CM, Chirumamilla R, Aoki S, Igarashi P, Wallingford JB, Carroll TJ. Wnt9b signaling regulates planar cell polarity and kidney tubule morphogenesis. *Nat Genet.* 2009;41:793–799.
56. Kispert A, Vainio S, Shen L, Rowitch DH, McMahon AP. Proteoglycans are required for maintenance of Wnt-11 expression in the ureter tips. *Development.* 1996;122:3627–3637.
57. Majumdar A, Vainio S, Kispert A, McMahon J, McMahon AP. Wnt11 and Ret/Gdnf pathways cooperate in regulating ureteric branching during metanephric kidney development. *Development.* 2003;130:3175–3185.
58. Nagy II, Xu Q, Naillat F, Ali N, Miinalainen I, Samoylenko A, Vainio SJ. Impairment of Wnt11 function leads to kidney tubular abnormalities and secondary glomerular cystogenesis. *BMC Dev Biol.* 2016;16:30.
59. Wang Y, Stokes A, Duan Z, Hui J, Xu Y, Chen Y, Chen HW, Lam K, Zhou CJ. LDL receptor-related protein 6 modulates Ret proto-oncogene signaling in renal development and cystic dysplasia. *J Am Soc Nephrol.* 2016;27:417–427.
60. Satow R, Chan TC, Asashima M. The role of *Xenopus* frizzled-8 in pronephric development. *Biochem Biophys Res Commun.* 2004;321:487–494.

61. Ye X, Wang Y, Rattner A, Nathans J. Genetic mosaic analysis reveals a major role for frizzled 4 and frizzled 8 in controlling ureteric growth in the developing kidney. *Development.* 2011;138:1161–1172.
62. Davies JA, Garrod DR. Induction of early stages of kidney tubule differentiation by lithium ions. *Dev Biol.* 1995;167:50–60.
63. Kuure S, Popsueva A, Jakobson M, Sainio K, Sariola H. Glycogen synthase kinase-3 inactivation and stabilization of beta-catenin induce nephron differentiation in isolated mouse and rat kidney mesenchymes. *J Am Soc Nephrol.* 2007;18:1130–1139.
64. Mohamed OA, Clarke HJ, Dufort D. Beta-catenin signaling marks the prospective site of primitive streak formation in the mouse embryo. *Dev Dyn.* 2004;231:416–424.
65. Park JS, Ma W, O'Brien LL, Chung E, Guo JJ, Cheng JG, Valerius MT, McMahon JA, Wong WH, McMahon AP. Six2 and Wnt regulate self-renewal and commitment of nephron progenitors through shared gene regulatory networks. *Dev Cell.* 2012;23:637–651.
66. Marose TD, Merkel CE, McMahon AP, Carroll TJ. Beta-catenin is necessary to keep cells of ureteric bud/Wolffian duct epithelium in a precursor state. *Dev Biol.* 2008;314:112–126.
67. Schmidt-Ott KM, Masckauchan TN, Chen X, Hirsh BJ, Sarkar A, Yang J, Paragas N, Wallace VA, Dufort D, Pavlidis P, Jagla B, Kitajewski J, Barasch J. Beta-catenin/TCF/Lef controls a differentiation-associated transcriptional program in renal epithelial progenitors. *Development.* 2007;134:3177–3190.
68. Grouls S, Iglesias DM, Wentzensen N, Moeller MJ, Bouchard M, Kemler R, Goodyer P, Niggli F, Grone HJ, Kriz W, Koesters R. Lineage specification of parietal epithelial cells requires beta-catenin/Wnt signaling. *J Am Soc Nephrol.* 2012;23:63–72.
69. Klein TJ, Mlodzik M. Planar cell polarization: an emerging model points in the right direction. *Annu Rev Cell Dev Biol.* 2005;21:155–176.
70. McNeill H. Planar cell polarity and the kidney. *J Am Soc Nephrol.* 2009;20:2104–2111.
71. Carroll TJ, Das A. Planar cell polarity in kidney development and disease. *Organogenesis.* 2011;7:180–190.
72. Schnell U, Carroll TJ. Planar cell polarity of the kidney. *Exp Cell Res.* 2016;343:258–266.
73. Papakrivopoulou E, Dean CH, Copp AJ, Long DA. Planar cell polarity and the kidney. *Nephrol Dial Transplant.* 2014;29:1320–1326.
74. Yates LL, Papakrivopoulou J, Long DA, Goggolidou P, Connolly JO, Woolf AS, Dean CH. The planar cell polarity gene Vangl2 is required for mammalian kidney-branching morphogenesis and glomerular maturation. *Hum Mol Genet.* 2010;19:4663–4676.
75. Rocque BL, Babayeva S, Li J, Leung V, Nezvitsky L, Cybulsky AV, Gros P, Torban E. Deficiency of the planar cell polarity protein Vangl2 in podocytes affects glomerular morphogenesis and increases susceptibility to injury. *J Am Soc Nephrol.* 2015;26:576–586.
76. Saburi S, Hester I, Fischer E, Pontoglio M, Eremina V, Gessler M, Quaggin SE, Harrison R, Mount R, McNeill H. Loss of Fat4 disrupts PCP signaling and oriented cell division and leads to cystic kidney disease. *Nat Genet.* 2008;40:1010–1015.
77. Miller RK, Canny SG, Jang CW, Cho K, Ji H, Wagner DS, Jones EA, Habas R, McCrea PD. Pronephric tubulogenesis requires Daam1-mediated planar cell polarity signaling. *J Am Soc Nephrol.* 2011;22:1654–1664.
78. Saadi-Kheddouci S, Berrebi D, Romagnolo B, Cluzeaud F, Peuchmaur M, Kahn A, Vandewalle A, Perret C. Early development of polycystic kidney disease in transgenic mice expressing an activated mutant of the beta-catenin gene. *Oncogene.* 2001;20:5972–5981.
79. Qian CN, Knol J, Igarashi P, Lin F, Zylstra U, Teh BT, Williams BO. Cystic renal neoplasia following conditional inactivation of apc in mouse renal tubular epithelium. *J Biol Chem.* 2005;280:3938–3945.

80. Wang W, Li F, Sun Y, Lei L, Zhou H, Lei T, Xia Y, Verkman AS, Yang B. Aquaporin-1 retards renal cyst development in polycystic kidney disease by inhibition of Wnt signaling. *FASEB J.* 2015;29:1551–1563.
81. Lancaster MA, Louie CM, Silhavy JL, Sintasath L, Decambre M, Nigam SK, Willert K, Gleeson JG. Impaired Wnt-beta-catenin signaling disrupts adult renal homeostasis and leads to cystic kidney ciliopathy. *Nat Med.* 2009;15:1046–1054.
82. Sugiyama N, Tsukiyama T, Yamaguchi TP, Yokoyama T. The canonical Wnt signaling pathway is not involved in renal cyst development in the kidneys of inv mutant mice. *Kidney Int.* 2011;79:957–965.
83. Miller MM, Iglesias DM, Zhang Z, Corsini R, Chu L, Murawski I, Gupta I, Somlo S, Germino GG, Goodyer PR. T-cell factor/beta-catenin activity is suppressed in two different models of autosomal dominant polycystic kidney disease. *Kidney Int.* 2011;80:146–153.
84. Harris PC, Torres VE. Polycystic kidney disease. *Annu Rev Med.* 2009;60:321–337.
85. Simons M, Walz G. Polycystic kidney disease: cell division without a c(l)ue? *Kidney Int.* 2006;70:854–864.
86. Fischer E, Legue E, Doyen A, Nato F, Nicolas JF, Torres V, Yaniv M, Pontoglio M. Defective planar cell polarity in polycystic kidney disease. *Nat Genet.* 2006;38: 21–23.
87. Simons M, Gloy J, Ganner A, Bullerkotte A, Bashkurov M, Kronig C, Schermer B, Benzing T, Cabello OA, Jenny A, Mlodzik M, Polok B, Driever W, Obara T, Walz G. Inversin, the gene product mutated in nephronophthisis type II, functions as a molecular switch between Wnt signaling pathways. *Nat Genet.* 2005;37:537–543.
88. Wuebken A, Schmidt-Ott KM. WNT/beta-catenin signaling in polycystic kidney disease. *Kidney Int.* 2011;80:135–138.
89. Lancaster MA, Gleeson JG. Cystic kidney disease: the role of Wnt signaling. *Trends Mol Med.* 2010;16:349–360.
90. Goggolidou P. Wnt and planar cell polarity signaling in cystic renal disease. *Organogenesis.* 2014;10:86–95.
91. Tan RJ, Zhou D, Zhou L, Liu Y. Wnt/beta-catenin signaling and kidney fibrosis. *Kidney Int Suppl (2011).* 2014;4:84–90.
92. Zhou D, Li Y, Lin L, Zhou L, Igarashi P, Liu Y. Tubule-specific ablation of endogenous beta-catenin aggravates acute kidney injury in mice. *Kidney Int.* 2012;82:537–547.
93. Ricci Z, Cruz DN, Ronco C. Classification and staging of acute kidney injury: beyond the RIFLE and AKIN criteria. *Nat Rev Nephrol.* 2011;7:201–208.
94. Bonventre JV, Yang L. Cellular pathophysiology of ischemic acute kidney injury. *J Clin Invest.* 2011;121:4210–4221.
95. Terada Y, Tanaka H, Okado T, Shimamura H, Inoshita S, Kuwahara M, Sasaki S. Expression and function of the developmental gene Wnt-4 during experimental acute renal failure in rats. *J Am Soc Nephrol.* 2003;14:1223–1233.
96. Lin SL, Li B, Rao S, Yeo EJ, Hudson TE, Nowlin BT, Pei H, Chen L, Zheng JJ, Carroll TJ, Pollard JW, McMahon AP, Lang RA, Duffield JS. Macrophage Wnt7b is critical for kidney repair and regeneration. *Proc Natl Acad Sci USA.* 2010;107:4194–4199.
97. Wang Z, Havasi A, Gall JM, Mao H, Schwartz JH, Borkan SC. Beta-catenin promotes survival of renal epithelial cells by inhibiting Bax. *J Am Soc Nephrol.* 2009;20: 1919–1928.
98. Kuncewitch M, Yang WL, Corbo L, Khader A, Nicastro J, Coppa GF, Wang P. WNT agonist decreases tissue damage and improves renal function after ischemia-reperfusion. *Shock.* 2015;43:268–275.
99. Xiao L, Zhou D, Tan RJ, Fu H, Zhou L, Hou FF, Liu Y. Sustained activation of Wnt/ beta-catenin signaling drives AKI to CKD progression. *J Am Soc Nephrol.* 2016;27: 1727–1740.

100. Kawakami T, Ren S, Duffield JS. Wnt signalling in kidney diseases: dual roles in renal injury and repair. *J Pathol.* 2013;229:221–231.
101. Maarouf OH, Ikeda Y, Humphreys BD. Wnt signaling in kidney tubulointerstitium during disease. *Histol Histopathol.* 2015;30:163–171.
102. DiRocco DP, Kobayashi A, Taketo MM, McMahon AP, Humphreys BD. Wnt4/beta-catenin signaling in medullary kidney myofibroblasts. *J Am Soc Nephrol.* 2013;24:1399–1412.
103. Zhou D, Fu H, Zhang L, Zhang K, Min Y, Xiao L, Lin L, Bastacky SI, Liu Y. Tubule-derived Wnts are required for fibroblast activation and kidney fibrosis. *J Am Soc Nephrol.* 2017;28:2322–2336.
104. Maarouf OH, Aravamudhan A, Rangarajan D, Kusaba T, Zhang V, Welborn J, Gauvin D, Hou X, Kramann R, Humphreys BD. Paracrine Wnt1 drives interstitial fibrosis without inflammation by tubulointerstitial cross-talk. *J Am Soc Nephrol.* 2016;27:781–790.
105. He W, Tan R, Dai C, Li Y, Wang D, Hao S, Kahn M, Liu Y. Plasminogen activator inhibitor-1 is a transcriptional target of the canonical pathway of Wnt/beta-catenin signaling. *J Biol Chem.* 2010;285:24665–24675.
106. He W, Tan RJ, Li Y, Wang D, Nie J, Hou FF, Liu Y. Matrix metalloproteinase-7 as a surrogate marker predicts renal Wnt/beta-catenin activity in CKD. *J Am Soc Nephrol.* 2012;23:294–304.
107. Zhou D, Tian Y, Sun L, Zhou L, Xiao L, Tan RJ, Tian J, Fu H, Hou FF, Liu Y. Matrix metalloproteinase-7 is a urinary biomarker and pathogenic mediator of kidney fibrosis. *J Am Soc Nephrol.* 2017;28:598–611.
108. Zhou L, Li Y, Hao S, Zhou D, Tan RJ, Nie J, Hou FF, Kahn M, Liu Y. Multiple genes of the renin-angiotensin system are novel targets of Wnt/beta-catenin signaling. *J Am Soc Nephrol.* 2015;26:107–120.
109. Zhao Y, Masiello D, McMillian M, Nguyen C, Wu Y, Melendez E, Smbatyan G, Kida A, He Y, Teo JL, Kahn M. CBP/catenin antagonist safely eliminates drug-resistant leukemia-initiating cells. *Oncogene.* 2016;35:3705–3717.
110. Henderson Jr WR, Chi EY, Ye X, Nguyen C, Tien YT, Zhou B, Borok Z, Knight DA, Kahn M. Inhibition of Wnt/beta-catenin/CREB binding protein (CBP) signaling reverses pulmonary fibrosis. *Proc Natl Acad Sci USA.* 2010;107:14309–14314.
111. Hao S, He W, Li Y, Ding H, Hou Y, Nie J, Hou FF, Kahn M, Liu Y. Targeted inhibition of beta-catenin/CBP signaling ameliorates renal interstitial fibrosis. *J Am Soc Nephrol.* 2011;22:1642–1653.
112. Hu MC, Kuro-o M, Moe OW. Klotho and chronic kidney disease. *Contrib Nephrol.* 2013;180:47–63.
113. Lu X, Hu MC. Klotho/FGF23 axis in chronic kidney disease and cardiovascular disease. *Kidney Dis.* 2017;3:15–23.
114. Zhou L, Li Y, Zhou D, Tan RJ, Liu Y. Loss of Klotho contributes to kidney injury by derepression of Wnt/beta-catenin signaling. *J Am Soc Nephrol.* 2013;24:771–785.
115. Satoh M, Nagasu H, Morita Y, Yamaguchi TP, Kanwar YS, Kashihara N. Klotho protects against mouse renal fibrosis by inhibiting Wnt signaling. *Am J Physiol Renal Physiol.* 2012;303:F1641–F1651.
116. Zhou L, Mo H, Miao J, Zhou D, Tan RJ, Hou FF, Liu Y. Klotho ameliorates kidney injury and fibrosis and normalizes blood pressure by targeting the renin-angiotensin system. *Am J Pathol.* 2015;185:3211–3223.
117. Heikkila E, Juhila J, Lassila M, Messing M, Perala N, Lehtonen E, Lehtonen S, Sjef Verbeek J, Holthofer H. Beta-catenin mediates adriamycin-induced albuminuria and podocyte injury in adult mouse kidneys. *Nephrol Dial Transplant.* 2010;25:2437–2446.
118. Kato H, Gruenwald A, Suh JH, Miner JH, Barisoni-Thomas L, Taketo MM, Faul C, Millar SE, Holzman LB, Susztak K. Wnt/beta-catenin pathway in podocytes integrates cell adhesion, differentiation and survival. *J Biol Chem.* 2011;286:26003–26015.

119. Wang D, Dai C, Li Y, Liu Y. Canonical Wnt/beta-catenin signaling mediates transforming growth factor-beta1-driven podocyte injury and proteinuria. *Kidney Int.* 2011;80:1159–1169.
120. Zhou L, Li Y, He W, Zhou D, Tan RJ, Nie J, Hou FF, Liu Y. Mutual antagonism of Wilms' tumor 1 and beta-catenin dictates podocyte health and disease. *J Am Soc Nephrol.* 2015;26:677–691.
121. He W, Kang YS, Dai C, Liu Y. Blockade of Wnt/beta-catenin signaling by paricalcitol ameliorates proteinuria and kidney injury. *J Am Soc Nephrol.* 2011;22:90–103.
122. Jiang L, Xu L, Song Y, Li J, Mao J, Zhao AZ, He W, Yang J, Dai C. Calmodulin-dependent protein kinase II/cAMP response element-binding protein/Wnt/beta-catenin signaling cascade regulates angiotensin II-induced podocyte injury and albuminuria. *J Biol Chem.* 2013;288:23368–23379.
123. Li C, Siragy HM. High glucose induces podocyte injury via enhanced (pro)renin receptor-Wnt-beta-catenin-snail signaling pathway. *PLoS One.* 2014;9:e89233.
124. Zhou L, Liu Y. Wnt/beta-catenin signaling and renin-angiotensin system in chronic kidney disease. *Curr Opin Nephrol Hypertens.* 2016;25:100–106.
125. Cox SN, Sallustio F, Serino G, Pontrelli P, Verrienti R, Pesce F, Torres DD, Ancona N, Stifanelli P, Zaza G, Schena FP. Altered modulation of WNT-beta-catenin and PI3K/Akt pathways in IgA nephropathy. *Kidney Int.* 2010;78:396–407.
126. Shkreli M, Sarin KY, Pech MF, Papeta N, Chang W, Brockman SA, Cheung P, Lee E, Kuhnert F, Olson JL, Kuo CJ, Gharavi AG, D'Agati VD, Artandi SE. Reversible cell-cycle entry in adult kidney podocytes through regulated control of telomerase and Wnt signaling. *Nat Med.* 2011;18:111–119.
127. Wang XD, Huang XF, Yan QR, Bao CD. Aberrant activation of the WNT/beta-catenin signaling pathway in lupus nephritis. *PLoS One.* 2014;9:e84852.
128. Niida A, Hiroko T, Kasai M, Furukawa Y, Nakamura Y, Suzuki Y, Sugano S, Akiyama T. DKK1, a negative regulator of Wnt signaling, is a target of the beta-catenin/TCF pathway. *Oncogene.* 2004;23:8520–8526.
129. Kruck S, Eyrich C, Scharpf M, Sievert KD, Fend F, Stenzl A, Bedke J. Impact of an altered Wnt1/beta-catenin expression on clinicopathology and prognosis in clear cell renal cell carcinoma. *Int J Mol Sci.* 2013;14:10944–10957.
130. Hsu RJ, Ho JY, Cha TL, Yu DS, Wu CL, Huang WP, Chu P, Chen YH, Chen JT, Yu CP. WNT10A plays an oncogenic role in renal cell carcinoma by activating WNT/beta-catenin pathway. *PLoS One.* 2012;7:e47649.
131. Kondratov AG, Kvasha SM, Stoliar LA, Romanenko AM, Zgonnyk YM, Gordiyuk VV, Kashuba EV, Rynditch AV, Zabarovsky ER, Kashuba VI. Alterations of the WNT7A gene in clear cell renal cell carcinomas. *PLoS One.* 2012;7:e47012.
132. Janssens N, Andries L, Janicot M, Perera T, Bakker A. Alteration of frizzled expression in renal cell carcinoma. *Tumour Biol.* 2004;25:161–171.
133. Xu R, Zeng S, Xie W, Sun C, Chen YL, Chen MJ, Zhang L. The expression and function of Frizzled-7 in human renal cell carcinoma. *Clin Transl Oncol.* 2016;18:269–276.
134. Saini S, Majid S, Dahiya R. The complex roles of Wnt antagonists in RCC. *Nat Rev Urol.* 2011;8:690–699.
135. Maiti S, Alam R, Amos CI, Huff V. Frequent association of beta-catenin and WT1 mutations in Wilms tumors. *Cancer Res.* 2000;60:6288–6292.
136. Major MB, Camp ND, Berndt JD, Yi X, Goldenberg SJ, Hubbert C, Biechele TL, Gingras AC, Zheng N, Maccoss MJ, Angers S, Moon RT. Wilms tumor suppressor WTX negatively regulates WNT/beta-catenin signaling. *Science.* 2007;316:1043–1046.

CHAPTER SEVEN

Wnt Signaling in Stem Cells and Cancer Stem Cells: A Tale of Two Coactivators

Michael Kahn[1]
Beckman Research Institute of the City of Hope, Duarte, CA, United States
[1] Corresponding author. E-mail address: mkahn@coh.org

Contents

Abstract

Wnt signaling in stem cells plays critical roles in development, normal adult physiology, and disease. In this chapter, we focus on the role of the Wnt signaling pathway in somatic stem cell biology and its critical role in normal tissue homeostasis and cancer. Wnt signaling can both maintain potency and initiate differentiation in somatic stem cells, depending on the cellular and environmental context. Based principally on studies from our lab, we will explain the dichotomous behavior of this signaling pathway in determining stem cell fate decisions, placing special emphasis on the

Progress in Molecular Biology and Translational Science, Volume 153
ISSN 1877-1173
http://dx.doi.org/10.1016/bs.pmbts.2017.11.007

interaction of β-catenin with either of the two highly homologous Kat3 coactivator proteins, CBP and p300. We will also discuss our results, both preclinical and clinical, demonstrating that small molecule modulators of the β-catenin/Kat3 coactivator interaction can be safely utilized to shift the balance between maintenance of potency and initiation of differentiation.

1. INTRODUCTION

By definition, a stem cell has the capacity to self-renew, that is, make at least one identical copy of itself at each mitotic event and also has the capacity to differentiate into more mature and less potent cells. Pluripotent stem cells are either embryonic stem (ES) cells, derived from embryonic inner cell mass or induced pluripotent stem (iPS) cells, that is, more differentiated cells that have been reprogrammed back to pluripotency.[1–3] Somatic stem cells (SSCs) are generally long lived tissue derived cells responsible for homeostatic maintenance and repair/regeneration after injury. Another specialized group of stem cells are germ stem cells, which are required for spermatogenesis and oogenesis. The significant interest elicited by ES cells and more recently to iPS cells is due to their inherent property of self-renewal pluripotency. Pluripotent cells possess the capacity to generate all cell types found in embryos, as well as adult organisms. This unique property is subsequently lost in SSCs. SSCs are therefore considered as being multipotent or unipotent, that is, capable of generating multiple differentiated cell types, however generally restricted to that of a particular tissue, organ, or physiological system (e.g., hematopoietic stem cells, neural stem cells, germ stem cells, etc.) in which they reside.

2. WNT SIGNALING PATHWAYS IN STEM CELL BIOLOGY

Wnt signaling is often considered to be divided into multiple arms. Three commonly delineated pathway subsets include: the canonical, the noncanonical planar cell polarity (PCP) pathway, and the noncanonical Wnt/calcium pathway. The central protein in the canonical pathway is β-catenin, whose cytoplasmic and nuclear levels are normally under very strict regulatory control. Wnt ligand binding to Frizzled (Fzd) receptors as well as LRP5/6 coreceptors (low density lipoprotein receptor-related protein 5/6) initiates an intracellular signaling cascade leading to the nuclear translocation of β-catenin. Cytoplasmic β-catenin, in the absence of

Wnt ligands, is targeted by a degradation complex composed of the tumor suppressor Adenomatous polyposis coli (APC), the scaffolding protein Axin and two constitutively activated kinases CK1α (casein kinase 1α) and GSK-3β (glycogen synthase kinase 3β).[4] In the canonical signaling cascade, β-catenin partners with members of the T-cell factor/lymphoid enhancer factor (TCF/LEF) family of transcription factors and recruits the general transcriptional apparatus including, the Kat3 coactivators, p300 (E1A binding protein p300) and/or CREB-binding protein (CBP), as well as other coactivator proteins, for example, Bcl9, TATA binding protein, SWI/SNF family members, and the mediator complex, to transcribe Wnt target genes and engender chromatin modifications.[5–8]

Although the nuclear translocation of β-catenin and subsequent β-catenin transcriptional processes are classically controlled by the canonical Wnt signaling cascade, in reality there are a number of alternative signaling pathways that can induce the nuclear translocation of β-catenin and its subsequent participation in transcription. These include the process of epithelial-to-mesenchymal transition (EMT), which involves the downregulation of E-cadherin, which normally binds to cytoplasmic β-catenin in a complex with α-catenin in normal differentiated epithelial cells.[9] A variety of receptor tyrosine kinases, including the receptors Met, Fer, Fyn, and Ron,[10] and nonreceptor tyrosine kinases, including Src[11] and Abl,[12] can also disrupt the interaction of E-cadherin with β-catenin, thereby activating β-catenin nuclear translocation and transcription. Additionally, various G protein-coupled receptors can activate β-catenin/TCF signaling, including prostaglandin E2 (PGE2),[13] prostaglandin F2 (PGF2α),[14] leukotriene D4 via the cysteinyl leukotriene receptors, as well as via the activation of an array of G proteins, such as Gα, Gα12, and Gα13.[15,16] Additionally, environmental conditions, such as hypoxia[17] and high glucose[18] can activate nuclear β-catenin signaling. These signals are subsequently integrated with signals from other key pathways including Notch, JAK/Stat, etc., providing nuclear β-catenin with an essential role in balancing self-renewal versus differentiation in SSCs.[17,19] Although few would argue that Wnt signaling is critical in stem cell biology; there is no consensus as to whether Wnt signaling is important for either maintenance of potency[5,20] or the differentiation of stem cells.[5] Wnt/β-catenin signaling clearly plays dichotomous roles in stem cell biology.[5,19–23]

β-Catenin is one member of a large family of armadillo proteins, consisting of three subfamilies: the p120 subfamily, the beta subfamily [comprising β-catenin and junction plakoglobin (also known as γ-catenin)], and the more distant alpha subfamily. Junction plakoglobin is involved in Wnt/catenin signaling, even in the absence of β-catenin.[24–26] Further adding to the complexity of Wnt/catenin transcriptional regulation in stem cells, β-catenin can bind to a wide array of transcription factors beyond TCF/LEF family members including forkhead box proteins (Foxo), nuclear receptors, Sox and Smad family members, etc., thereby modulating a plethora of downstream biological processes including maintenance of pluripotency/multipotency, differentiation, and EMT.[27]

The noncanonical PCP and noncanonical Wnt/calcium pathways are termed "β-catenin-independent pathways," which coexist and interact with the canonical Wnt pathway. The PCP pathway is characterized by Wnt ligand binding to Fzd receptors and activation of small GTPases, such as RhoA (Ras homolog gene family member A), RAC (Ras-related C3 botulinum toxin substrate), and Cdc42 (cell division control protein 42), via recruitment and activation of Dishevelled (Dvl).[28] The PCP pathway affects the cytoskeleton and triggers the transcriptional activation of target genes responsible for cell adhesion and migration.[29] The Wnt/calcium pathway is characterized by engagement of Wnt ligands with Fzd receptors and or RYK or ROR (alternative receptors) enhancing cell migration and Wnt canonical pathway inhibition through the management of intracellular calcium flux and activation of calmodulin kinase II (CaMK2), Jun kinase (JNK), and protein kinase C (PKC).[30] Although these three Wnt pathways are separately delineated for convenience, in reality Wnt signaling involves the integration of all three pathways.[31–33]

3. EMBRYONIC AND INDUCED PLURIPOTENT STEM CELLS

The field of regenerative medicine hinges on advances in stem cell biology and a significant focus to date has been placed on pluripotent stem cells, both ES and iPS cells. These pluripotent stem cells are capable of differentiating into any cell type in the body and retain the property of self-renewal, hence providing in theory an unlimited resource. Therefore, understanding the mechanisms controlling the maintenance of self-renewal and somatic cell reprogramming is critical for the safe and effective utilization of pluripotent stem cells in clinical applications. A number of intrinsic factors are known to regulate pluripotency in ES cells including the transcription factors octamer-binding transcription factor-4 (Oct4), sex determining region-Y box-2 (Sox2), and Nanog. Ground breaking research by the

Yamanaka lab determined that ectopic expression of the four transcription factors, Oct4, Sox2, Klf4 (Kruppel-like factor 4), and c-Myc, can reprogram somatic fibroblasts to pluripotent iPS cells, which have been shown to contribute to chimeric mice that are germline competent.[1]

Somatic cell reprogramming has enabled the generation of patient-specific pluripotent cells without destroying blastocysts, thereby circumventing important ethical and practical (i.e., rejection) concerns. iPS cell derivation originally required the viral expression of the four transcription factors, that is, Oct4, Sox2, Klf4, and c-Myc, which had substantial drawbacks in regards to clinical usage. However, more recent efforts to avoid genetic manipulation have successfully led researchers to further define the signaling pathways involved in cellular reprogramming with the aim of replacing ectopic gene expression with nonintegrating vectors and removable vectors, proteins, or small molecules.[34–37] Wnt/β-catenin signaling in particular has been highlighted as one of the critical pathways in the reprogramming process.[38]

Wnt signaling regulates cell–cell interactions during embryogenesis and plays major roles at various stages of development including, early trophoblast lineage commitment and development, blastocyst activation, implantation, and chorionallantois fusion.[39] Deletion of β-catenin results in a peri-implantation lethal phenotype in knockout mice, however other studies indicate that lack of both maternal and zygotic β-catenin does not impair blastocyst development.[40,41] Additionally, inhibition of the β-catenin/TCF complex by small molecule inhibitors or overexpression of Dkk1, an extracellular negative regulator of Wnt signaling, did not negatively affect blastocyst formation.[42] These results have been interpreted to mean that Wnt signaling may contribute through noncanonical signaling, such as the Ca^{2+}-dependent pathway at the preimplantation stage of embryonic development.[43] It has been demonstrated that the increase in Ca^{2+} signaling in response to Wnt-5a and Fz-2/4 results in intracellular Ca^{2+} release with subsequent activation of the Ca^{2+}-sensitive enzymes Ca^{2+}-calmodulin-dependent protein kinase II (CamKII) and PKC.[44,45] Interestingly, these Ca^{2+} dependent enzymes are both known to phosphorylate p300 serine 89, thereby enhancing the p300/β-catenin interaction and promoting differentiation in ES and SSCs.[46,47]

ES cells are derived from the inner cell mass of blastocysts. However, the molecular mechanisms controlling pluripotency and differentiation of ES cells are still not fully understood despite extensive research. Pluripotency in ES cell is regulated by a combination of exogenous factors. The maintenance of murine ES (mES) cells in culture is achieved through leukemia inhibitory factor (LIF)-dependent as well as LIF-independent mechanisms. Although Wnt signaling has been demonstrated to enable the maintenance of stemness without LIF, Wnt supplementation is not a widely accepted protocol to maintain ES cell self-renewal.[48,49] The major reason being that Wnt signaling has also been shown to initiate the differentiation of ES cells. The role of Wnt signaling in stem cells is relatively complex and has engendered significant controversy. As briefly discussed earlier, Wnt signaling is essential for organogenesis and also contributes to the differentiation of ES cells.

Wnt3a promotes, in a LEF-1-dependent manner, trophectoderm formation in ES cells by inducing Cdx2 expression, a critical transcription factors in early trophoblast development.[50] A number of reports have also suggested that activation of the Wnt pathway in mES cells induces both mesodermal and endodermal differentiation, including the expression of high levels of mesodermal and endodermal markers, for example, Brachyury-T, Flk-1, Foxa2, Lhx1, and AFP. In sharp contrast, increasing evidence suggests that Wnt signaling is also involved in stem cell maintenance by modulating levels of intrinsic pluripotency factors, for example, Oct4, Nanog, and Sox2.[5,23,51] For example, overexpression of Wnt1 attenuated neural differentiation in mES cells.[52] β-Catenin deficient mES cells display strong downregulation of several stem cell pluripotency genes including Rex-1, dppa-4, and dppa-5, with upregulation of the neural differentiation marker Oct-6.[51] Wnt3a supplementation has been demonstrated to maintain the pluripotency of mES cells.[48,49] GSK3β inhibitors, such as lithium chloride, BIO, or CHIR99021, via stabilization of β-catenin allow for the maintenance of both mouse and human ES cell self-renewal.[23,52] Austin Smith's group reported in 2008 that a combination of the MEK inhibitor, PD0325901, and the GSK3β inhibitor, CHIR99021, sustains mES cell self-renewal while treatment with CHIR99021 alone induced nonectoderm differentiation.[53] Dominant-negative ΔNhLef1-expressing Stat3 null mES demonstrated that β-catenin mediated self-renewal was independent of LIF/Stat3.[53] Furthermore, mutations affecting the APC protein thereby activating the Wnt pathway inhibited mES cell differentiation in vitro.[54] Although Wnt/β-catenin signaling can contribute to ES cell self-renewal, significant controversy exists as Wnt activation in ES cells can also initiate differentiation. For example, Dravid *et al.* showed that recombinant Wnt3a stimulated not only hES cell proliferation, but also differentiation and further showed that canonical Wnt/β-catenin/TCF mediated signaling was minimal in undifferentiated hES cells, but greatly upregulated during Wnt induced differentiation.[55] Ogawa *et al.* demonstrated that constitutively activated mutant β-catenin required LIF supplementation to maintain mES cell self-renewal.[56] Conversely, activated β-catenin maintained the stemness of mES cells and Nanog expression for at least 6 days after withdrawal of LIF.[57] Takao *et al.* further demonstrated that stabilized β-catenin physically interacts with Oct4 thereby upregulating Nanog expression in mES cells.[57] These contradictory results have sparked significant controversy in regards to the role of Wnt/β-catenin signaling in stem cells. β-catenin does not directly bind to DNA. However, as previously mentioned, it binds and interacts with a large number of transcription factors beyond members of the TCF/LEF family, including members of the Smad, Sox, Foxo, and Nuclear Receptor families etc., as well as a number of coactivators including CBP, p300, Bcl9, and MLL1.[27] In this regard, the classical readouts for canonical Wnt activation (e.g., TOPFLASH) do not encompass all of the potential transcriptional effects of nuclear β-catenin.

Studies with cancer cells and knockout cells have indicated that in the case of deletion of β-catenin, other forms of catenin (e.g., γ-catenin/plakoglobin) can at least partially compensate.[24] Mammals possess four TCF family member genes, TCF1, TCF3, TCF4, and LEF1. TCF3 is the most abundantly expressed of these in ES cells. β-Catenin/LEF1 transcription was shown to initiate trophoblast lineage differentiation in mES cells,[50] whereas TCF3 was responsible for Nanog promoter specific repression.[58] The repressive role of TCF3 in mES cells was further demonstrated by genome –wide TCF3 ChIP–ChIP analysis.[59] This demonstrated that Tcf3, Oct4, and Nanog cooccupy more than 1000 promoter regions including their own promoters and Tcf3 knockdown in mES cells resulted in higher transcript levels for Oct4, Sox2, and Nanog.[59]

Furthermore, dominant-negative TCF expression did not impair Nanog and Oct4 levels in GSK-3α/β double knockout mES that possess abundant cytoplasmic and nuclear β-catenin.[60] Taken in total, these results suggest that β-catenin/TCFs complexes are involved in stem cell differentiation, whereas a TCF independent β-catenin nuclear complex is required to maintain ES cell self-renewal (Fig. 1).

4. SOMATIC STEM CELLS (SSC)

The pioneering work of Siminovitch *et al.* on mouse bone marrow stem cells provided the first concrete evidence for the existence of SSCs.[61] Subsequently SSCs have been identified in many organs and tissues,

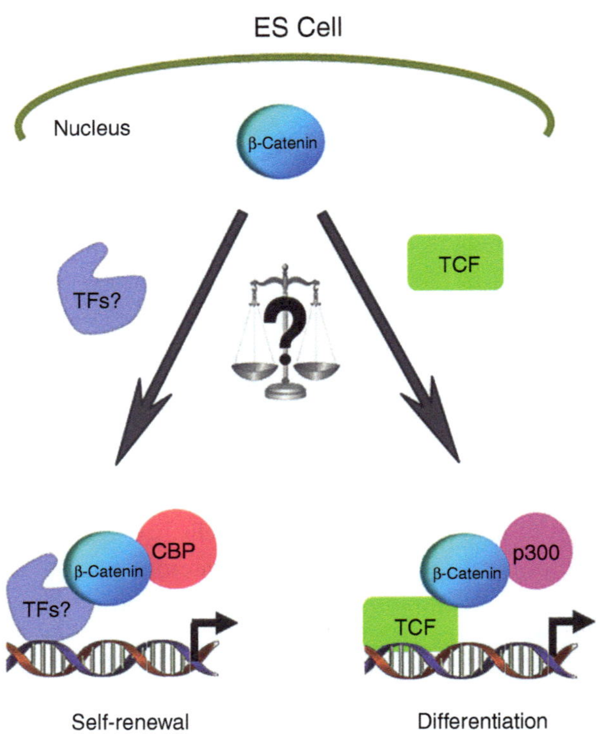

Fig. 1 The differential roles of the Kat3 coactivators CBP (Kat3A) and p300 (Kat3B) in the regulation of embryonic stem (ES) cell self-renewal versus differentiation. In ES cells, it is known that β-catenin recruits and partners with T-cell factor (TCF) and p300 to initiate the differentiation process. However, recruitment of TCF does not seem to be necessary to maintain ES cell self-renewal. *CBP*, CREB-binding protein; *TFs*, transcription factors.

including liver,[62] gut,[63] lung,[64] heart,[65] and CNS.[66] SSCs are critical in homeostatic maintenance, as well as regeneration and repair in adult tissues.[67] The ability of stem cells and progenitors to switch between different modes of cell division, that is, symmetric and asymmetric, is critical for the hierarchical maintenance of tissue organization to limit the accumulation of somatic mutations.[68,69] Symmetric cell division, can be further subdivided into differentiative or nondifferentiative symmetric division, thereby producing either two identical daughter cells with reduced differentiation potential and a higher degree of specification, or two daughter cells without changes in differentiation potential, thereby increasing the pool of stem cells respectively (Fig. 2). On the other hand, asymmetric division results in the production of two distinct daughter cells: one retaining the

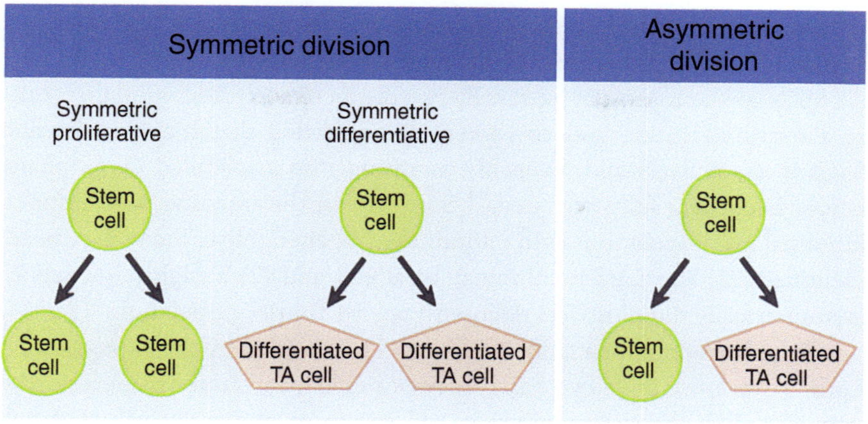

Fig. 2 A stem cell can undergo symmetric or asymmetric division. There are two modes of symmetric division. In symmetric proliferative division, the two daughter cells remain as stem cells in their niche. Stem cells can also undergo symmetric differentiative division where both daughter cells leave the niche and initiate differentiation. Both symmetric proliferative, as well as symmetric differentiative are thought to be detrimental to normal long lived somatic stem cells (SSCs), as they result in an increase in the number of DNA mutations accumulated in the stem cell pool (via symmetric proliferative divisions), or lead to premature exhaustion of the stem cell pool (via symmetric differentiative divisions). An asymmetric division produces progeny with different cell fates—one daughter cell that remains a stem cell and the other daughter cell that undergoes differentiation to become a transient amplifying (TA) cell.

characteristics of the parental stem cell, the other entering differentiation and exiting the stem cell niche to become a transient amplifying (TA) cell (Fig. 2). The Wnt signaling pathway has emerged as a pivotal player in the specification and maintenance of SSCs lineages and has been shown to have an important role in multiple stem cell compartments in a wide array of tissues and organs.[70–76] Involvement of Wnt/β-catenin in the "self-renewal" of a variety of stem cell populations has been previously reviewed.[77–79] For example, in 1998 the Clevers' laboratory reported that elimination of Tcf4 in mice resulted in the complete absence of the stem cell compartment in the small intestine.[80]

Canonical Wnt signaling cooperates with BMP and Notch signaling in the intestinal stem cell niche to control stem cell self-renewal.[74] The small intestine is organized into villi (apical) and crypts (basal). Intestinal stem cells (ISC) and Paneth cells (PC) require Wnt-signaling for establishment and maintenance [81,82] and the loss of positive Wnt regulators, for example, TCF4 or β-catenin, or overexpression of negative Wnt regulators, for example, Dickkopf1 (Dkk1), dramatically decreases stem cell capacity.[80,83]

SSCs appear to exist in minimally two distinct states based upon their ease of activation. Rando and coworkers demonstrated that highly quiescent SSCs in G_0 actively and reversibly transition into a so-called G_{Alert} phase before becoming fully activated.[84] Similarly, in the intestine, two distinct intestinal ISC populations with multilineage regenerating capacity have been identified: (1) the Lgr5+ columnar basal cell and (2) a highly quiescent, asymmetrically dividing, less radiosensitive +4 Bmi1+ population.[85]

Crypt basis columnar cells, which can be identified by the expression of Lgr5,[86,87] continuously cycle and are responsible for sustained tissue homeostasis.[82,83] PC, through expression and secretion of various Wnt related factors (e.g., Wnt3, 6, 9b, Fzd 4, 6, 7, Lrp5, Sfrp5),[82,88] are an important source of Wnt ligands crucial in ISC maintenance and depletion of PC's leads to a decrease in the number of ISCs.[89] Wnt signaling is also critical for Sox9 expression, which is critical for PC lineage commitment.[90,91] Due to the traits listed previously and their mode of chromosomal segregation,[92] the +4 population appears to behave more like bona fide long-lived quiescent SSCs. However, these populations can apparently interconvert and both possess the capacity to generate all intestinal epithelial lineages.[93] Similar situations exist in a variety of other stem cell populations including the mammary and lung epithelium,[94] as well as the hematopoietic system.[95]

Regulation of hematopoietic stem/progenitors, as well as lineage commitment of progenitors during hematopoiesis is highly Wnt dependent and Wnt3a has been implicated in HSC self-renewal and proliferation.[73,77,96] Survivin expression, which we previously demonstrated to be a Wnt/CBP/β-catenin regulated gene,[97] is important during hematopoiesis and is prominently upregulated in $CD34^+$ hematopoietic stem/progenitor cells upon growth factor treatment.[98] Hematopoietic progenitors deficient in survivin expression show defects in erythroid and megakaryocytic formation.[99]

Wnt signaling has also been implicated in mammary gland development as well as breast cancer development.[75,100] Overexpression of Axin[101] or loss of Lef1 inhibits mammary precursor cell differentiation.[102] The bilayered mammary epithelium consists of luminal cells ($CK8^+$, $Muc1^+$) and basal cells ($CK5^+$, $p63^+$). Basal cells have been shown to express both obligate canonical Wnt signaling receptors Lrp5 and 6.[103] Ductal mammary stem cells comprise a sub-population of basal epithelial cells, which are capable of regenerating cleared mammary fat pads.[104] Knockout of *Lrp5*[105] and loss of function of *Lrp6*[106] demonstrated significantly reduced regenerative capacity in this cellular compartment and impaired gland branching, suggestive of impaired stem cell functionality.

Finally, Wnt-activity has also been implicated in neuronal stem cell biology.[71,107] The small molecule tankyrase inhibitor XAV939 stabilizes Axin2 and thereby amplifies Wnt/β-catenin negative feedback signaling leading to accelerated differentiation of spinal cord stem cells and improved myelination after hypoxic and demyelinating injury.[108] Specific inhibition of the CBP/β-catenin interaction in neuronal stem/progenitors can rescue differentiation defects in a cell-based model of Alzheimer's disease.[109]

5. SYMMETRIC OR ASYMMETRIC DIVISION: THAT IS THE QUESTION

Long-term SSC are highly quiescent for the majority of their lifetime perhaps dividing only once every few months[110] or less.[111] However, under both normal homeostatic conditions and more frequently after insult or injury quiescent SSC enter cell cycle and undergo mitosis to give rise to two daughter cells. As described earlier, when they divide they can do so in two basic modes, that is, symmetrically or asymmetrically. Ideally, an asymmetric balance is maintained, whereby one of the daughter cells remains in its niche as a stem cell, while the other daughter initiates differentiation to maintain tissue homeostasis (Fig. 2). This asymmetric balance however is not always maintained. SSC symmetric divisions, whether symmetric nondifferentiative division, where both daughter cells remain as stem cells in their niche or symmetric differentiative divisions, where both cells leave the niche, thereby losing their "stemness" (Fig. 2) are presumed to be deleterious to the normal long-lived stem cell population. Symmetric nondifferentiative divisions increase the number of DNA mutations accumulated in the SSC pool, whereas symmetric differentiative divisions lead to premature exhaustion of the stem cell pool. Although still controversial, a potential rationale for SSC to preferentially undergo asymmetric cell divisions stems from the Cairn's "immortal strand hypothesis."[112] Simply stated, during an asymmetric mitosis, the stem cell desires to retain the original uncopied strand of DNA and pass on the duplicated strand that contain multiple copy errors to the differentiated daughter cell. Thereby, the total number of DNA mutations that accumulate in the long-lived SSC population can be minimized. This symmetric/asymmetric decision process likely becomes corrupted with age due to a variety of factors including accumulating mutations in the SSC pool, chronic or acute injury, reversion of differentiated daughters to SSCs, serum factors, and SSC senescence. This is likely the most critical decision a stem cell makes in adult organisms and likely underlies problems associated with aging, including a host of diseases associated with aging, for example, cancer, neurodegeneration, and decreased tissue maintenance/homeostasis (as in wound healing, sarcopenia, fibrosis, and osteoporosis).

6. CANCER STEM CELLS (CSC)

The "dark side" of the immortality of SSCs is that they can be corrupted into so-termed cancer stem cells [CSCs or cancer-initiating cell (CIC) populations]. Drug resistance, disease relapse, and metastasis remain major obstacles in the treatment of cancer. CSC share the properties of self-renewal and pluripotency with their normal SSC counterparts and appear to be the culprits associated with cancer initiation, metastasis, and disease relapse.[20] This is not a new concept though as almost 150 years ago Cohnheim *et al.* proposed that cancer might arise from a rare population of cells with stem cell-like properties.[113] The CSC concept proposes that a small population of tumor cells provide for the long-term maintenance of the tumor, whereas the bulk of the tumor consists of rapidly proliferating and partially differentiated cells. CSC self-renew,[114] actively express telomerase[115] and express multidrug resistance pathways.[116,117] CSCs are also relatively quiescent, however they produce rapidly dividing TA cells, which form the bulk of the tumor. John Dick and coworkers were the first to isolate CSC, in this case leukemic stem/initiating cells (LSC/LIC), from bulk acute myeloid leukemia cells in 1997.[118] Over the past decade, a large number of studies have also identified CSC in solid tumors, including brain,[119] melanoma,[120] breast,[121] liver,[122] pancreatic,[123] colon cancer,[124] as well as other organ systems.[125] CSC may emerge from normal SSC via genetic alterations acquired during DNA replication, via various insults and/or from microenvironmental factors.[20] Thus a significant effort in cancer research over the past decade has been focused on both prospectively identifying CSCs and the development of therapeutic strategies to safely eliminate this cell population without deleterious effects to the normal SSC populations.

The Wnt pathway is a key player in the maintenance of SSC in a wide array of tissues and organs including the intestines, heart, blood, brain, and mammary gland.[126] Therefore, it is not surprising that aberrant regulation of Wnt signaling is a recurrent theme in cancer biology.[127,128] In 1991 it was discovered that mutations in the tumor suppressor gene APC were associated with >80% of sporadic colorectal cancers via aberrant Wnt signaling activation.[129,130] Furthermore, APC is the most frequently mutated gene in human cancers.[131,132] Mutations in the Wnt pathway or silencing of negative regulators of the pathway are not restricted to colon cancer. Loss-of-function mutations in Axin have been found in hepatocellular carcinomas, oncogenic β-catenin mutations, first described in colon cancer and melanoma,[133] and also in a wide variety of solid tumors,[134] including hepatocellular carcinomas,[135] thyroid tumors,[136] and ovarian endometrioid adenocarcinomas.[137]

Thus, development of therapeutic approaches to target the Wnt pathway has engendered substantial effort. Several factors have delayed progress however, including the enormous complexity of this pathway.[138] Further complexity is encountered when targeting transcriptionally competent β-catenin. Beyond members of the TCF/LEF family, β-catenin can bind to a broad spectrum of transcription factor families.[27] β-catenin transcriptional regulation modulates a number of downstream processes including pluripotency, EMT, oxidative stress, metabolism, and lineage commitment.[27] Successful therapeutic manipulation of Wnt/catenin transcription to target endogenous "stemness," both normal and cancerous, offers enormous promise, however, it requires significant precision to provide the desired effects without inducing deleterious ones (e.g., depletion of the normal SSC population or increasing somatic mutations within the endogenous SSC population).[139]

7. SSCs AND CSCs; MORE SIMILARITIES THAN DIFFERENCES

It appears that the similarities between normal adult SSCs and CSCs far outweigh their differences thus complicating the safe targeting of CSCs. CSCs and SSCs express similar "stemness markers" and exhibit highly similar cellular behaviors. Both CSCs and SSCs appear to reside in the same specialized niches within tissues or organs, and in fact can compete with one another for the limited space within the niche. The same signaling pathways regulating SSC maintenance (i.e., Wnt, Notch, Hedeghog, TGFβ/BMP, JAK/Stat, Hippo, FGF/MAPK/PI3K) are also critical in the regulation of CSCs.[140–142] Multiple points of intersection and cross-talk, including feedback and feed-forward loops connecting these signaling cascades regulate "stemness" in both CSCs and SSCs. How is all of this information integrated to decide a stem cell's fate: to exit quiescence and subsequently divide asymmetrically or symmetrically, be it in normal SSC or CSC? From a therapeutic point of view, can we safely pharmacologically manipulate the endogenous stem cell population to eliminate CSCs without deleterious effects on SSCs? This would constitute a significant breakthrough in treating malignancies as well as to potentially ameliorate the aging process.

8. AN EXTREMELY CRITICAL DECISION

Symmetric versus asymmetric cell division is essentially a simple binary decision process. However, the stem cell in the niche undergoing mitosis must integrate an enormously complex array of information from its microenvironment (e.g., oxygen levels, nutrient levels, circadian cycles, nervous system innervation, growth factors, adhesion molecules, kinase cascades, cell–cell contacts, etc.) to decide to divide symmetrically or asymmetrically (Fig. 3). Interestingly, CSCs seem to have a preference for symmetric over asymmetric divisions compared to SSCs. This appears to be one fundamental difference between CSCs and normal SSCs. For example, breast cancer stem cells with p53 mutations preferentially undergo symmetric divisions.[143] Loss of function of the tumor suppressor PTEN leads to premature exhaustion of the normal HSC population due to increased symmetric differentiative divisions, whereas the leukemic stem cell population expands due to increased symmetric non-differentiative divisions.[144] Indirect perturbation of Notch signaling via genetic activation of Hedgehog signaling causes an increase in neural stem cell symmetric divisions.[145] The decision to preferentially undergo symmetric non-differentiative versus symmetric differentiative divisions respectively, appears to be another intrinsic difference between CSCs and SSCs carrying critical genetic mutations (e.g., p53, p73, and PTEN). This would provide a potential mechanism to stochastically eliminate mutated defective SSCs prior to the accumulation of additional deleterious mutations. Clearly, multiple signaling cascades can affect the stem cell's ultimate decision to become activated and divide either asymmetrically or symmetrically.

9. DIFFERENTIAL COACTIVATOR MODULATION

To generate a transcriptionally active complex, β-catenin must recruit one of the two Kat3 transcriptional coactivators, cAMP response element binding protein [CBP or its closely related homolog, p300 (E1A-binding

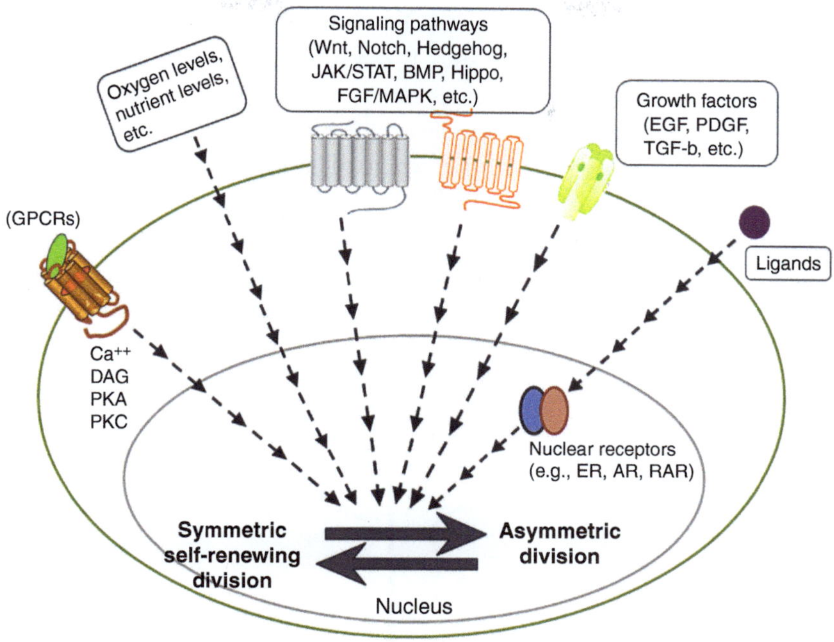

Fig. 3 A somatic stem cell or cancer stem cell (CSCs) in the niche has to integrate a plethora of signals from multiple pathways (i.e., Wnt, Notch, Hedhehog, etc.) as well as inputs from external sources, such as nutrient and oxygen levels, and then decides whether to remain quiescent, exit quiescence and subsequently divide either symmetrically or asymmetrically. This figure illustrates our model demonstrating that this decision is transcriptionally regulated after the integration from all of these signals. *GPCRs*, G-protein–coupled receptors.

protein, 300 kDa)].[5,7] CBP and p300 interact with hundreds of proteins, including the basal transcriptional apparatus, for example, TATA binding protein, RNA polymerase II, SWI/SNF, and Mediator complex members, in their roles as master orchestrators of transcription. They have long been considered largely redundant due to their high degree of protein sequence identity and even greater homology. However, accumulating data has demonstrated that CBP and p300 are not redundant and play definitive and unique roles both in vitro and in vivo.[146–148] From a library of 5000 secondary structure peptide mimetics, which we designed and synthesized, our lab identified ICG-001 in a cell-based TopFlash reporter gene assay in SW480 colon carcinoma cells. ICG-001 had an IC_{50} value of 3 μM. We subsequently demonstrated via deletion analysis that ICG-001 bound specifically and with very high affinity (~1 nM) to the N-terminus of the coactivator

CBP,[149,150] without binding to p300. We found that selectively antagonizing the interaction between CBP and β-catenin leads to the initiation of a differentiation program in a wide variety of stem/progenitor cells.[151,152] Subsequently, we developed a model of differential coactivator usage, which highlights the distinct roles of the coactivators CBP and p300 in catenin-mediated transcription (Fig. 4).[46] The differential utilization of either CBP or p300 by catenin is the first decision that guides a stem cell to either maintain potency or initiate a differentiative transcriptional program, respectively. We have subsequently identified several small molecules including

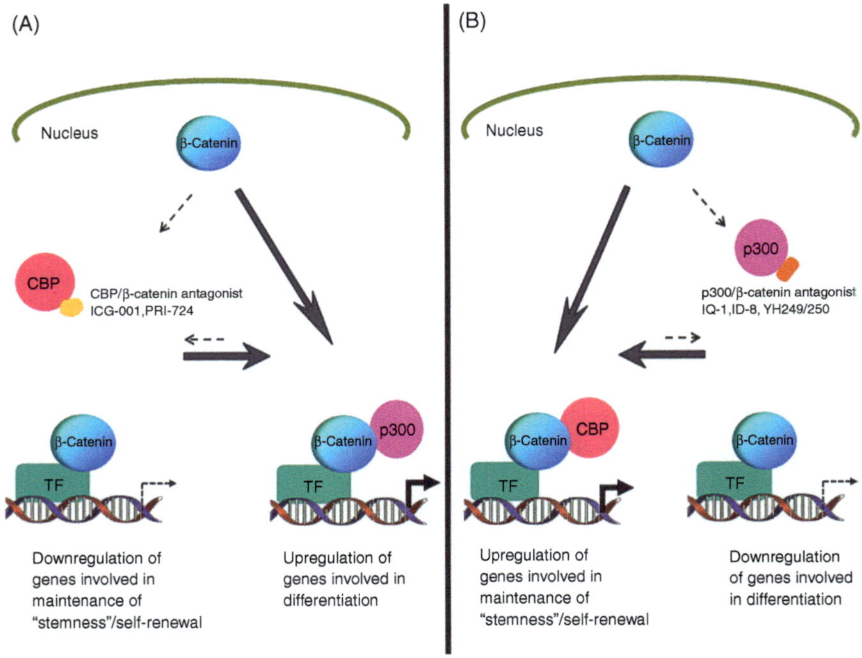

Fig. 4 Depending on differential Kat3 coactivator recruitment, nuclear β-catenin can transcriptionally direct either a self-renewal/proliferative pathway or initiate a differentiative pathway. (A) ICG-001 and PRI-724 are CBP/β-catenin antagonists that specifically bind to the N-terminus of CBP (Kat3A) and prevent CBP from partnering with β-catenin. This leads to enhanced usage of p300 by β-catenin and results in the upregulation of expression of genes involved in initiating differentiation. (B) IQ-1 and ID-8 indirectly and YH249/250, p300 (Kat3B) binding, direct p300/β-catenin antagonists, specifically prevent the interaction between p300 and β-catenin. This leads to enhanced usage of CBP by β-catenin. β-catenin partnering with CBP as its coactivator results in the expression of genes that are important in self-renewal/proliferation being upregulated.

IQ-1, ID-8, and more recently the specific direct p300/catenin antagonists YH249/250 that selectively block the p300/catenin interaction, thereby increasing the CBP/catenin interaction and resulting in enhancement of symmetric divisions and the maintenance of potency in a variety of stem cell populations (ES, iPS, and SSC), in both mouse and human cells (Fig. 4).[46,68,151,153]

10. PHARMACOLOGICALLY MANIPULATING SSC AND CSC

We have extensively examined the therapeutic potential of selectively pharmacologically antagonizing the CBP/β-catenin interaction in a variety of preclinical tumor and injury repair models. We and others have demonstrated that specifically antagonizing the CBP/β-catenin interaction allows safe elimination of drug-resistant CSCs, via forced differentiation, without deleterious effects on the normal endogenous stem cell populations.[154–158] CBP/catenin antagonists have also demonstrated efficacy in a variety of injury models including pulmonary, dermal, myocardial, hepatic and renal fibrosis, myocardial infarction, and neurodegeneration.[109,159–164] The beneficial effects of CBP/catenin antagonists in these models are apparently due to enhanced asymmetric differentiation of SSCs and progenitors thereby accelerating repair.[68] The differential effects of CBP/catenin antagonists on CSCs versus normal SSCs (i.e., forced differentiation and elimination vs. differentiation and enhanced repair without depletion, respectively) are cell intrinsic. CBP/catenin antagonists utilize the intrinsic propensity of CSCs to increase their number of symmetric divisions at the expense of asymmetric divisions, whereas normal long term repopulating SSCs preferentially divide asymmetrically (Fig. 5A).[68] When CSC are treated with CBP/catenin antagonists, CSCs will stochastically be cleared from their niche via symmetric differentiative divisions (Fig. 5B), whereas SSCs that divide asymmetrically will always retain one of the dividing daughter cells in the stem cell niche (Fig. 5A). This cell intrinsic fundamental difference between SSCs and CSCs provides a unique opportunity to therapeutically target and eliminate CSC, as well as enhance the repair potential of normal SSCs, without damaging the normal endogenous stem cell populations.[68,154,155,164,165]

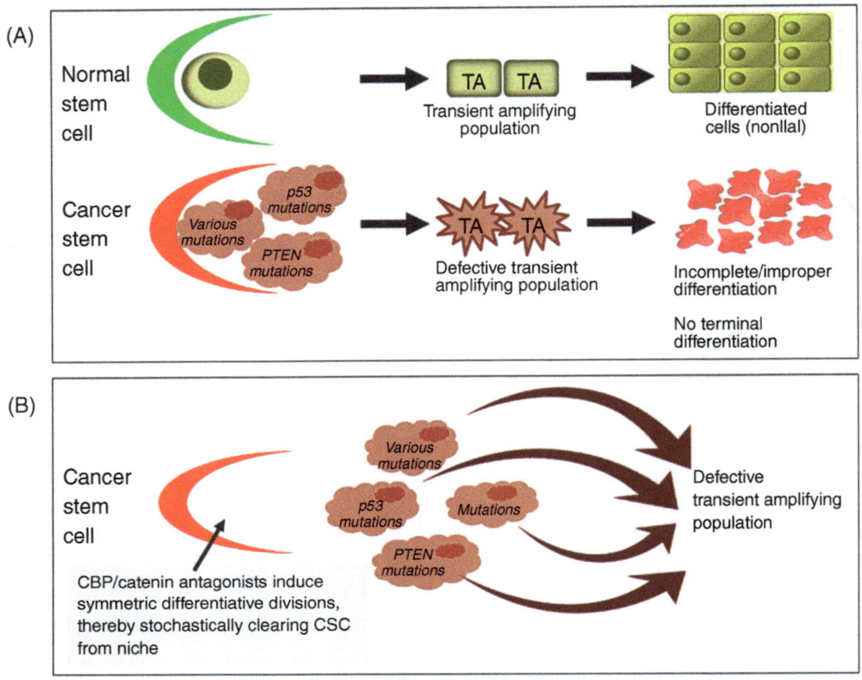

Fig. 5 (A) Normal somatic stem cells favor asymmetric division resulting in a daughter stem cell that remains in the niche and a transient amplifying population that goes on to differentiate into specialized cells. CSCs undergo both symmetric proliferative as well as asymmetric divisions. Symmetric proliferative divisions lead to an increase in CSCs in the niche. Asymmetric divisions in CSCs are also detrimental as this mode of division results not only in the maintenance of CSCs in the niche, but also to a defective transient amplifying population that undergoes rapid proliferation and incomplete/improper differentiation, failing to terminally differentiate. (B) CSCs preferentially undergo symmetric divisions. Treatment with CBP/catenin antagonists, forces CSCs in the niche to undergo symmetric differentiative divisions. Forced symmetric differentiative divisions result in stochastic elimination of cancer stem cells from their niche. Normal stem cells when treated with a CBP/catenin antagonist continue to undergo asymmetric divisions, and therefore the niche is never depleted of normal somatic stem cells.

11. CBP/CATENIN ANTAGONISTS IN THE CLINIC

Significant concerns about specificity arise when using small molecule inhibitors and in particular one that targets the coactivator protein CBP, as it has as many as 500 molecular partners, including a vast number of transcription factors.[165] However, both preclinically and even more importantly in the clinic, specific CBP/catenin antagonists (i.e., ICG-001 or the second-generation CBP/catenin antagonist PRI-724), have proven to be quite safe. Although at first blush perhaps very surprising, a few salient features are worth mentioning. First, ICG-001/PRI-724 have extremely high biochemical selectivity for their molecular target CBP. Second, these small molecules only disrupt a small subset of all CBP interactions. Third, the differences between the N-terminal binding domain in CBP for ICG-001/PRI-724 and the corresponding domain in p300, which will be discussed in more detail later in this chapter, contribute to the unique selectivity and safety of these small molecules.

PRI-724 (IC_{50} ~150 nM), which is a close structural analog of ICG-001, was developed by Prism Pharma. PRI-724 was extremely safe in preclinical Investigational New Drug enabling toxicology studies, with the no adverse event level for PRI-724 being 120 mg/kg/day in dogs given 28-days of continuous infusion. The results of the first in man phase Ia safety study in subjects with solid tumors were reported at the American Society for Clinical Oncology in June 2013.[166] PRI-724 had a very acceptable toxicity profile with dose escalation from 40 to 1280 $mg/m^2/day$ with 7 days of continuous intravenous infusion. Downregulation of the biomarker survivin/BIRC5 with increased expression of the differentiation antigen CK20 in circulating tumor cells strongly correlated with increasing plasma concentrations of drug.[166] Additional oncology trials,[167,168] as well as a trial for HCV-induced hepatic fibrosis with PRI-724 were initiated.

As described earlier, CBP/catenin antagonists have demonstrated efficacy in a wide variety of preclinical injury models. Given the attractive safety profile of these agents, additional clinical trials with CBP/catenin antagonists are anticipated in the future.

12. CBP/CATENIN ANTAGONISTS: TARGETING CSCs ACHILLES' HEEL?

The ability to safely target the drug resistant CSC population remains an extremely vexing problem that we must be able to overcome to dramatically change overall outcome when treating malignancies. In this respect, CBP/catenin antagonists appear to be truly unique in their ability to target drug-resistant CSCs in a wide range of tumors that seemingly have little in common with respect to mutational drivers, mechanisms associated with malignancy, that is, genetic mutations or epigenetic changes, or cell of origin.[154,155] Even more critically, CBP/catenin antagonists do so without damaging the normal endogenous stem cell population and even ameliorate toxicity associated with chemotherapy.[5,68] What endows CBP/catenin antagonists with this unique profile? Essentially immortal SSCs utilize quiescence to provide a safeguard to preserve their functionality by limiting damage to the cell caused by mitochondrial oxidative phosphorylation, DNA damage, and uncontrolled cell cycle entry and exhaustion of the stem cell pool via symmetric differentiative divisions.[169–172] Preferential use of glycolysis rather than oxidative phosphorylation, despite the inefficiency in regards to ATP generation,[173] has been demonstrated in both SSCs and CSCs. Switching from glycolysis to oxidative phosphorylation is associated with activation of quiescent SSC and the initiation of differentiation.[174]

The evolution of vertebrates, beginning roughly 450 million years ago, initiated a new lifestyle requiring critical adaptations. These included long-term homeostatic maintenance and tissue repair, which necessitated the advent of SSCs and their corresponding niches to maintain a relatively quiescent anaerobic metabolic state as opposed to their more proliferative aerobic-differentiated daughter cells, in order to protect the integrity of the genetic material in the stem cell pool.[110] This additionally required a high fidelity mechanism to ensure the maintenance of "stemness" in one daughter cell, while allowing for the initiation of a differentiative program in the other daughter. The Kat3 coactivators evolved after undergoing gene duplication apparently just prior to the vertebrate radiation. CBP and p300 are extremely large proteins encoded over 33 and 31 exons respectively. CBP and p300 retain an extremely high degree of identity, up to 93% at the amino acid level, particularly over a large central core that includes the CH1, KIX, Bromodomain, and CH2 and CH3 regions, despite having diverged over 450 million years ago (Fig. 6).[68] The least conserved region between the two Kat3 coactivators, with only 66% identity, is the extreme N-terminal region, to which both β- and γ-catenin and the small molecules ICG-001/PRI-724 bind. However, the N-terminal regions within each orthologous group are highly conserved for at least the past 100 million years of evolution. Human and mouse CBP are 98% identical at the amino acid level within this region.[68]

Logically, over millions of years of evolution, one would assume that "naturally occurring" CBP/catenin antagonists would have evolved that could insure the asymmetric differentiation of the highly quiescent SSC pool. In this regard, it is interesting that the very amino-termini of both CBP and p300 appear to be a nexus for the integration of many signal transduction pathways. All-trans retinoic acid (ATRA) is a very effective

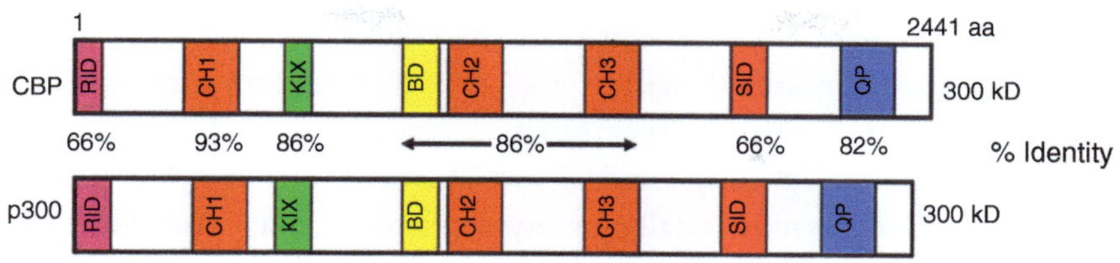

Fig. 6 Schematic representation of the homology between the two Kat3 coactivators: CBP (Kat3A) and p300 (Kat3B). There is an extremely high degree of identity at the amino acid level. CBP/catenin antagonists bind to the very amino terminal region of CBP, by far the most divergent region between these two Kat3 coactivators. *BD*, bromodomain; *CH1–3*, cysteine and histidine-rich regions 1–3; *KIX* domain, binding site of CREB; *QP*, glutamine- and proline-rich domain; *RID*, receptor-interacting domain; *SID*, steroid receptor coactivator-1 interaction domain.

therapeutic agent for the treatment of acute promyelocytic leukemia. Similarly to ICG-001, ATRA does not kill the malignant cells but induces them to differentiate. Vitamin D occupies a prominent position in cancer prevention. Both ATRA and Vitamin D, via their respective transcriptional complexes (RAR/RXR and VDR/RXR), are recruited via a highly conserved LXXLL sequence that is present in the N-termini of both CBP and p300, thereby antagonizing aberrant Wnt signaling via competition for catenin binding to the N-terminus of CBP. ATRA and vitamin D however, are not "pure antagonists" of CBP/catenin signaling as they also have distinct sets of gene expression that they drive through their respective nuclear receptors. A number of nuclear receptors, have also demonstrated the ability to regulate potency in a variety of stem cell populations, in a manner similar to what we have observed with specific CBP/catenin or p300/catenin antagonists. PPARδ agonists improved HSC maintenance, presumably via CBP/catenin antagonism and increased asymmetric division. Therefore, nuclear receptor ligands can behave as CBP/catenin antagonists partially phenocopying the direct antagonists ICG-001 and PRI-724. There are several important differences however. Small molecule CBP/catenin antagonists directly bind to CBP and do not require any protein cofactors (e.g., RAR/RXR). They are also pure CBP antagonists as they have no agonistic activity per se and they allow for stochastic differentiation (i.e., nondeterministic). On the other hand, ATRA or vitamin D, after antagonizing the CBP/catenin interaction, bias lineage commitment via p300-dependent agonistic properties. This p300-dependent lineage biasing is associated with the teratogenic effects of high concentrations of ATRA on embryonic development, which are not seen in mice treated with ICG-001 in utero.

There are a large number of serine and threonine residues that can be posttranslationally modified within the first 111 amino acid residues of CBP and p300. We believe they function as a nexus for the integration of an array of kinase signaling cascades that determine the critical stem cell decision to divide symmetrically or asymmetrically, via controlling the balance between the CBP/catenin and the p300/catenin interaction.[68]

Therefore, CBP/catenin antagonists potentially can safely target a common "Achilles' Heel" in CSC by forcing CSC to undergo symmetric differentiative divisions, whereas the normal SSC population is maintained in its niche via asymmetric division. Resistance to CBP/catenin antagonists appears to be quite unusual, likely due to this regions critical role in stem cell maintenance, in that escape mutations were apparently not observed in the clinic with PRI-724. Interestingly, the fundamental nature of differential coactivator usage by catenin is already manifested at the first cellular decision point in mammalian biology (i.e., at the 8 cell stage of embryogenesis) and we have proposed that it is carried through all stem cell populations in vertebrates throughout their life span.[68]

13. AGING AND AGING STEM CELLS AND WNT SIGNALING

Our aging population is creating major health and economic issues for our society. The efficiency and the fidelity of our bodies homeostatic and repair processes decline with age. This could be due to a decrease in SSC populations; however, it appears rather than a loss of SSC populations, a decrease in the efficiency of SSC to serve as a regenerative pool during homeostasis and repair decreases with age. Parabiosis experiments, which connect the circulatory systems of aged and young mice, demonstrated the presence of factors in young blood that can restore many of the features of youth, including faster recovery and decreased fibrosis after muscle injury, and increased neurogenesis and memory consolidation in aged mice.[175,176] These studies further highlight that rather than an issue of endogenous stem cell deficiency, the problem is due to either an extrinsic gain of deleterious and/or a dimunition of beneficial circulating factors that regulate the activation of our SSC. A number of mouse models of premature aging have demonstrated an increase in Wnt signaling.[177–179] Additionally, serum complement factor C1q, whose concentration increases with age, enhances Wnt signaling and decreases skeletal muscle regeneration.[180] This decreased efficiency in homeostasis and repair, we propose arises from defective SSC activation associated with an increase in the number of symmetric renewing divisions at the expense of asymmetric divisions in the SSC pool. Furthermore, this is due to an increase in CBP/catenin transcription at the expense of the p300/catenin dependent transcription. This hypothesis nicely fits with epidemiologic data demonstrating the significant increased incidence, after the age of 50, of developing cancer, fibrosis, or neurodegeneration.[181] Increased SSC symmetric versus asymmetric division with age could be influenced by a variety of factors including genetics (both germline and somatic variations), various insults and life style decisions, including infection, xenobiotics and pollutants, diet and caloric intake, and radiation, for example, UV and X-ray. In combination, these factors bias the equilibrium between CBP/catenin and p300/catenin-driven transcription leading to an increase in CBP/catenin-driven processes and an increase in symmetric versus asymmetric divisions in SSCs populations (Fig. 7). Interestingly, clonal hematopoiesis with somatic mutations was recently found in approximately 10% of persons 65 years or older, whereas in only 1% of those younger than

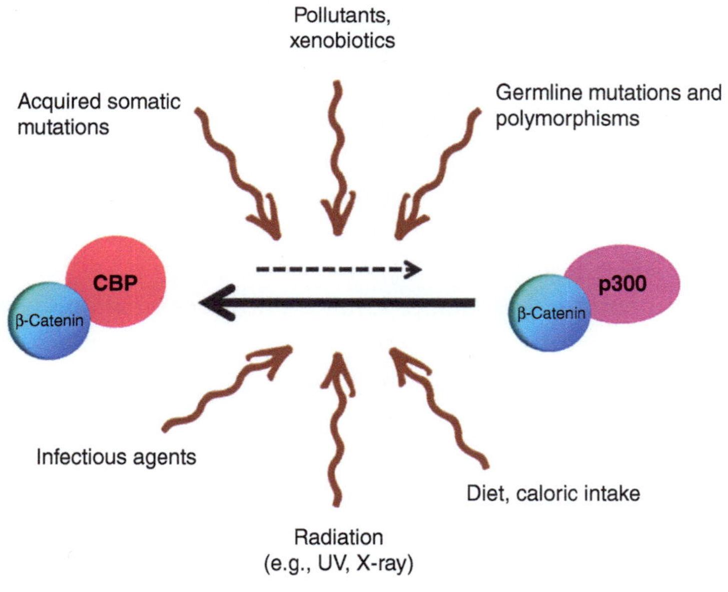

Fig. 7 An increase in the percentage of somatic stem cells undergoing symmetric versus asymmetric division could be influenced by a many factors including genetics, various insults, and external environmental factors. These factors affect the balance between CBP/catenin-dependent and p300/catenin-dependent transcription. With aging, there is a bias toward CBP/catenin-driven processes.

50 years of age[182] and appears to represent "preleukemic clones" that originated in the hematopoietic stem cell population. If the deleterious effects of aging are associated with an imbalance in Kat3/catenin coactivator usage and in particular biased CBP/catenin transcription, then selective small molecule CBP/catenin antagonists could ameliorate this and potentially provide prophylaxis against many of the common diseases of aging (e.g., cancer, fibrosis, neurodegeneration, etc.).[183]

14. KAT3A/CBP AND KAT3B/P300 EVOLUTIONARY RATIONALE

The vertebrate lineage ushered in a new lifestyle with complex body patterns and a relatively long-lived adult stage. In order to successfully accommodate this "life style" change, a mechanism for long-term homeostatic maintenance and tissue repair was required. The advent of long lived quiescent SSCs utilizing an "anaerobic" metabolic state as opposed to their more proliferative "aerobic"-differentiated daughter cells provided a mechanism to insure the integrity of the genetic material in SSCs.[184] A high fidelity mechanism to ensure the proper maintenance of "stemness" in one daughter cell, while also initiating a differentiative program in the other daughter cell was also required. We proposed that this gene duplication occurred in conjunction with the expansion of the nuclear receptor and Wnt signaling families and solved the problem of the requirement for a high fidelity control mechanism to control asymmetric SSC division and guide subsequent differential lineage commitment.[68]

Yet, symmetric versus asymmetric mitotic divisions significantly predate the vertebrate radiation. Furthermore, the Wnt signaling and the nuclear receptor pathways appear to be inventions of the first multicellular animals[185,186] and found in all animals including nematodes, flies, and vertebrates. Wnt signaling plays a critical role in regulating symmetric versus asymmetric division in both flies and nematodes,[187] clearly predating Kat3 gene duplication.

In *Caenorhabditis elegans* (*C. elegans*), multiple gene duplication events led to at least four β-catenin homologs: bar-1, hmp-2, wrm-1, and sys-1. Bar-1 is utilized in the canonical Wnt/catenin (Wβ) pathway in conjunction with the TCF homolog pop-1,[188,189] that is generally conserved among animals including vertebrates. *C. elegans*, unlike vertebrates and flies, also has a Wnt/catenin asymmetry (WβA) pathway. This system utilizes pop-1 in conjunction with the divergent (by sequence but not function) β-catenin-like molecule sys-1.[190] Sys-1 has no known orthologs outside the nematode family. It appears as though the WβA asymmetry pathway evolved specifically in *C. elegans*. Two additional β-catenin-like molecules, hmp-2 and wrm-1 in *C. elegans* play major roles in adhesion and regulation of TCF export, respectively.[191] The WβA pathway controls key asymmetric divisions at multiple stages of development in *C. elegans*.[191] The role of sys-1 in asymmetric cell division is an example of subfunctionalization, in which the multiple roles of the β-catenin/armadillo family in other animals has been divided among the different catenins in nematodes.[192] Asymmetric division in the ancestral bilaterian (i.e., *C. elegans*) was likely controlled via a Wnt/catenin/TCF-like pathway and the single Kat3 coactivator gene still found today. However, duplication of this ancestral coactivator thereby generating CBP and p300 provided a more robust mechanism to regulate asymmetric stem cell division. Cooperation between Wnt and nuclear receptor signaling, many of which are associated with metabolism, including (RAR, Vit D, PPAR, FXR, and others) can regulate SSC quiescence and also drive lineage commitment to accommodate longer lived, more complex organismal development. Multiple nuclear receptor ligands via their respective transcriptional complexes have been shown to either antagonize[193] or synergize with Wnt/β-catenin signaling.[194] We previously proposed and have subsequently experimentally validated, a model to explain these divergent outcomes of nuclear receptors and Wnt signaling. We recently demonstrated, using CRISPR/CAS9 editing in P19 cells, that a unique highly evolutionarily conserved N-terminal 27 bp/9aa deletion in CBP between the β-catenin-binding region (DELI-sequence) and the nuclear receptor binding sequence (LXXLL) is the key to tight control over antagonism versus synergy between Wnt and nuclear receptor signaling (Fig. 8).[68] Furthermore, the increased fidelity afforded by this new mechanism was essential to enhance the fidelity of SSCs that would be critical for increasing longevity (i.e., *C. elegans* a few weeks versus vertebrates that can live up to 100 years or more) in vertebrate populations. The closest orthologs of the Kat3 coactivators CBP and p300 in other species (including Nejire in *Drosophila* and cbp-1 in *C. elegans*) contain a significantly longer "spacer" sequence between the β-catenin and nuclear receptor binding motifs than exists in p300. The ancestral Kat3 coactivator could thus bind β-catenin (with or without nuclear receptor binding) to drive proliferation and differentiation following either symmetric or asymmetric cell division, with the possibility for either antagonistic or synergistic interactions with nuclear receptor signaling. The evolution of Kat3A/CBP in vertebrates, allowed for nuclear receptor binding to purely antagonize the β-catenin/CBP

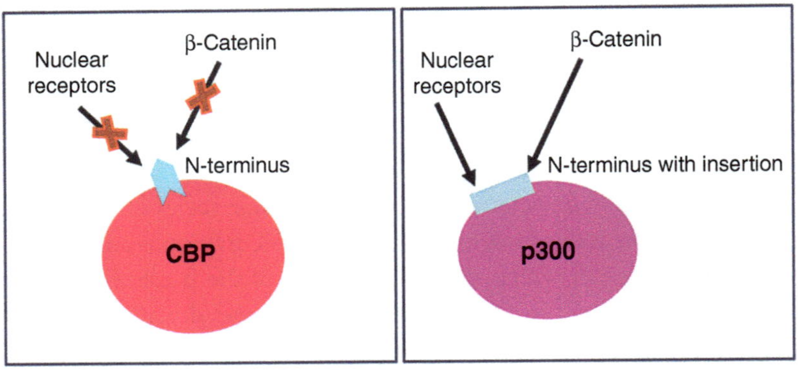

Fig. 8 Model depicting the different modes of interaction between nuclear receptors and β-catenin on CBP (antagonistic) and p300 (potentially synergistic). Nuclear receptors and β-catenin can transcriptionally interact with each other via the N-terminus of the Kat3 coactivator. This interaction can either be synergistic or antagonistic. The N-terminus of CBP contains an evolutionary conserved deletion of 27 base pairs/9 amino acids between the β-catenin binding region and the nuclear receptor binding sequence that is not present in p300. This deletion leads to a "built-in" antagonism, whereby only one entity—either β-catenin or nuclear receptor—can bind the N-terminus of CBP at any one time. On the other hand, on p300, nuclear receptors and β-catenin can potentially simultaneously bind to the N-terminus, leading to a synergistic mode of interaction.

interaction. This was a key evolutionary innovation that provided a high fidelity mechanism for the control of asymmetric mitosis in SSC. Thereby, one daughter can maintain (using a CBP-regulated program) its quiescent, multipotent stem cell status in its niche, while the other daughter generates via p300-regulated differentiation a rapidly symmetrically dividing TA population. This enabled tighter control over SSCs during normal tissue homeostasis and repair in longer-lived organisms.

15. CONCLUSIONS AND FUTURE PROSPECTS

Small molecule direct CBP/catenin selective antagonists are unique in their ability to specifically induce stem cell differentiation and do not require any protein cofactors (as in the case of ATRA and RAR/RXR). Pure direct small molecule CBP antagonists have no agonistic activity via p300 coactivation (as in the case of ATRA and RAR/RXR) and thus allow for stochastic differentiation, that is, nondeterministic, nonlineage biasing. They provide a safe pharmacological mechanism to manipulate our endogenous SSC populations. They do not deplete the normal endogenous long-lived SSC pool that is preferentially undergoing asymmetric division, however they offer the potential to eliminate CSC or even potentially pre-CSC[195] via forced symmetric differentiation. We believe that selective CBP/catenin antagonists will have a substantial impact on diseases of aging, that is, cancer, fibrosis, and neurodegeneration that are associated with defects in our SSC. We hope that these concepts will be further evaluated in the clinic in the not too distant future.

ACKNOWLEDGMENTS

I would like to thank Dr. Jia-Ling Teo for editorial assistance in preparing this chapter. Support from the USC Norris Comprehensive Cancer Center Support Grant P30 CA014089, NIH 1R01CA166161-01A1, and NIH 1R01 HL112638-01 is also gratefully acknowledged.

REFERENCES

1. Takahashi K, Yamanaka S. Induced pluripotent stem cells in medicine and biology. *Development.* 2013;140(12):2457–2461.
2. Nakagawa M, Yamanaka S. Reprogramming of somatic cells to pluripotency. *Adv Exp Med Biol.* 2010;695:215–224.
3. Izpisúa Belmonte JC, Ellis J, Hochedlinger K, Yamanaka S. Induced pluripotent stem cells and reprogramming: seeing the science through the hype. *Nat Rev Genet.* 2009;10(12):878–883.
4. Nakamura T, Hamada F, Ishidate T, et al. Axin, an inhibitor of the Wnt signalling pathway, interacts with beta-catenin, GSK-3beta and APC and reduces the beta-catenin level. *Genes Cells.* 1998;3(6):395–403.
5. Teo JL, Kahn M. The Wnt signaling pathway in cellular proliferation and differentiation: a tale of two coactivators. *Adv Drug Deliv Rev.* 2010;62(12):1149–1155.
6. Mosimann C, Hausmann G, Basler K. Beta-catenin hits chromatin: regulation of Wnt target gene activation. *Nat Rev Mol Cell Biol.* 2009;10(4):276–286.
7. Moon RT. Wnt/beta-catenin pathway. *Sci STKE.* 2005;2005(271):cm1.
8. Veeman MT, Axelrod JD, Moon RT. A second canon. Functions and mechanisms of beta-catenin-independent Wnt signaling. *Dev Cell.* 2003;5(3):367–377.
9. Onder TT, Gupta PB, Mani SA, Yang J, Lander ES, Weinberg RA. Loss of E-cadherin promotes metastasis via multiple downstream transcriptional pathways. *Cancer Res.* 2008;68(10):3645–3654.
10. Wagh PK, Gray JK, Zinser GM, et al. β-catenin is required for Ron receptor induced mammary tumorigenesis. *Oncogene.* 2011;30(34):3694–3704.
11. Coluccia AM, Benati D, Dekhil H, De Filippo A, Lan C, Gambacorti-Passerini C. SKI-606 decreases growth and motility of colorectal cancer cells by preventing pp60 (c-Src)-dependent tyrosine phosphorylation of β-catenin and its nuclear signaling. *Cancer Res.* 2006;66(4):2279–2286.
12. Ress A, Moelling K. Bcr interferes with β-catenin-Tcf1 interaction. *FEBS Lett.* 2006;580(5):1227–1230.
13. Fujino H, West KA, Regan JW. Phosphorylation of glycogen synthase kinase-3 and stimulation of T-cell factor signaling following activation of EP2 and EP4 prostanoid receptors by prostaglandin E2. *J Biol Chem.* 2002;277(4):2614–2619.

14. Fujino H, Regan JW. FP prostanoid receptor activation of a T-cell factor/β-catenin signaling pathway. *J Biol Chem.* 2001;276(16):12489–12492.
15. Liu X, Rubin JS, Kimmel AR. Rapid, Wnt-induced changes in GSK3β associations that regulate β-catenin stabilization are mediated by Gα proteins. *Curr Biol.* 2005;15 (22):1989–1997.
16. Meigs TE, Fields TA, McKee DD, Casey PJ. Interaction of Gα12 and Gα13 with the cytoplasmic domain of cadherin provides a mechanism for β-catenin release. *Proc Natl Acad Sci USA.* 2001;98(2):519–524.
17. Mazumdar J, O'Brien WT, Johnson RS, et al. O2 regulates stem cells through Wnt/β-catenin signalling. *Nat Cell Biol.* 2010;12(10):1007–1013.
18. Chocarro-Calvo A, García-Martínez JM, Ardila-González S, De la Vieja A, García-Jiménez C. Glucose-induced β-catenin acetylation enhances Wnt signaling in cancer. *Mol Cell.* 2013;49(3):474–486.
19. Kida A, Kahn M. Hypoxia selects for a quiescent, CML stem/leukemia initiating-like population dependent on CBP/catenin transcription. *Curr Mol Pharmacol.* 2013;6 (3):204–210.
20. Reya T, Morrison SJ, Clarke MF, Weissman IL. Stem cells, cancer, and cancer stem cells. *Nature.* 2001;414(6859):105–111.
21. Brembeck FH, Rosario M, Birchmeier W. Balancing cell adhesion and Wnt signaling, the key role of beta-catenin. *Curr Opin Genet Dev.* 2006;16(1):51–59.
22. van VW, Le NH, Helvensteijn W, et al. beta-catenin tyrosine 654 phosphorylation increases Wnt signalling and intestinal tumorigenesis. *Gut.* 2011;60(9):1204–1212.
23. Sato N, Meijer L, Skaltsounis L, Greengard P, Brivanlou AH. Maintenance of pluripotency in human and mouse embryonic stem cells through activation of Wnt signaling by a pharmacological GSK-3-specific inhibitor. *Nat Med.* 2004;10(1):55–63.
24. Kim YM, Ma H, Oehler VG, et al. The γ-catenin/CBP complex maintains survivin transcription in β-catenin deficient/depleted cancer cells. *Curr Cancer Drug Targets.* 2011;11(2):213–225.
25. Maeda O, Usami N, Kondo M, et al. Plakoglobin (γ-catenin) has TCF/LEF family-dependent transcriptional activity in β-catenin deficient cell line. *Oncogene.* 2004;23 (4):964–972.
26. Jeannet G, Scheller M, Scarpellino L, et al. Long-term, multilineage hematopoiesis occurs in the combined absence of β-catenin and γ-catenin. *Blood.* 2008;111 (1):142–149.
27. Le NH, Franken P, Fodde R. Tumour-stroma interactions in colorectal cancer: converging on β-catenin activation and cancer stemness. *Br J Cancer.* 2008;98(12): 1886–1893.
28. Lai SL, Chien AJ, Moon RT. Wnt/Fz signaling and the cytoskeleton: potential roles in tumorigenesis. *Cell Res.* 2009;19(5):532–545.
29. Yamamoto S, Nishimura O, Misaki K, et al. Cthrc1 selectively activates the planar cell polarity pathway of Wnt signaling by stabilizing the Wnt-receptor complex. *Dev Cell.* 2008;15(1):23–36.
30. van Amerongen R, Nusse R. Towards an integrated view of Wnt signaling in development. *Development.* 2009;136(19):3205–3214.
31. Moon RT, Kohn AD, De Ferrari GV, Kaykas A. WNT and beta-catenin signalling: diseases and therapies. *Nat Rev Genet.* 2004;5(9):691–701.
32. Thrasivoulou C, Millar M, Ahmed A. Activation of intracellular calcium by multiple Wnt ligands and translocation of beta-catenin into the nucleus: a convergent model of Wnt/Ca2+ and Wnt/beta-catenin pathways. *J Biol Chem.* 2013;288(50): 35651–35659.
33. Florian MC, Nattamai KJ, Dörr K, et al. A canonical to non-canonical Wnt signalling switch in haematopoietic stem-cell ageing. *Nature.* 2013;503(7476):392–396.

34. Uhlin E, Rönnholm H, Day K, et al. Derivation of human iPS cell lines from monozygotic twins in defined and xeno free conditions. *Stem Cell Res.* 2017;18:22–25.
35. Kang PJ, Moon JH, Yoon BS, et al. Reprogramming of mouse somatic cells into pluripotent stem-like cells using a combination of small molecules. *Biomaterials.* 2014;35(26):7336–7345.
36. Pasha Z, Haider HKh. Ashraf M. Efficient non-viral reprogramming of myoblasts to stemness with a single small molecule to generate cardiac progenitor cells. *PLoS One.* 2011;6(8):e23667.
37. Marson A, Foreman R, Chevalier B, et al. Wnt signaling promotes reprogramming of somatic cells to pluripotency. *Cell Stem Cell.* 2008;3(2):132–135.
38. Merrill BJ. Develop-WNTs in somatic cell reprogramming. *Cell Stem Cell.* 2008;3(5):465–466.
39. Sonderegger S, Pollheimer J, Knofler M. Wnt signalling in implantation, decidualisation and placental differentiation–review. *Placenta.* 2010;31(10):839–847.
40. De Vries WN, Evsikov AV, Haac BE, et al. Maternal beta-catenin and E-cadherin in mouse development. *Development.* 2004;131(18):4435–4445.
41. Haegel H, Larue L, Ohsugi M, Fedorov L, Herrenknecht K, Kemler R. Lack of beta-catenin affects mouse development at gastrulation. *Development.* 1995;121(11):3529–3537.
42. Xie H, Tranguch S, Jia X, et al. Inactivation of nuclear Wnt-beta-catenin signaling limits blastocyst competency for implantation. *Development.* 2008;135(4):717–727.
43. Chen Q, Zhang Y, Lu J, et al. Embryo-uterine cross-talk during implantation: the role of Wnt signaling. *Mol Hum Reprod.* 2009;15(4):215–221.
44. Slusarski DC, Corces VG, Moon RT. Interaction of Wnt and a Frizzled homologue triggers G-protein-linked phosphatidylinositol signalling. *Nature.* 1997;390(6658):410–413.
45. Kuhl M, Sheldahl LC, Malbon CC, Moon RT. Ca(2+)/calmodulin-dependent protein kinase II is stimulated by Wnt and Frizzled homologs and promotes ventral cell fates in Xenopus. *J Biol Chem.* 2000;275(17):12701–12711.
46. Miyabayashi T, Teo JL, Yamamoto M, McMillan M, Nguyen C, Kahn M. Wnt/beta-catenin/CBP signaling maintains long-term murine embryonic stem cell pluripotency. *Proc Natl Acad Sci USA.* 2007;104(13):5668–5673.
47. Rieger ME, Zhou B, Solomon N, et al. Differential β-catenin/coactivator interactions regulate adult stem/progenitors. *J Biol Chem.* 2016;291(12):6569–6582.
48. Singla DK, Schneider DJ, LeWinter MM, Sobel BE. Wnt3a but not wnt11 supports self-renewal of embryonic stem cells. *Biochem Biophys Res Commun.* 2006;345(2):789–795.
49. Hao J, Li TG, Qi X, Zhao DF, Zhao GQ. WNT/beta-catenin pathway up-regulates Stat3 and converges on LIF to prevent differentiation of mouse embryonic stem cells. *Dev Biol.* 2006;290(1):81–91.
50. He S, Pant D, Schiffmacher A, Meece A, Keefer CL. Lymphoid enhancer factor 1-mediated Wnt signaling promotes the initiation of trophoblast lineage differentiation in mouse embryonic stem cells. *Stem Cells.* 2008;26(4):842–849.
51. Anton R, Kestler HA, Kuhl M. Beta-catenin signaling contributes to stemness and regulates early differentiation in murine embryonic stem cells. *FEBS Lett.* 2007;581(27):5247–5254.
52. Aubert J, Dunstan H, Chambers I, Smith A. Functional gene screening in embryonic stem cells implicates Wnt antagonism in neural differentiation. *Nat Biotechnol.* 2002;20(12):1240–1245.
53. Ying QL, Wray J, Nichols J, et al. The ground state of embryonic stem cell self-renewal. *Nature.* 2008;453(7194):519–523.

54. Kielman MF, Rindapaa M, Gaspar C, et al. Apc modulates embryonic stem-cell differentiation by controlling the dosage of beta-catenin signaling. *Nat Genet.* 2002;32 (4):594–605.
55. Dravid G, Ye Z, Hammond H, et al. Defining the role of Wnt/beta-catenin signaling in the survival, proliferation, and self-renewal of human embryonic stem cells. *Stem Cells.* 2005;23(10):1489–1501.
56. Ogawa K, Nishinakamura R, Iwamatsu Y, Shimosato D, Niwa H. Synergistic action of Wnt and LIF in maintaining pluripotency of mouse ES cells. *Biochem Biophys Res Commun.* 2006;343(1):159–166.
57. Takao Y, Yokota T, Koide H. Beta-catenin up-regulates Nanog expression through interaction with Oct-3/4 in embryonic stem cells. *Biochem Biophys Res Commun.* 2007;353(3):699–705.
58. Pereira L, Yi F, Merrill BJ. Repression of Nanog gene transcription by Tcf3 limits embryonic stem cell self-renewal. *Mol Cell Biol.* 2006;26(20):7479–7491.
59. Cole MF, Johnstone SE, Newman JJ, Kagey MH, Young RA. Tcf3 is an integral component of the core regulatory circuitry of embryonic stem cells. *Genes Dev.* 2008;22(6):746–755.
60. Kelly KF, Ng DY, Jayakumaran G, Wood GA, Koide H, Doble BW. Beta-Catenin enhances Oct-4 activity and reinforces pluripotency through a TCF-independent mechanism. *Cell Stem Cell.* 2011;8(2):214–227.
61. Siminovitch L, McCulloch EA, Till JE. The distribution of colony-forming cells among spleen colonies. *J Cell Comp Physiol.* 1963;62:327–336.
62. Stachelscheid H, Urbaniak T, Ring A, Spengler B, Gerlach JC, Zeilinger K. Isolation and characterization of adult human liver progenitors from ischemic liver tissue derived from therapeutic hepatectomies. *Tissue Eng Part A.* 2009;15(7):1633–1643.
63. Lin SA, Barker N. Gastrointestinal stem cells in self-renewal and cancer. *J Gastroenterol.* 2011;46(9):1039–1055.
64. Leeman KT, Fillmore CM, Kim CF. Lung stem and progenitor cells in tissue homeostasis and disease. *Curr Top Dev Biol.* 2014;107:207–233.
65. Anversa P, Leri A. Innate regeneration in the aging heart: healing from within. *Mayo Clin Proc.* 2013;88(8):871–883.
66. Williams A. Central nervous system regeneration—where are we? *QJM.* 2014;107 (5):335–339.
67. Blanpain C, Simons BD. Unravelling stem cell dynamics by lineage tracing. *Nat Rev Mol Cell Biol.* 2013;14(8):489–502.
68. Thomas PD, Kahn M. Kat3 coactivators in somatic stem cells and cancer stem cells: biological roles, evolution, and pharmacologic manipulation. *Cell Biol Toxicol.* 2016;32 (1):61–81.
69. Derényi I, Szöllősi GJ. Hierarchical tissue organization as a general mechanism to limit the accumulation of somatic mutations. *Nat Commun.* 2017;8:14545.
70. Boras-Granic K, Wysolmerski JJ. Wnt signaling in breast organogenesis. *Organogenesis.* 2008;4(2):116–122.
71. Inestrosa NC, Arenas E. Emerging roles of Wnts in the adult nervous system. *Nat Rev Neurosci.* 2010;11(2):77–86.
72. Naito AT, Shiojima I, Akazawa H, et al. Developmental stage-specific biphasic roles of Wnt/beta-catenin signaling in cardiomyogenesis and hematopoiesis. *Proc Natl Acad Sci USA.* 2006;103(52):19812–19817.
73. Reya T, Duncan AW, Ailles L, et al. A role for Wnt signalling in self-renewal of haematopoietic stem cells. *Nature.* 2003;423(6938):409–414.
74. Clarke AR. Wnt signalling in the mouse intestine. *Oncogene.* 2006;25(57):7512–7521.
75. Brennan KR, Brown AM. Wnt proteins in mammary development and cancer. *J Mammary Gland Biol Neoplasia.* 2004;9(2):119–131.

76. Zechner D, Fujita Y, Hulsken J, et al. Beta-Catenin signals regulate cell growth and the balance between progenitor cell expansion and differentiation in the nervous system. *Dev Biol.* 2003;258(2):406–418.
77. Fleming HE, Janzen V, Lo Celso C, et al. Wnt signaling in the niche enforces hematopoietic stem cell quiescence and is necessary to preserve self-renewal in vivo. *Cell Stem Cell.* 2008;2(3):274–283.
78. Ahn JY, Park G, Shim JS, Lee JW, Oh IH. Intramarrow injection of beta-catenin-activated, but not naive mesenchymal stromal cells stimulates self-renewal of hematopoietic stem cells in bone marrow. *Exp Mol Med.* 2010;42(2):122–131.
79. Kim JA, Kang YJ, Park G, et al. Identification of a stroma-mediated Wnt/beta-catenin signal promoting self-renewal of hematopoietic stem cells in the stem cell niche. *Stem Cell.* 2009;27(6):1318–1329.
80. Korinek V, Barker N, Moerer P, et al. Depletion of epithelial stem-cell compartments in the small intestine of mice lacking Tcf-4. *Nature Genet.* 1989;19(4):379–383.
81. Cheng H, Leblond CP. Origin, differentiation and renewal of the four main epithelial cell types in the mouse small intestine. V. unitarian theory of the origin of the four epithelial cell types. *Am J Anat.* 1974;141(4):537–561.
82. Farin HF, van Es JH, Clevers H. Redundant sources of Wnt regulate intestinal stem cells and promote formation of Paneth cells. *Gastroenterol.* 2012;143(6):1518–1529.
83. Pinto D, Gregorieff A, Begthel H, Clevers H. Canonical Wnt signals are essential for homeostasis of the intestinal epithelium. *Genes Dev.* 2003;17(14):1709–1713.
84. Rodgers JT, King KY, Brett JO, et al. mTORC1 controls the adaptive transition of quiescent stem cells from G0 to G(Alert). *Nature.* 2014;510(7505):393–396.
85. Barker N, van Es JH, Kuipers J, et al. Identification of stem cells in small intestine and colon by marker gene Lgr5. *Nature.* 2007;449(7165):1003–1007.
86. Koo BK, Clevers H. Stem cells marked by the R-spondin receptor LGR5. *Gastroenterol.* 2014;147(2):289–302.
87. Muñoz J, Stange DE, Schepers AG, et al. The Lgr5 intestinal stem cell signature: robust expression of proposed quiescent '+4' cell markers. *EMBO J.* 2012;31(14): 3079–3091.
88. Ootani A, Li X, Sangiorgi E, et al. Sustained in vitro intestinal epithelial culture within a Wnt-dependent stem cell niche. *Nat Med.* 2009;15(6):701–706.
89. Sato T, van Es JH, Snippert HJ, et al. Paneth cells constitute the niche for Lgr5 stem cells in intestinal crypts. *Nature.* 2011;469(7330):415–418.
90. Blache P, van de Wetering M, Duluc I, et al. SOX9 is an intestine crypt transcription factor, is regulated by the Wnt pathway, and represses the CDX2 and MUC2 genes. *J Cell Biol.* 2004;166(1):37–47.
91. Bastide P, Darido C, Pannequin J, et al. Sox9 regulates cell proliferation and is required for Paneth cell differentiation in the intestinal epithelium. *J Cell Biol.* 2007;178(4): 635–648.
92. Potten CS, Hume WJ, Reid P, Cairns J. The segregation of DNA in epithelial stem cells. *Cell.* 1978;15(3):899–906.
93. Ritsma L, Ellenbroek SI, Zomer A, et al. Intestinal crypt homeostasis revealed at single-stem-cell level by in vivo live imaging. *Nature.* 2014;507(7492):362–365.
94. Blanpain C, Fuchs E. Stem cell plasticity. Plasticity of epithelial stem cells in tissue regeneration. *Science.* 2014;344(6189):1242281.
95. Oh IH, Humphries RK. Concise review: multidimensional regulation of the hematopoietic stem cell state. *Stem Cells.* 2012;30(1):82–88.
96. Vijayaragavan K, Szabo E, Bossé M, Ramos-Mejia V, Moon RT, Bhatia M. Noncanonical Wnt signaling orchestrates early developmental events toward hematopoietic cell fate from human embryonic stem cells. *Cell Stem Cell.* 2009;4 (3):248–262.

97. Ma H, Nguyen C, Lee KS, Kahn M. Differential roles for the coactivators CBP and p300 on TCF/beta-catenin-mediated survivin gene expression. *Oncogene.* 2005;24 (22):3619–3631.
98. Carter BZ, Qiu Y, Huang X, et al. Survivin is highly expressed in CD34(+)38(-) leukemic stem/progenitor cells and predicts poor clinical outcomes in AML. *Blood.* 2012;120(1):173–180.
99. Fukuda S, Pelus LM. Elevation of Survivin levels by hematopoietic growth factors occurs in quiescent CD34+ hematopoietic stem and progenitor cells before cell cycle entry. *Cell Cycle.* 2002;1(5):322–326.
100. Yu QC, Verheyen EM, Zeng YA. Mammary development and breast cancer: a Wnt perspective. *Cancers.* 2016;8:65.
101. Hsu W, Shakya R, Costantini F. Impaired mammary gland and lymphoid development caused by inducible expression of Axin in transgenic mice. *J Cell Biol.* 2001;155 (6):1055–1064.
102. van Genderen C, Okamura RM, Fariñas I, et al. Development of several organs that require inductive epithelial-mesenchymal interactions is impaired in LEF-1-deficient mice. *Genes Dev.* 1994;8(22):2691–2703.
103. Goel S, Chin EN, Fakhraldeen SA, Berry SM, Beebe DJ, Alexander CM. Both LRP5 and LRP6 receptors are required to respond to physiological Wnt ligands in mammary epithelial cells and fibroblasts. *J Biol Chem.* 2012;287(20):16454–16466.
104. Plaks V, Brenot A, Lawson DA, et al. Lgr5-expressing cells are sufficient and necessary for postnatal mammary gland organogenesis. *Cell Rep.* 2013;3(1):70–78.
105. Badders NM, Goel S, Clark RJ, et al. The Wnt receptor, Lrp5, is expressed by mouse mammary stem cells and is required to maintain the basal lineage. *PLoS One.* 2009;4(8): e6594.
106. Lindvall C, Zylstra CR, Evans N, et al. The Wnt co-receptor Lrp6 is required for normal mouse mammary gland development. *PLoS One.* 2009;4(6):e5813.
107. L'episcopo F, Serapide MF, Tirolo C, et al. A Wnt1 regulated Frizzled-1/β-catenin signaling pathway as a candidate regulatory circuit controlling mesencephalic dopaminergic neuron-astrocyte crosstalk: therapeutical relevance for neuron survival and neuroprotection. *Mol Neurodegener.* 2011;6:49.
108. Fancy SP, Harrington EP, Yuen TJ, et al. Axin2 as regulatory and therapeutic target in newborn brain injury and remyelination. *Nat Neurosci.* 2011;14(8):1009–1016.
109. Teo JL, Ma H, Nguyen C, Lam C, Kahn M. Specific inhibition of CBP/beta-catenin interaction rescues defects in neuronal differentiation caused by a presenilin-1 mutation. *Proc Natl Acad Sci USA.* 2005;102(34):12171–12176.
110. Foudi A, Hochedlinger K, Van Buren D, et al. Analysis of histone 2B-GFP retention reveals slowly cycling hematopoietic stem cells. *Nat Biotechnol.* 2009;27(1):84–90.
111. Baker AM, Cereser B, Melton S, et al. Quantification of crypt and stem cell evolution in the normal and neoplastic human colon. *Cell Rep.* 2014;8(4):940–947.
112. Cairns J. Mutation selection and the natural history of cancer. *Nature.* 1975;255 (5505):197–200.
113. Cohnheim J. Ueber entzuendung und eiterung [about inflammation and suppuration]. *Path Anat Physiol Klin Med.* 1867;40(1–2):1–79.
114. O'Brien CA, Kreso A, Jamieson CH. Cancer stem cells and self-renewal. *Clin Cancer Res.* 2010;16(12):3113–3120.
115. Armanios M, Greider CW. Telomerase and cancer stem cells. *Cold Spring Harb Symp Quant Biol.* 2005;70:205–208.
116. Clevers H. The cancer stem cell: premises, promises and challenges. *Nat Med.* 2011;17 (3):313–319.
117. Visvader JE, Lindeman GJ. Stem cells and cancer—the promise and puzzles. *Mol Oncol.* 2010;4(5):369–372.

118. Bonnet D, Dick JE. Human acute myeloid leukemia is organized as a hierarchy that originates from a primitive hematopoietic cell. *Nat Med.* 1997;3(7):730–737.
119. Singh SK, Hawkins C, Clarke ID, et al. Identification of human brain tumour initiating cells. *Nature.* 2004;432(7015):396–401.
120. Fang D, Nguyen TK, Leishear K, et al. A tumorigenic subpopulation with stem cell properties in melanomas. *Cancer Res.* 2005;65(20):9328–9337.
121. Dontu G, El-Ashry D, Wicha MS. Breast cancer, stem/progenitor cells and the estrogen receptor. *Trends Endocrinol Metab.* 2004;15(5):193–197.
122. Ma S, Chan KW, Hu L, et al. Identification and characterization of tumorigenic liver cancer stem/progenitor cells. *Gastroenterol.* 2007;132(7):2542–2556.
123. Li C, Heidt DG, Dalerba P, et al. Identification of pancreatic cancer stem cells. *Cancer Res.* 2007;67(3):1030–1037.
124. O'Brien CA, Pollett A, Gallinger S, Dick JE. A human colon cancer cell capable of initiating tumour growth in immunodeficient mice. *Nature.* 2007;445(7123):106–110.
125. Holland JD, Klaus A, Garratt AN, et al. Wnt signaling in stem and cancer stem cells. *Curr Opin Cell Biol.* 2013;25(2):254–264.
126. Kühl SJ, Kühl M. On the role of Wnt/β-catenin signaling in stem cells. *Biochim Biophys Acta.* 2013;1830(2):2297–2306.
127. Anastas JN, Moon RT. WNT signalling pathways as therapeutic targets in cancer. *Nat Rev Cancer.* 2013;13(1):11–26.
128. Polakis P. Drugging Wnt signalling in cancer. *EMBO J.* 2012;31(15):2737–2746.
129. Groden J, Thliveris A, Samowitz W, et al. Identification and characterization of the familial adenomatous polyposis coli gene. *Cell.* 1991;66(3):589–600.
130. Kinzler KW, Nilbert MC, Su LK, et al. Identification of FAP locus genes from chromosome 5q21. *Science.* 1991;253(5020):661–665.
131. Clements WM, Lowy AM, Grode J. Adenomatous polyposis coli/beta-catenin interaction and downstream targets: altered gene expression in gastrointestinal tumors. *Clin Colorectal Cancer.* 2003;3(2):113–120.
132. Drier Y, Lawrence MS, Carter SL, et al. Somatic rearrangements across cancer reveal classes of samples with distinct patterns of DNA breakage and rearrangement-induced hypermutability. *Genome Res.* 2013;23(2):228–235.
133. Rubinfeld B, Robbins O, El-Gamil M, Albert I, Porfiri E, Polakis P. Stabilization of beta-catenin by genetic defects in melanoma cell lines. *Science.* 1997;275(5307): 1790–1792.
134. Herr P, Hausmann G, Basler K. WNT secretion and signalling in human disease. *Trends Mol Med.* 2012;18(8):483–493.
135. Han ZG. Functional genomic studies: insights into the pathogenesis of liver cancer. *Annu Rev Genomics Hum Genet.* 2012;13:171–205.
136. Sastre-Perona A, Santisteban P. Role of the wnt pathway in thyroid cancer. *Front Endocrinol.* 2012;3:31.
137. Ratcliffe TA, Monk BJ, Planutis K, Holcombe RF. Wnt signaling in ovarian tumorigenesis. *Int J Gynecol Cancer.* 2008;18(5):954–962.
138. Niehrs C. The complex world of WNT receptor signalling. *Nat Rev Mol Cell Biol.* 2012;13(12):767–779.
139. Takahashi-Yanaga F, Kahn M. Targeting Wnt signaling: can we safely eradicate cancer stem cells? *Clin Cancer Res.* 2010;16(12):3153–3162.
140. LaBarge M. The difficulty of targeting cancer stem cell niches. *Clin Cancer Res.* 2010;16 (12):3121–3129.
141. Merchant AA, Matsui W. Targeting Hedgehog—a cancer stem cell pathway. *Clin Cancer Res.* 2010;16(12):3130–3140.
142. Pannuti A, Foreman K, Rizzo P, et al. Targeting Notch to target cancer stem cells. *Clin Cancer Res.* 2010;16(12):3141–3152.

143. Cicalese A, Bonizzi G, Pasi CE, et al. The tumor suppressor p53 regulates polarity of self-renewing divisions in mammary stem cells. *Cell.* 2009;138(6):1083–1095.
144. Lee JY, Nakada D, Yilmaz OH, et al. mTOR activation induces tumor suppressors that inhibit leukemogenesis and deplete hematopoietic stem cells after Pten deletion. *Cell Stem Cell.* 2010;7(5):593–605.
145. Ferent J, Cochard L, Faure H, et al. Genetic activation of Hedgehog signaling unbalances the rate of neural stem cell renewal by increasing symmetric divisions. *Stem Cell Rep.* 2014;3(2):312–323.
146. Kung AL, Rebel VI, Bronson RT, et al. Gene dose-dependent control of hematopoiesis and hematologic tumor suppression by CBP. *Genes Dev.* 2000;14(3):272–277.
147. Yamauchi T, Oike Y, Kamon J, et al. Increased insulin sensitivity despite lipodystrophy in Crebbp heterozygous mice. *Nat Genet.* 2002;30(2):221–226.
148. Roth JF, Shikama N, Henzen C, et al. Differential role of p300 and CBP acetyltransferase during myogenesis: p300 acts upstream of MyoD and Myf5. *EMBO J.* 2003;22 (19):5186–5196.
149. McMillan M, Kahn M. Investigating Wnt signaling: a chemogenomic safari. *Drug Discov Today.* 2005;10(21):1467–1474.
150. Emami KH, Nguyen C, Ma H, et al. A small molecule inhibitor of beta-catenin/CREB-binding protein transcription [corrected]. *Proc Natl Acad Sci USA.* 2004;101 (34):12682–12687.
151. Hasegawa K, Yasuda SY, Teo JL, et al. Wnt signaling orchestration with a small molecule DYRK inhibitor provides longterm xeno-free human pluripotent cell expansion. *Stem Cells Transl Med.* 2012;1(1):18–28.
152. Banerjee ER, Laflamme MA, Papayannopoulou T, et al. Human embryonic stem cells differentiated to lung lineage-specific cells ameliorate pulmonary fibrosis in a xenograft transplant mouse model. *PLoS One.* 2012;7(3):e33165.
153. Higuchi Y, Nguyen C, Yasuda S, et al. Specific direct small molecule p300/β-Catenin antagonists maintain stem cell potency. *Curr Mol Pharm.* 2016;9(3):272–279.
154. Zhao Y, Masiello D, McMillan M, et al. CBP/catenin antagonist safely eliminates drug-resistant leukemia-initiating cells. *Oncogene.* 2016;35(28):3705–3717.
155. Gang EJ, Hsieh YT, Pham J, et al. Small-molecule inhibition of CBP/catenin interactions eliminates drug-resistant clones in acute lymphoblastic leukemia. *Oncogene.* 2014;33(17):2169–2178.
156. Wend P, Fang L, Zhu Q, et al. Wnt/β-catenin signalling induces MLL to create epigenetic changes in salivary gland tumours. *EMBO J.* 2013;32(14):1977–1989.
157. Chan KC, Chan LS, Ip JC, et al. Therapeutic targeting of CBP/β-catenin signaling reduces cancer stem-like population and synergistically suppresses growth of EBV-positive nasopharyngeal carcinoma cells with cisplatin. *Sci Rep.* 2015;5:9979.
158. Mavila N, James D, Utley S, et al. Fibroblast growth factor receptor-mediated activation of AKT-β-catenin-CBP pathway regulates survival and proliferation of murine hepatoblasts and hepatic tumor. *PLoS One.* 2012;7(11):e50401.
159. Blyszczuk P, Müller-Edenborn B, Valenta T, et al. Transforming growth factor-β-dependent Wnt secretion controls myofibroblast formation and myocardial fibrosis progression in experimental autoimmune myocarditis. *Eur Heart J.* 2017;38 (18):1413–1425.
160. Osawa Y, Oboki K, Imamura J, et al. Inhibition of Cyclic Adenosine Monophosphate (cAMP)-response element-binding protein (CREB)-binding protein (CBP)/β-catenin initiating stem cells reduces liver fibrosis in mice. *EBioMedicine.* 2015;2(11): 1751–1758.
161. Sasaki T, Hwang H, Nguyen C, et al. The small molecule Wnt signaling modulator ICG-001 improves contractile function in chronically infarcted rat myocardium. *PLoS One.* 2013;8(9):e75010.

162. Beyer C, Reichert H, Akan H, et al. Blockade of canonical Wnt signalling ameliorates experimental dermal fibrosis. *Ann Rheum Dis.* 2013;72(7):1255–1258.
163. Hao S, He W, Li Y, et al. Targeted inhibition of β-catenin/CBP signaling ameliorates renal interstitial fibrosis. *J Am Soc Nephrol.* 2011;22(9):1642–1653.
164. Henderson Jr WR, Chi EY, Ye X, et al. Inhibition of Wnt/betacatenin/CREB binding protein (CBP) signaling reverses pulmonary fibrosis. *Proc Natl Acad Sci USA.* 2010;107 (32):14309–14314.
165. Ring A, Kim YM, Kahn M. Wnt/catenin signaling in adult stem cell physiology and disease. *Stem Cell Rev.* 2014;10(4):512–525.
166. El-Khoueiry AB, Ning Y, Yang D, et al. A phase I first-in-human study of PRI-724 in patients (pts) with advanced solid tumors. *J Clin Oncol.* 2013;31(15 suppl):2501.
167. Zhou HS, Carter BZ, Andreeff M. Bone marrow niche-mediated survival of leukemia stem cells in acute myeloid leukemia: Yin and Yang. *Cancer Biol Med.* 2016;13(2): 248–259.
168. McWilliams RR, Ko AH, Chiorean EG, et al. A phase Ib dose-escalation study of PRI-724, a CBP/beta-catenin modulator, plus gemcitabine (GEM) in patients with advanced pancreatic adenocarcinoma (APC) as second-line therapy after FOLFIRINOX or FOLFOX. *J Clin Oncol.* 2015;33(15 suppl):e15270.
169. Weerkamp F, van Dongen JJ, Staal FJ. Notch and Wnt signaling in T-lymphocyte development and acute lymphoblastic leukemia. *Leukemia.* 2006;20(7):1197–1205.
170. Belmonte M, Hoofd C, Weng AP, Giambra V. Targeting leukemia stem cells: which pathways drive self-renewal activity in T-cell acute lymphoblastic leukemia? *Curr Oncol.* 2016;23(1):34–41.
171. Guo Z, Dose M, Kovalovsky D, et al. Beta-catenin stabilization stalls the transition from double-positive to singlepositive stage and predisposes thymocytes to malignant transformation. *Blood.* 2007;109(12):5463–5472.
172. Guo W, Lasky JL, Chang CJ, et al. Multi-genetic events collaboratively contribute to Pten-null leukaemia stem-cell formation. *Nature.* 2008;453(7194):529–533.
173. Giambra V, Jenkins CE, Lam SH, et al. Leukemia stem cells in T-ALL require active Hif1alpha and Wnt signaling. *Blood.* 2015;125(25):3917–3927.
174. Ng OH, Erbilgin Y, Firtina S, et al. Deregulated WNT signaling in childhood T-cell acute lymphoblastic leukemia. *Blood Cancer J.* 2014;4:e192.
175. Laviano A. Young blood. *N Engl J Med.* 2014;371(6):573–575.
176. Sinha M, Jang YC, Oh J, et al. Restoring systemic GDF11 levels reverses age-related dysfunction in mouse skeletal muscle. *Science.* 2014;344(6184):649–652.
177. Hernandez L, Roux KJ, Wong ES, et al. Functional coupling between the extracellular matrix and nuclear lamina by Wnt signaling in progeria. *Dev Cell.* 2010;19(3):413–425.
178. Liu H, Fergusson MM, Castilho RM, et al. Augmented Wnt signaling in a mammalian model of accelerated aging. *Science.* 2007;317(5839):803–806.
179. Brack AS, Conboy MJ, Roy S, et al. Increased Wnt signaling during aging alters muscle stem cell fate and increases fibrosis. *Science.* 2007;317(5839):807–810.
180. Naito AT, Sumida T, Nomura S, et al. Complement C1q activates canonical Wnt signaling and promotes aging-related phenotypes. *Cell.* 2012;149(6):1298–1313.
181. Tomasetti C, Vogelstein B. Cancer etiology. Variation in cancer risk among tissues can be explained by the number of stem cell divisions. *Science.* 2015;347(6217):78–81.
182. Genovese G, Kähler AK, Handsaker RE, et al. Clonal hematopoiesis and blood-cancer risk inferred from blood DNA sequence. *N Engl J Med.* 2014;371(26):2477–2487.
183. Kim YM, Gang EJ, Kahn M. CBP/catenin antagonists: targeting LSCs' Achilles' heel. *Exp Hematol.* 2017;52:1–11.
184. Trosko JE, Kang KS. Evolution of energy metabolism, stem cells and cancer stem cells: how the Warburg and Barker hypotheses might be linked. *Int J Stem Cells.* 2012;5(1):39–56.

185. Holstein TW. The evolution of the Wnt pathway. *Cold Spring Harb Perspect Biol.* 2012;4 (7):a007922.
186. Markov GV, Laudet V. Origin and evolution of the ligand-binding ability of nuclear receptors. *Mol Cell Endocrinol.* 2011;334(1–2):21–30.
187. Lu B, Jan LY, Jan YN. Asymmetric cell division: lessons fromflies and worms. *Curr Opin Genet Dev.* 1998;8(4):392–399.
188. Lin R, Thompson S, Priess JR. Pop-1 encodes an HMG box protein required for the specification of a mesoderm precursor in early C. elegans embryos. *Cell.* 1995;83 (4):599–609.
189. Thorpe CJ, Schlesinger A, Carter JC, et al. Wnt signaling polarizes an early *C. elegans* blastomere to distinguish endoderm from mesoderm. *Cell.* 1997;90(4):695–705.
190. Liu J, Phillips BT, Amaya MF, et al. The *C. elegans* SYS-1 protein is a bona fide β-catenin. *Dev Cell.* 2008;14(5):751–761.
191. Phillips BT, Kimble J. A new look at TCF and beta-catenin through the lens of a divergent *C. elegans* Wnt pathway. *Dev Cell.* 2009;17(1):27–34.
192. McCrea PD, Gottardi CJ. Beyond beta-catenin: prospects for a larger catenin network in the nucleus. *Nat Rev Mol Cell Biol.* 2016;17(1):55–64.
193. Dillard AC, Lane MA. Retinol decreases beta-catenin protein levels in retinoic acid-resistant colon cancer cell lines. *Mol Carcinog.* 2007;46:315–329.
194. Szeto W, Jiang W, Tice DA, et al. Overexpression of the retinoic acid-responsive gene Stra6 in human cancers and its synergistic induction by Wnt-1 and retinoic acid. *Cancer Res.* 2001;61:4197–4205.
195. Koeffler HP, Leong G. Preleukemia: one name, many meanings. *Leukemia.* 2017;31 (3):534–542.

FURTHER READING

Arany Z, Sellers WR, Livingston DM, et al. E1A-associated p300 and CREB-associated CBP belong to a conserved family of coactivators. *Cell.* 1994;77(6):799–800.

Eckner R, Ewen ME, Newsome D, et al. Molecular cloning and functional analysis of the adenovirus E1A-associated 300-kD protein (p300) reveals a protein with properties of a transcriptional adaptor. *Genes Dev.* 1994;8(8):869–884.

Mi H, Muruganujan A, Thomas PD. PANTHER in 2013: modeling the evolution of gene function, and other gene attributes, in the context of phylogenetic trees. *Nucleic Acids Res.* 2013;41:D377–D886.

CHAPTER EIGHT

Chemical Modulation of WNT Signaling in Cancer

Li-shu Zhang, Lawrence Lum[1]
University of Texas Southwestern Medical Center, Dallas, TX, United States
[1] Corresponding author. E-mail address: Lawrence.lum@utsouthwestern.edu

Contents

Abstract

Genetically based observations stemming from defects in development and in regeneration form the foundation of our understanding regarding how the secreted WNT proteins control coordinated cell fate decision-making in adult tissues. At the same time, our anticipation of potential benefits and unwanted toxicities associated with candidate anticancer agents targeting WNT signal transduction are also reliant upon this blueprint of WNT-associated physiology. Despite the long established role of WNT signaling in cancer, the emergence of WNT signaling as a suppressor of immunological attack in melanoma reveals an unanticipated anticancer potential in targeting WNT signaling. Here we review the literature associated with WNT signaling in cancer and discuss potential challenges that may be associated with the chemical attack of

Progress in Molecular Biology and Translational Science, Volume 153
ISSN 1877-1173
http://dx.doi.org/10.1016/bs.pmbts.2017.11.008

this important cellular process in achieving therapeutic goals. Although a number of small molecules targeting WNT signaling are introduced here, we center our discussion on antagonists of the WNT acyltransferase porcupine (PORCN) given the recent entry of two candidate molecules in clinical testing.

1. INTRODUCTION

The 19 secreted WNT ligands constitute a major component in the repertoire of cell-to-cell signaling molecules that are essential to metazoan development and adult tissue regeneration. The endoplasmic reticulum-localized acyltransferase porcupine (PORCN) is essential to the lipidation of all 19 WNT proteins[1] (Fig. 1). In the absence of lipidation, WNT proteins are unable to engage the polytopic Wntless (WLS) chaperone protein and are thus sequestered in the secretory pathway.[2,3] PORCN is the founding member of the membrane-bound O-acyltransferase (MBOAT) gene family[4] and is unique among family members for using a monounsaturated fatty acid as a fatty acyl donor.[5]

Responses to WNTs are broadly categorized into those that induce activity of the transcriptional coactivator CTNNB1 (also known as β-catenin) and those that elicit other forms of cellular responses including those involving GTPases and G proteins.[6,7] Although CTNNB1 activation entails WNT-induced dimerization of a member of the LDL receptor protein (LRP) family and a Frizzled (FZD) family member at the membrane,[8] our understanding of the outcomes as a consequence of combinatorial WNT protein exposure remains limited. The threshold for WNT responses are also set by two transmembrane ubiquitin ligases (RNF43 and ZNRF3) that control the abundance of LRP and FZD receptors on the cell surface.[9–11] In turn, the secreted R-spondin (RSPO) proteins regulate the cell surface abundance of these transmembrane ubiquitin ligases[9,10] by a mechanism that entails their direct interaction.[9,12]

CTNNB1 levels and activity is regulated by large protein complex scaffolded by the AXIN and APC proteins[13] (Fig. 1). In the absence of WNT receptor activation, phosphorylation of CTNNB1 by members of the casein kinase (CSNK) and glycogen synthase kinase (GSK) found in the AXIN/APC complex promotes CTNNB1 ubiquitinylation and subsequent destruction by the proteasome. Activation of FZD/LRP receptor complexes on the other hand recruits AXIN/APC into a membrane-associated

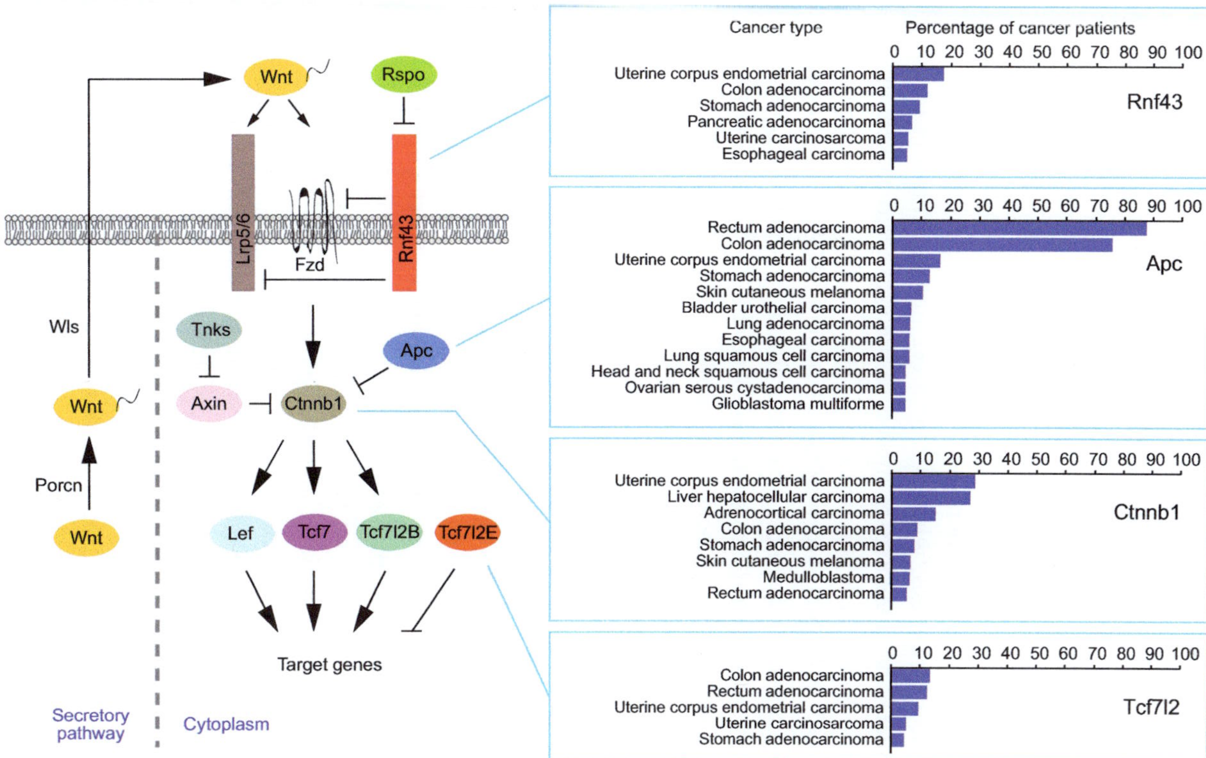

Fig. 1 Mutations in WNT pathway components found in solid tumors. Left: The Wnt signaling pathway includes 19 Wnt ligands that require porcupine (PORCN)-mediated lipidation and WLS for secretion, 10 Frizzled (FZD) receptors and 2 LDL-related protein (LRP) coreceptors, and 4 transcriptional regulators (TCF/LEF). The RSPO-RNF43 signaling axis regulates the turnover of LRP/FZD receptor complexes. APC and AXIN scaffolds a β-catenin destruction complex. TNKS enzymes promote AXIN destruction by an ADP-ribosylation-dependent mechanism. Right: Solid tumor types rank-ordered based on the frequency of pathway component mutations identified using The Cancer Genome Atlas (TCGA). Cancer types with mutations greater than ~5% incidence are shown. *The results shown here are in part based upon data generated by Genomic Data Commons Data Portal (portal.gdc.cancer.gov) managed by the National Cancer Institute. Data was retrieved on July 7, 2017.*

signaling complex thereby suppressing GSK3β activity within the complex.[13] The abundance of AXIN is a rate-limiting determinant of CTNNB1 metabolism particularly in cells with compromised APC activity.[14] The poly (ADP ribose) polymerase (PARP) family members tankyrases 1 and 2 (TNKS1 and TNKS2) promote AXIN protein destruction by a direct ADP-ribosylation-dependent mechanism (Fig. 1). Thus, activity of TNKS enzymes in addition to those of RNF43 and ZNRF3 set thresholds for cellular responsiveness to WNT ligands.

Mutations in signal transduction pathway components in the cancer cell-of-origin, or tumor microenvironment-derived sources of ligands frequently contribute to cancer biogenesis; whereas cancer cell intrinsic mutations in pathway components have seen some success as predictive biomarkers of anticancer agent response, such as in the case of Patched (PTCH) receptor mutations for small molecules targeting the Hedgehog (HH) pathway in basal cell carcinoma,[15] attempts to successfully target tumor--microenvironment interactions have been less successful.[16–18] Nevertheless more recent insights into how cancer cells communicate with other cells within a tumor or with stromal cells may give way to viable approaches for exploiting these interactions for therapeutic goals.[19–21]

The recent success of agents that disrupt immune cell suppression by tumors has galvanized research focused on a nonresidential component of the tumor microenvironment—the immune cells. Although studies focused on WNT signaling in blood cells have significantly contributed to our understanding of how WNT molecules transduce signals, their roles in the regeneration of epithelial cells and in cancer have largely fueled the development of small molecules targeting this pathway. The success of agents targeting immune "checkpoints" and the recent attribution of WNT signaling to the suppression of immune responses suggests a greater need to understand WNT signaling in immune cells. Indeed, the existence of many WNTs and the potential complexity of the responses in various immune cells to combinatorial exposure to these different WNTs constitute a code that remains mostly unsolved. Indeed, the role of WNT signaling in hematopoiesis and immune responses remains unclear but has important implications for the use of WNT inhibitors in the context of immunooncology.

WNT signaling dependent and independent of CTNNB1 participation appears to be important in immune surveillance and immune system development.[22,23] Small molecules targeting the WNT acyltransferase PORCN are able to inhibit both forms of WNT signaling.[24,25] Accounting also for the recent entry of two PORCN inhibitors into clinical testing, this chapter will focus on the potential utility of targeting PORCN in cancer and for eliciting more robust immune responses to tumors. At the same time, we discuss potential challenges to these therapeutic strategies posed by the systemic suppression of WNT signaling within adult tissues.

2. THE DEVELOPMENT OF PORCN INHIBITORS AS WNT SIGNALING ANTAGONISM

The frequent exploitation of WNT signaling for achieving unfettered cell growth underlies the sustained investment by academic and private research enterprises in the development of WNT pathway specific antagonists as anticancer agents (Fig. 1; Table 1). The primary goal in most of these endeavors is the realization of small molecules that disable deviant β-catenin-dependent transcriptional responses induced by loss of APC in >80% of colorectal cancers (CRCs) and gain of CTNNB1 function in ~27% of hepatocellular carcinoma (HCC).[26,27] Although to date no molecule with such activity has advanced into clinical practice, a better understanding of the mechanistic underpinnings of WNT signaling, as well as chemically tractable pathway components has emerged from these efforts.[28]

First-in-class chemicals that disrupt PORCN activity are the products of large chemical library screens in cultured cells focused on identifying novel intervention points in WNT signaling.[24,32] Subsequent medicinal chemistry efforts to develop PORCN inhibitors with drug-like properties revealed PORCN to be an attractive target given its capacity for accommodating inhibitors with picomolar activity and selectivity among evaluated members of the membrane bound O-acyltransferase (MBOAT) gene family. The basis of this chemical selectivity potentially lies in the strong preference of PORCN for monounsaturated Δ9 fatty acyl donors as compared to other MBOAT family members, such as Hedgehog acyltransferase (HHAT) and Ghrelin O-acyltransferase (GOAT) that lipidate protein substrates with saturated fatty acids.[5]

The ability to chemically disable the production of WNT ligands is associated with new opportunities for disabling cancer-promoting signals provided by the tumor microenvironment or from mutations that increase cancer cell responsiveness to ligands. For example, deleterious mutations in RNF43 and ZNRF3 result in increased cell surface expression of FZD/LRP6 receptors and elevated WNT signaling in a number of cancer types, including CRC, pancreatic ductal carcinoma (PDAC), and endometrial

Table 1 Summary of Wnt Inhibitors Discussed in the Chapter

Agents	Classes	Targets	Diseases	References
Wnt974	Chemical	PORCN	PDAC	29
			Ovarian cancer	30
			Lung cancers	20,31
			HNSCC	32
			Heart injury	33
C59	Chemical	PORCN	CC	34
			Melanoma	35
ETC-159	Chemical	PORCN	CRC	36
GNF-6231	Chemical	PORCN	Heart injury	37
G007-LK	Chemical	TNKS	CRC	38,39
XAV939	Chemical	TNKS	HCC	40
SSTC3	Chemical	CSNK1A1	CRC	41
ICG-001	Chemical	CTNNB1/CBP interaction	AML	42
			CC	34
PKF115-584	Chemical	CTNNB1/TCF interaction	MM	43
Indomethacin	Chemical	PTGS2	CML	44
IgG-2919	Antibody	FZD5/8	PDAC	45
anti-RSPO3	Antibody	RSPO3	CRC	46
OMP-18R5	Antibody	FZD1/2/5/7/8	Multiple cancers	47
OMP-131R10	Antibody	RSPO3	Multiple cancers	47
OMP-54F28	Decoy receptor	Multiple WNTs	Multiple cancers	47

AML, Acute myeloid leukemia; CC, cholangiocarcinoma; CML, chronic myeloid leukemia; CRC, colorectal cancer; HCC, hepatocellular carcinoma; HNSCC, head and neck squamous cell carcinoma; MM, multiple myeloma; PDAC, pancreatic ductal adenocarcinoma.

cancer.[9,10,48] The addition of PORCN inhibitors is able to reduce tumor cell growth in models of pancreatic ductal adenocarcinomas with *RNF43* mutations.[29] Confirming the on-target activity of PORCN inhibitors in these contexts is the tumor suppressing activity of an antibody targeting FZD5 in pancreatic ductal carcinoma.[45] In a similar vein, recurrent cancer-associated chromosomal translocations that result in the altered production of RSPO proteins are associated with 10% of colorectal cancer.[49–51] Cancerous cells harboring such mutations derived from the ovary or from the gut exhibited reduced growth when exposed to either a PORCN inhibitors[30] or an RSPO antibody,[46] respectively.

Despite the new therapeutic opportunities made available by PORCN inhibitors or biologics targeting extracellular components of WNT signaling, such as antibodies targeting RSPO or FZD, the call for agents capable of blocking *APC* mutation-induced CTNNB1 signaling remains unanswered. On the other hand, a number of cancers harboring other mutations in pathway components, including RPSO fusion proteins and loss of RNF43 is currently being evaluated with two PORCN-targeted clinical candidates—WNT974 (Novartis) and ETC-1922159 (D3).[28,52] Although not yet in testing either in combination with other agents or in a monotherapeutic setting, preclinical studies have shown response to PORCN inhibitors, including $BRAF^{V600E}$ and $KRAS^{G12D}$; $TRP53^{-/-}$ induced lung tumors[20,31] and human head and neck squamous cell carcinoma.[32] Notably absent is any report of studies focused on evaluating the in vivo activity of PORCN inhibitors on preclinical models of HCC, which is associated with frequent mutations in *CTNNB1*. Despite the shared end result of increased β-catenin transcriptional activity stemming from both *APC* and *CTNNB1* mutations, the ability of WNT ligands to influence β-catenin activity in each context remains unclear.

3. NORMAL TISSUE RESPONSES TO TRANSIENT WNT SIGNALING SUPPRESSION

3.1 Gut Toxicity and Chemical Inhibition of WNT Signaling

A number of genetically based observations suggest that the lining of the gut would be particularly sensitive to loss of WNT ligand production thereby imposing a dose limitation for the therapeutic use of PORCN inhibitors.[53] However, two observations suggest that a therapeutic window can be found with these agents. First, complete loss of cells expressing $LGR5^+$, a marker of WNT/β-catenin signaling stem cells in the gut, disrupts regeneration of the gut lining following injury (in this case induced by diphtheria toxin) but not homeostatic renewal[54] suggesting that chemical suppression of WNT signaling could be tolerated, but must be monitored. The source of this resilience may be another stem cell population marked by BMI1, a gene encoding a ring finger protein that functions in the polycomb group complex.[54,55] Second, in preclinical models there appears to be tolerance for systemic exposure to PORCN inhibitors at concentrations that are able to blunt cancerous cell growth or other nonvital WNT-dependent signaling events (e. g., hair regrowth).[32,33,56] The tolerability to PORCN inhibitors in contrast to another class of WNT inhibitors targeting the Tankyrase (TNKS) enzymes may lie in the polypharmacology of TNKS inhibitors, which appear to indirectly target β-catenin by controlling levels of the destruction complex scaffolding proteins AXIN1/2 but also the angiomotin (AMOT) proteins that in turn regulate HIPPO pathway gene transcription[57–60] and the cell polarity gene PALS.[57] The HIPPO transcriptional effector Yes Associated Protein 1 (YAP1), which is suppressed by AMOT family proteins is essential to gut epithelium repair following injury and in CRC initiation in mice.[61] Although Phase I clinical data for PORCN have yet to be disclosed, the entry of at least two PORCN inhibitors into clinical testing suggests that at doses sufficient to exhibit activity against certain types of tumors such compounds can be tolerated.[28,52] At the same time, clinical management protocols that avoid prolonged systemic exposure to PORCN inhibitors could be used to bypass gut toxicity associated with long-term suppression of WNT signaling.

3.2 Taste Sensation and WNT Signaling

The loss or alteration of taste sensation (ageusia and dysgeusia, respectively) is an unfortunate and common phenomenon associated with the use of chemotherapeutic agents.[62] Although oftentimes considered a "necessary evil" akin to hair loss that is associated with the management of more immediate pathophysiological crises stemming from cancer progression, the loss of normal taste sensation can influence protocol compliance in the clinical testing of cancer agents and in patients with diseases managed with FDA-approved drugs.[63] In more extreme instances, this could result in complications associated with malnutrition in patients. WNT signaling is essential to the formation of fungiform papilla and taste bud development and their innervation.[64–66] The level of WNT signaling likely controls the relative distribution of the three cell types found in taste buds from keratin 5 and 14-expressing proliferative basal keratinocytes that are found adjacent to taste buds.[67] In adults, the average half-life of a taste bud in mammals is 11 days with the receptor cells types having different life spans.[68] Our anticipation of how PORCN inhibitors might affect taste could be informed from human experience with agents targeting HH signaling, a pathway that is essential for taste bud renewal in adult tissues.[69] Loss of the HH pathway transcriptional regulation by manipulating the expression of levels of GLI1, 2, or 3 in epithelial cell populations of taste papillae results in the loss of taste bud cells.[69] The majority of patients on the Smoothened (SMO) antagonist Vismodegib report dysgeusia thus also suggesting HH signaling retains an important role in adult taste bud homeostasis.[70] The reversibility of some genetically induced loss of taste bud cells in mice suggests that at least some chemical effects associated with taste aberrations are also reversible.[69] Whereas any effects on taste bud renewal in patients treated with PORCN inhibitors will likely be an inevitable consequence of on-target drug activity, the druggability of PORCN and the importance of WNT signaling in taste bud renewal suggests that inadvertent disruption of PORCN activity by other chemotherapeutic agents could be identified using WNT-related assays and then neutralized using chemistry.

3.3 Hematopoiesis and WNT Signaling

Another rapidly dividing adult tissue that oftentimes sees unwanted effects with chemotherapeutic agents is blood. Given the essential role of WNT signaling in many aspects of development, much of what we know regarding the role of WNT signaling in hematopoiesis has been garnered from developing animals or hematopoietic stem cells (HSCs) derived from fetal liver.[71] Differences between developing and adults tissues likely account for some challenges associated with the use of observations from such studies in modeling the role of WNT signaling in adult blood cells.[72] In addition, dose-dependent variability in WNT signaling loss based on the use of different genetic approaches has also hindered the ability to gain consensus among researchers regarding the role of WNT signaling in adult hematopoeisis.[73] We frame our discussion regarding the potential impact of PORCN inhibitors on adult HSC function upon a few studies that interrogate WNT signaling in adult animals. Elimination of PORCN expression using an inducible gene knock-out strategy in hematopoietic cells or prolonged exposure to a PORCN inhibitor failed to elicit changes in hematopoiesis suggesting that WNT signaling is not essential for this function.[74] These observations are consistent with the absence of hematopoietic defects in animals lacking CTNNB1 in HSCs.[75] On the other hand, the previous assignment for a role of WNT3A in hematopoeisis may be due to the use of fetal liver derived HSCs and not adult cells from bone.[76] The gain of CTNNB1 activity in HSCs due to the introduction of truncating mutations in APC suggest that TCF/LEF proteins can contribute to hematopoiesis in adult tissues. Yet, these observations fail to model the potential action of

PORCN inhibitors in HSCs given these compounds would suppress WNT controlled CTNNB1 signaling and likely would not impact non-WNT signaling related functions controlled by APC.[73]

Despite some controversy regarding the role of Wnt signaling in adult hematopoiesis, there is general consensus that the production of blood cells during stress or following injury (so called emergency hematopoiesis) is dependent upon WNT signaling.[77–80] Studies using animals with either loss and gain of CTNNB1 activity in HSCs point to an essential role of WNT/CTNNB1 signaling in the formation of granulocyte-macrophage progenitor (GMP) clusters following HSC ablation with the thymidine synthesis inhibitor 5 fluorouracil (5-FU).[81] These GMP clusters appear to be essential for replenishing the myeloid cells following catastrophic loss. Whether PORCN antagonists are able to recapitulate the effects of *CTNNB1* loss is an important question to address particularly in cancer patients managed with a WNT pathway antagonists and who often see catastrophic loss of HSCs as a consequence of other chemotherapeutic agents. Other genetically based observations that supports a role of WNT ligand-mediated signaling in emergency hematopoiesis include the inability of bone marrow cells lacking the WNT receptor Frizzled 6 (FZD6) to competitively repopulate blood cells in lethally irradiated animals.[82] Thus, the effects of WNT pathway antagonists, such as PORCN inhibitors particularly on HSC counts in postchemotherapy patients should be monitored.

4. WNT INHIBITORS AND THEIR POTENTIAL UTILITY IN SOLID TUMORS

The prevalence of mutations in known WNT pathway components in a number of solid tumor types, including CRC, HCC, and PDAC has revealed the importance of WNT signaling in cell growth regulation in certain tissues (Fig. 1). Despite a shared impact of these mutations on TCF/LEF-mediated transcriptional regulation, different tumor types appear to be enriched for certain mutations thus suggesting either a tissue-specific susceptibility of certain genomic regions to mutagenesis, or tissue-specific roles of these genes either in WNT signaling or other cellular processes. For example, >80% of CRCs harbor mutations in *APC*, whereas nearly no mutations in the same gene are found in HCC[26,27] (Fig. 1). On the other hand, gain-of-function mutations in *CTNNB1* are found in 27% of HCC cases and only 5% of CRC cases. In this section, we discuss major target cancer types that constitute litmus tests for the anticancer utility of WNT pathway antagonists.

4.1 Countering the Cancer-Promoting Effects of RNF43 and ZNRF3 Mutations

The type I transmembrane RNF43 and ZNRF3 ubiquitin ligases control the turnover of WNT receptors in a number of tissues and thus control WNT ligand-dependent signaling.[9–11] Unlike the majority of CRCs, traditional serrated adenomas of the gut are not associated with *APC* mutations but rather *RNF43* mutations coupled with microsatellite instability.[50,83–85] Other cancer types that also see common mutations in *RNF43* include pancreatic ductal carcinoma,[29,86] endometrial cancer,[84] and mucinous tumors of the ovary[87] (Fig. 1). In addition to showing responsiveness in mouse to PORCN inhibitors, pancreatic cancer cells with *RNF43* mutations are also responsive to antibodies targeting FZD5.[45] These antibodies at high doses also result in loss of gut epithelium integrity. The similarity in gut phenotype to that observed in mice lacking another FZD receptor (*FZD7*) reveals nonredundant roles of FZD proteins in gut stem cells.[88] Indeed, the code for WNT/FZD interaction specificities (19 WNTs with 10FZDs) remains incomplete although recent efforts have begun to make important strides toward addressing this gap in knowledge.[89] Despite some challenges associated with bone toxicity, the multi-FZD neutralizing antibody OMP-18R5 (vantictumab) and the WNT decoy receptor protein OMP-54F28 (Fzd8-Fc; ipafricept) may also be useful against RNF43-associated cancers.[47]

4.2 A Long-Standing Challenge: APC Mutations and Colorectal Cancer

Studies using genetic approaches have hinted that eliminating CTNNB1/TCF activity could be effective at slowing the progression of CRC induced by loss of APC[13] (Fig. 1). Restoration of APC function in gut epithelial cells following transformation (loss of APC and TP53 function and gain of KRAS activity) is able to restore crypt homeostasis and normal levels of WNT signaling in mice.[90] The long-standing hypothesis that targeting CTNNB1/TCF signaling in cells with compromised *APC* function would disable deviant cell growth has now also been chemically evaluated. An inhibitor of the TNKS enzymes (G007-LK) is able to block CTNNB1 activity and gut polyposis in *APC* compromised mice and in xenografted CRC cells harboring truncation mutations in APC.[38,39] The antigrowth activity of TNKS inhibitors appears to be most robust in CRC cells lacking all seven CTNNB1 binding repeats in APC thus providing a potential biomarker of drug sensitivity.[91] TNKS inhibitors also induce gut tissue toxicity at concentrations required to inhibit tumor growth thereby rendering these molecules potentially difficult to develop for clinical use.[92,93] Thus, current strategies for targeting CTNNB1 activity using TNKS appear to exert anticancer activity but also exhibit toxicities that could be addressed by eliminating their polypharmacology. At the same time, a report that some APC mutations may continue to sustain ligand-dependent WNT pathway activity could provide a rationale for pursuing therapeutic opportunity for PORCN inhibitors in some CRCs.[94]

Other approaches in preclinical evaluation that may neutralize *APC* mutation-induced β-catenin activity and that may bypass the toxicity associated with TNKS inhibitors include small molecule activation of CK1α, a kinase that promotes CTNNB1 degradation, targeting metabolic changes induced by the expression of truncated APC proteins, or promoting read-through of the premature termination codons found in *APC* using the aminogylcoside geneticin.[41,92,95] In the case of CK1α agonists, the absence of effect on the gut is similar to that seen in preclinical studies of PORCN inhibitors, which have seen success in entering clinical testing. Further studies will be required to see if these agents, such as PORCN inhibitors at higher concentrations also result in similar drug toxicities and whether these can be tolerated in human patients. Notably, the gut toxicities induced by WNT inhibitors appear to be reversible.[93]

The four members of the R-spondin (RSPO) gene family constitute a family of secreted signaling molecules that enhance WNT signaling in development and intestinal homeostasis by in part engaging a receptor complex consisting of RNF43 or ZNRF3 and the stem cell marker LGR5, a gene controlled by WNT/CTNNB1 signaling.[9,10] RSPOs induce the cell surface clearance of RNF43/ZNRF3, thereby facilitating WNT receptor abundance and WNT-ligand mediated signaling.[10,48] Despite its recognition as cell surface molecule, RNF43 may also have an intracellular function in suppressing WNT signaling by promoting nuclear membrane sequestration of TCF7L2 (also known as TCF4).[96] An antibody targeting anti-RSPO3 (OMP-131R10) currently in clinical testing[47] may thus exert activity against deviant growth of *APC* mutant cells by increasing the nuclear membrane localized RNF43 thereby suppressing the activity of TCF7L2 isoforms and potentially other TCF/LEF family members.[97]

4.3 Cancer-Associated Mutations in CTNNB1

Akin to the regenerative role of WNT signaling in the gut epithelium, WNT signaling is essential to the regeneration of hepatocytes.[11,98] Consistent with the importance of this pathway in liver homeostasis, ~35% of liver hepatocellular tumors are associated with mutations in either *CTNNB1* or *AXIN1* (Fig. 1). In *CTNNB1*, these mutations typically disrupt DNA sequence encoding the GSK3β phosphorylation site that promotes CTNNB1 turnover.[99] Given the shared effects of *APC* and *CTNNB1* mutations on promoting deviant TCF/CTNNB1 activity, a common assumption is that the activity of WNT antagonists in *APC* mutant cells would mirror that in cells with mutations in *CTNNB1*; whereas mutations that result in C-terminally truncated APC protein have variable effects on both CTNNB1 subcellular distribution and protein abundance, mutations that disrupt the CTNNB1 destruction sequence in theory would only affect protein turnover but leave intact mechanism that regulate its nuclear abundance. Indeed, TNKS inhibitors exert activity in HCC cell lines with annotated *CTNNB1* mutations.[40] Although in vitro studies suggest that CRC cells with a CTNNB1 mutation may still rely on ligand-dependent WNT signaling for growth, in vivo evaluation of PORCN inhibitors in liver cancers with *CTNNB1* mutations remains untested.[94] Approximately 10%–20% of primary liver cancers are cholangiocarcinomas (CC; bile duct cancers), a deadly disease with a limited understanding of its genetic etiology.[100] The reliance of CCs on increased activity of WNT7B and WNT10 appears to be a vulnerability to PORCN inhibitors, as well as another class of WNT inhibitors targeting the interaction between CTNNB1 and the transcriptional coactivator CBP.[34] *CTNNB1* mutations are also frequently found in the WNT subtype of medulloblastomas (86%)[101] (Fig. 1).

4.4 TCF7L2 and Cancer

The DNA binding protein TCF7L2 (TCF4) has dual roles in WNT-dependent signaling as both positive and negative transcriptional regulator—functions that are dependent on the presence of isoform-specific protein domains that mediate interactions with various regulatory proteins.[102,103] Initial assigned a role as a facilitator of tumorigenesis in the gut, genetically based observations in cultured cells and in vivo have since suggested a primary role as a tumor suppressor.[103,104] The broad distribution of CRC-associated mutations found in the coding sequence *TCF7L2* support a frequent loss of function in cancer and also its assignment as a tumor suppressor in this disease (Fig. 1). The absence of a CTNNB1 binding domain in TCF7L2 fusion proteins generated by chromosomal translocations in CRC also speaks to loss of function events in the case of *TCF7L2*.[105] Typically found in tumors with mutations in other WNT pathway components is unclear whether loss of *TCF7L2* function merely increases the activity of other positively acting TCF/LEF family members or confers new transcriptional dependencies that cannot be neutralized with WNT antagonists.[26]

4.5 Other Genetic Indications of Deviant WNT Signaling

A disproportionate investment in monotherapeutic cancer management strategies is a reasonable front line approach to securing a clinical development path for most candidate agents. On the other hand, codependencies of traditional tumor drivers with WNT signaling may provide combinatorial therapeutic rationale for WNT antagonists. For example, *RNF43* mutations are frequently associated with *BRAF* mutations in colonic serrated adenomas with one study reporting 87% of microsatellite instability tumors with mutations found in both genes.[83,106] The ability of a PORCN inhibitor to disrupt lung tumor growth in mice induced by BRAFV600E by blunting the expression of the TCF/LEF target gene and protooncogene MYC suggests that other well-characterized tumor drivers may be similarly dependent upon WNT-driven MYC expression.[31]

Chromosomal rearrangements that give rise to fusion proteins that incorporate RSPO genes (EIF3E-RSPO2 and PTPRK-RSPO3) are frequently associated with traditional serrated adenomas.[49–51] Cancerous cells with RSPO fusion proteins are growth-sensitive to RSPO antibodies[46] and PORCN inhibitors[36] suggesting that the mutational status of RSPO genes will likely be strong drug efficacy prognosticators for agents targeting WNT and RSPO proteins despite their distinct roles in gut stem cell renewal.[107]

The identification of mutations in the *CTNNB1* promoter region in breast cancer suggests that likely other cancer types will also exploit deviant WNT signaling for cell growth and that noncoding regions should be more deeply mined for predictive markers of WNT inhibitor responses.[108] Indeed, the identification of many mutations in noncoding DNA sequences regulating the expression of telomerase (TERT), for example, has reinvigorated efforts to target this hallmark of cancer.[109] Yet, given the absence of alterations in the coding sequence of WNT pathway components associated with changes in gene regulatory elements, chemical-genetic interactions previously established for WNT pathway inhibitors using landmark mutations in *APC* and *CTNNB1*, for example, would have to be reevaluated for their anticancer potential.

5. WNT INHIBITORS AND THEIR POTENTIAL UTILITY IN LIQUID TUMORS

In contrast to solid tumors, blood cell cancers offer few genetically based clues in human tissue that directly associate deviant activation of WNT signaling with aberrant cell growth. Yet, genetic observations in mice indicate a collaborative role between well-established drivers of chronic myeloid leukemia (CML) or acute myeloid leukemia (AML) and CTNNB1 in influencing drug efficacy or leukemic stem cell (LSC) renewal, respectively.[44,110] Efforts to chemically evaluate the therapeutic relevance of these observations have included the testing of COX-2 inhibitors in a BCR-ABL driven mouse model of CML.[44] However, any enthusiasm for the use of COX-2 inhibitors in WNT-related cancers generated from the mixed results of clinical trials evaluating Celecoxib in familial adenomatous polyposis (FAP) patients[111] have been offset by the adverse cardiovascular events seen with at least one COX-2 inhibitor (Rofecoxib). Yet another approach undergoing clinical testing as an antileukemic strategy entails disrupting Src associated with mitosis 68K protein (Sam68) activity within the WNT pathway in several forms of hematological malignancies.[42,47] Based on the success of this study, other WNT antagonists programs with the goal of disabling leukemia may be initiated.

PORCN inhibitors or biologics targeting extracellularly localized components of WNT signaling may be useful in leukemias if ligand dependencies can be first established in these diseases. Although a contribution of CTNNB1 in murine models of leukomagenesis has been established, the role of CTNNB1, WNT ligands, and WNT receptors in cancers of blood cells remains unclear.[81,110] Self-sufficiency with respect to WNT signaling appears to be important in the colonization of distal tissue in other cancers, such as breast- and lung cancer.[112,113] In these cases, the abandonment of niche provided WNT signaling made possible by cell-autonomously produced WNT ligands may be an important step toward metastasis. This may be also true in cancers of blood cells as well. Indeed, intrinsic WNT signaling has been reported in both multiple myeloma (MM) and AML.[114,115] In the case of the MM, a molecule targeting CTNNB1/TCF interactions (PKF115-584) prolongs survival in a mouse model of the disease.[43]

6. WNT AND IMMUNOONCOLOGY

Long-standing efforts to enlist immune cells to attack cancerous cells have seen recent successes with the discovery of agents that disengage immune regulatory check-point mechanisms, such as those controlled by CTLA-4 and PD1/PDL1.[116] Dendritic cells presenting cancer-associated antigens contribute to the ability of $CD8^+$ T lymphocytes to initiate and sustain an attack on cancerous cells.[117,118] A number of extracellular signaling and cell adhesion molecules, including WNT ligands, Toll-like receptors (TLRs), and E-cadherin, appear to converge on CTNNB1-dependent regulation of a tolerogenic state in dendritic cells (DCs), which are professional antigen presenting cells[119] (Fig. 2). Indeed, genetically based ablation of LRP5/6 from DCs results in heightened immune response to transplanted melanoma cells.[120] At least in the case of melanoma, cancer cell-autonomous WNT signaling appears to be a barrier to immune cell infiltration and activation.[121,122] In the study by Spranger *et al.*, an impressive 48% of nonimmune cell infiltrated tumors have a gain of WNT signaling, a caveat is that the gene set (with the exception of the cMYC, which is frequently found elevated in cancer) used to define a signature of WNT signaling and to stratify melanomas (*EFNB3*, *APC2*, *TCF1*, *c-MYC*, *TCF12*, *VEGFA*) are not well established WNT pathway target genes. WNT signaling may also be involved in suppressing inflammatory responses in other cancer types given that activation of β-catenin in dendritic cells regulates immunity versus tolerance in the intestine.[123]

Nevertheless, based on studies using genetically engineered mouse models or xenotransplantation approaches, WNT pathway antagonists may be useful for suppressing a tolerogenic state in the host[121,122] (Fig. 2). Of the different classes of WNT inhibitors that have been identified, PORCN inhibitors may be particularly useful if previous studies reporting WNT-mediated interactions between cellular components of the cancer microenvironments and DCs can be confirmed in vivo.[124–129] Indeed, a PORCN inhibitor is able to increase immunogenic responses in mice hosting transplanted human melanoma cells.[35] Furthermore, toxicities associated with long-term exposure to PORCN inhibitors will likely be mitigated when codelivered with an immune checkpoint inhibitor which are routinely administered once every two weeks.[130] However, given the potential role of WNT signaling in $CD8^+$ lymphocyte production and persistence,

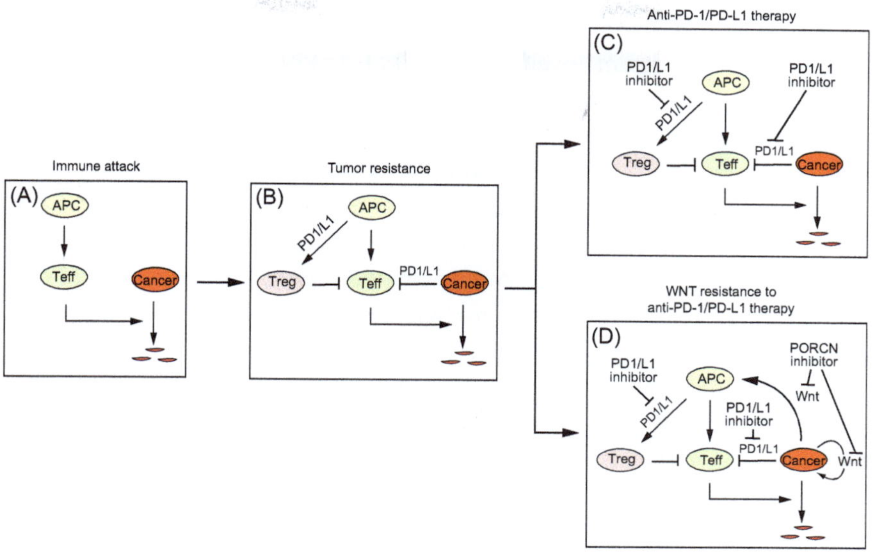

Fig. 2 Eliciting a durable response to immune checkpoint inhibitors by disabling Wnt signaling. (A) Antigen presenting cells (APC) present cancer cell-associated peptides to T cells, including cytotoxic T cells and T helper cells. Activated T cells further differentiate and proliferate into T effector (Teff) cells to counteract cancer cells through cytotoxic T lymphocyte killing or/and antibody-dependent cell-mediated cytotoxicity. (B) Expressing the ligand of programmed cell death 1 (PD-1) receptor PD-L1 in tumors or APCs induces immune tolerance by engaging PD-1 in Treg or Teffs. (C) Inhibitors that disrupt PD1/PD-L1 interactions counter immune tolerance. (D) WNT ligands produced by cancerous cells either signaling autonomously or to APCs promote Treg formation either by promoting a tolerogenic state in APCs or by suppressing APC recruitment thus suppressing the formation of Teffs. In either case, the efficacy of PD1/PD-L1 inhibitors is blunted. Inhibitors of WNT signaling particularly those targeting PORCN could be used to reverse this process.

systemic use of PORCN inhibitors may also have unintentional consequences that may counteract durable anticancer effects using this strategy.[22,131,132]

7. LOOKING AHEAD

Solid tumors particularly those that originate in the gut have provided the major framework and motivation for developing agents that target WNT signaling. Indeed, the spectra of mutations in CRC typically found today with the help of genomic sequencing would have provided sufficient clues as to assemble our current inventory of WNT/CTNNB1 signaling components. Yet, this focus on the potential benefits of suppressing WNT signaling in solid tumors has also to some extent disenfranchised the pursuit of other utilities for chemical modulators for WNT signaling particularly in cases where genetic evidence is not as readily available as in solid tumors. For example, the epigenetically driven transcriptional storms that rewire cellular signaling pathways commonly found in blood tumors challenges our ability to identify prognostic indicators for when such inhibitors might be effectively deployed.

Perhaps an equally important lesson learned from years devoted to drugging the WNT pathway is how to avoid its unintentional disruption by small molecules. As in the case of SMO in the HH pathway, PORCN appears to be highly druggable with determinants that enable selective chemical targeting.[5] Potentially unwanted side effects due to PORCN inhibition, such as dysgeusia could be engineered out of FDA approved drugs or avoided at the preclinical drug development phase by eliminating pharmacophores that engage PORCN active site features. Here, access to a structure, such as in the case of SMO[133] or a stronger understanding of consensus chemical features that support the selective targeting of the PORCN active site[5] would be helpful. Incorporating robust assays for PORCN or SMO activity during medicinal chemistry due diligence may also be useful. Altering parameters in the cycling of the SMO inhibitor Vismodegib exposure (e.g., 12 weeks on; 8 weeks off; 12 weeks on, compared to 24 weeks on; 8 weeks off; 8 weeks on) appears to improve protocol adherence and to reduce adverse effects with improvement in BCC tumor shrinkage in the former.[134]

After many decades of observing the ravages of deviant WNT signaling on adult tissue homeostasis, we are now poised with precision reagents in hand to test the hypothesis in humans that disabling certain WNT pathway components would provide therapeutic benefits. Unclear at the moment is how these reagents would be best deployed in disease (e.g., monotherapeutic approaches based on predictive biomarkers or in combination with immune checkpoint inhibitors). At the same time, some regenerative medicine utilities premised upon disrupting WNT signaling might require short systemic exposure to antagonists therefore bypassing some of the toxicity issues discussed here. For example, a brief exposure to PORCN inhibitors following a myocardial infarction event may be sufficient to blunt fibrotic responses and improve heart function after injury.[33,37] What is already evident is that the lessons learned from both regenerative medicine and anticancer efforts using WNT pathway modulators will be pivotal to our ability to safely use such agents for therapeutic ends.

ACKNOWLEDGEMENTS

We would like to thank Drs. James Kim and Jesung Moon for their insightful comments. This work was supported in part by the Welch Foundation (I-1665) and CPRIT (RP130212). Research reported in this publication was supported by the National Cancer Institute of the National Institutes of Health under Award Number R01CA168761, R01CA196851, and P50-CA70907 (Minna). The content is solely the responsibility of the authors and does not necessarily represent the official views of the National Institutes of Health.

REFERENCES

1. Gross JC, Boutros M. Secretion and extracellular space travel of Wnt proteins. *Curr Opin in Genet Dev.* 2013;23(4):385–390.
2. Banziger C, Soldini D, Schutt C, Zipperlen P, Hausmann G, Basler K. Wntless, a conserved membrane protein dedicated to the secretion of Wnt proteins from signaling cells. *Cell.* 2006;125(3):509–522.
3. Bartscherer K, Pelte N, Ingelfinger D, Boutros M. Secretion of Wnt ligands requires Evi, a conserved transmembrane protein. *Cell.* 2006;125(3):523–533.
4. Hofmann K. A superfamily of membrane-bound O-acyltransferases with implications for wnt signaling. *Trends Biochem Sci.* 2000;25(3):111–112.
5. Tuladhar R, Lum L. Fatty acyl donor selectivity in membrane bound O-acyltransferases and communal cell fate decision-making. *Biochem Soc Trans.* 2015;43(2):235–239.
6. Angers S, Moon RT. Proximal events in Wnt signal transduction. *Nat Rev.* 2009;10 (7):468–477.
7. van Amerongen R. Alternative Wnt pathways and receptors. *Cold Spring Harb Perspect Biol.* 2012;4(10), a007914.
8. MacDonald BT, He X. Frizzled and LRP5/6 receptors for Wnt/beta-catenin signaling. *Cold Spring Harb Perspect Biol.* 2012;4(12), a007880.
9. Hao HX, Xie Y, Zhang Y, et al. ZNRF3 promotes Wnt receptor turnover in an R-spondin-sensitive manner. *Nature.* 2012;485(7397):195–200.
10. Koo BK, Spit M, Jordens I, et al. Tumour suppressor RNF43 is a stem-cell E3 ligase that induces endocytosis of Wnt receptors. *Nature.* 2012;488(7413):665–669.
11. Planas-Paz L, Orsini V, Boulter L, et al. The RSPO-LGR4/5-ZNRF3/RNF43 module controls liver zonation and size. *Nat Cell Biol.* 2016;18(5):467–479.
12. Zebisch M, Jones EY. Crystal structure of R-spondin 2 in complex with the ectodomains of its receptors LGR5 and ZNRF3. *J Struct Biol.* 2015;191(2):149–155.
13. Nusse R, Clevers H. Wnt/beta-catenin signaling, disease, and emerging therapeutic modalities. *Cell.* 2017;169(6):985–999.
14. Lee E, Salic A, Kruger R, Heinrich R, Kirschner MW. The roles of APC and Axin derived from experimental and theoretical analysis of the Wnt pathway. *PLoS Biol.* 2003;1(1):E10.
15. Basset-Seguin N, Sharpe HJ, de Sauvage FJ. Efficacy of Hedgehog pathway inhibitors in Basal cell carcinoma. *Mol Cancer Ther.* 2015;14(3):633–641.
16. Berlin J, Bendell JC, Hart LL, et al. A randomized phase II trial of vismodegib versus placebo with FOLFOX or FOLFIRI and bevacizumab in patients with previously untreated metastatic colorectal cancer. *Clin Cancer Res.* 2013;19(1):258–267.

17. Catenacci DV, Junttila MR, Karrison T, et al. Randomized phase Ib/II study of gemcitabine plus placebo or vismodegib, a Hedgehog pathway inhibitor, in patients with metastatic pancreatic cancer. *J Clin Oncol.* 2015;33(36):4284–4292.
18. Kaye SB, Fehrenbacher L, Holloway R, et al. A phase II, randomized, placebo-controlled study of vismodegib as maintenance therapy in patients with ovarian cancer in second or third complete remission. *Clin Cancer Res.* 2012;18(23):6509–6518.
19. Lim JS, Ibaseta A, Fischer MM, et al. Intratumoural heterogeneity generated by Notch signalling promotes small-cell lung cancer. *Nature.* 2017;545(7654):360–364.
20. Tammela T, Sanchez-Rivera FJ, Cetinbas NM, et al. A Wnt-producing niche drives proliferative potential and progression in lung adenocarcinoma. *Nature.* 2017;545 (7654):355–359.
21. Zhao C, Cai S, Shin K, et al. Stromal Gli2 activity coordinates a niche signaling program for mammary epithelial stem cells. *Science.* 2017;356(6335).
22. Staal FJ, Arens R. Wnt signaling as master regulator of T-lymphocyte responses: implications for transplant therapy. *Transplantation.* 2016;100(12):2584–2592.
23. Staal FJ, Luis TC, Tiemessen MM. WNT signalling in the immune system: WNT is spreading its wings. *Nat Rev Immunol.* 2008;8(8):581–593.
24. Chen B, Dodge ME, Tang W, et al. Small molecule-mediated disruption of Wnt-dependent signaling in tissue regeneration and cancer. *Nat Chem Biol.* 2009;5 (2):100–107.
25. Najdi R, Proffitt K, Sprowl S, et al. A uniform human Wnt expression library reveals a shared secretory pathway and unique signaling activities. *Differentiation.* 2012;84 (2):203–213.
26. Cancer Genome Atlas Network.Comprehensive molecular characterization of human colon and rectal cancer. *Nature.* 2012;487(7407):330–337.
27. Cancer Genome Atlas Network.Comprehensive and integrative genomic characterization of hepatocellular carcinoma. *Cell.* 2017;169(7):1327–1341.
28. Lum L, Chen C. Chemical disruption of Wnt-dependent cell fate decision-making mechanisms in cancer and regenerative medicine. *Curr Med Chem.* 2015;22 (35):4091–4103.
29. Jiang X, Hao HX, Growney JD, et al. Inactivating mutations of RNF43 confer Wnt dependency in pancreatic ductal adenocarcinoma. *Proc Natl Acad Sci USA.* 2013;110 (31):12649–12654.
30. Boone JD, Arend RC, Johnston BE, et al. Targeting the Wnt/beta-catenin pathway in primary ovarian cancer with the porcupine inhibitor WNT974. *Lab Invest.* 2016;96 (2):249–259.
31. Juan J, Muraguchi T, Iezza G, Sears RC, McMahon M. Diminished WNT → beta-catenin → c-MYC signaling is a barrier for malignant progression of BRAFV600E-induced lung tumors. *Genes Dev.* 2014;28(6):561–575.
32. Liu J, Pan S, Hsieh MH, et al. Targeting Wnt-driven cancer through the inhibition of porcupine by LGK974. *Proc Natl Acad Sci USA.* 2013;110(50):20224–20229.
33. Moon J, Zhou H, Zhang LS, et al. Blockade to pathological remodeling of infarcted heart tissue using a porcupine antagonist. *Proc Natl Acad Sci USA.* 2017;114 (7):1649–1654.
34. Boulter L, Guest RV, Kendall TJ, et al. WNT signaling drives cholangiocarcinoma growth and can be pharmacologically inhibited. *J Clin Invest.* 2015;125 (3):1269–1285.
35. Holtzhausen A, Zhao F, Evans KS, et al. Melanoma-derived Wnt5a promotes local dendritic-cell expression of IDO and immunotolerance: opportunities for pharmacologic enhancement of immunotherapy. *Cancer Immunol Res.* 2015;3(9):1082–1095.
36. Madan B, Ke Z, Harmston N, et al. Wnt addiction of genetically defined cancers reversed by PORCN inhibition. *Oncogene.* 2016;35(17):2197–2207.

37. Bastakoty D, Saraswati S, Joshi P, et al. Temporary, systemic inhibition of the WNT/beta-catenin pathway promotes regenerative cardiac repair following myocardial infarct. *Cell Stem Cells Regen Med.* 2016; 2(2). Available from: https://doi.org/10.16966/2472-6990.111
38. Lau T, Chan E, Callow M, et al. A novel tankyrase small-molecule inhibitor suppresses APC mutation-driven colorectal tumor growth. *Cancer Res.* 2013;73(10):3132–3144.
39. Waaler J, Machon O, Tumova L, et al. A novel tankyrase inhibitor decreases canonical Wnt signaling in colon carcinoma cells and reduces tumor growth in conditional APC mutant mice. *Cancer Res.* 2012;72(11):2822–2832.
40. Ma L, Wang X, Jia T, Wei W, Chua MS, So S. Tankyrase inhibitors attenuate WNT/beta-catenin signaling and inhibit growth of hepatocellular carcinoma cells. *Oncotarget.* 2015;6(28):25390–25401.
41. Li B, Orton D, Neitzel LR, et al. Differential abundance of CK1alpha provides selectivity for pharmacological CK1alpha activators to target WNT-dependent tumors. *Sci Signal.* 2017;10(485), eaak9916.
42. Benoit YD, Mitchell RR, Risueno RM, et al. Sam68 allows selective targeting of human cancer stem cells. *Cell Chem Biol.* 2017;24(7):833–844. e839.
43. Sukhdeo K, Mani M, Zhang Y, et al. Targeting the beta-catenin/TCF transcriptional complex in the treatment of multiple myeloma. *Proc Natl Acad Sci USA.* 2007;104(18):7516–7521.
44. Heidel FH, Bullinger L, Feng Z, et al. Genetic and pharmacologic inhibition of beta-catenin targets imatinib-resistant leukemia stem cells in CML. *Cell stem cell.* 2012;10(4):412–424.
45. Steinhart Z, Pavlovic Z, Chandrashekhar M, et al. Genome-wide CRISPR screens reveal a Wnt-FZD5 signaling circuit as a druggable vulnerability of RNF43-mutant pancreatic tumors. *Nat Med.* 2017;23(1):60–68.
46. Storm EE, Durinck S, de Sousa e Melo F, et al. Targeting PTPRK-RSPO3 colon tumours promotes differentiation and loss of stem-cell function. *Nature.* 2016;529(7584):97–100.
47. Tai D, Wells K, Arcaroli J, et al. Targeting the WNT signaling pathway in cancer therapeutics. *Oncologist.* 2015;20(10):1189–1198.
48. Hao HX, Jiang X, Cong F. Control of Wnt receptor turnover by R-spondin-ZNRF3/RNF43 signaling module and its dysregulation in cancer. *Cancers.* 2016;8(6.):E54.
49. Han T, Schatoff EM, Murphy C, et al. R-spondin chromosome rearrangements drive Wnt-dependent tumour initiation and maintenance in the intestine. *Nat Commun.* 2017;8:15945.
50. Sekine S, Yamashita S, Tanabe T, et al. Frequent PTPRK-RSPO3 fusions and RNF43 mutations in colorectal traditional serrated adenoma. *J Pathol.* 2016;239(2):133–138.
51. Seshagiri S, Stawiski EW, Durinck S, et al. Recurrent R-spondin fusions in colon cancer. *Nature.* 2012;488(7413):660–664.
52. Pai SG, Carneiro BA, Mota JM, et al. Wnt/beta-catenin pathway: modulating anticancer immune response. *J Hematol Oncol.* 2017;10(1):101.
53. Kretzschmar K, Clevers H. Wnt/beta-catenin signaling in adult mammalian epithelial stem cells. *Dev Biol.* 2017;428(2):273–282.
54. Tian H, Biehs B, Warming S, et al. A reserve stem cell population in small intestine renders Lgr5-positive cells dispensable. *Nature.* 2011;478(7368):255–259.
55. Yan KS, Chia LA, Li X, et al. The intestinal stem cell markers Bmi1 and Lgr5 identify two functionally distinct populations. *Proc Natl Acad Sci USA.* 2012;109(2):466–471.
56. Proffitt KD, Madan B, Ke Z, et al. Pharmacological inhibition of the Wnt acyltransferase PORCN prevents growth of WNT-driven mammary cancer. *Cancer Res.* 2013;73(2):502–507.

57. Campbell CI, Samavarchi-Tehrani P, Barrios-Rodiles M, Datti A, Gingras AC, Wrana JL. The RNF146 and tankyrase pathway maintains the junctional crumbs complex through regulation of angiomotin. *J Cell Sci.* 2016;129(18):3396–3411.
58. Troilo A, Benson EK, Esposito D, et al. Angiomotin stabilization by tankyrase inhibitors antagonizes constitutive TEAD-dependent transcription and proliferation of human tumor cells with Hippo pathway core component mutations. *Oncotarget.* 2016;7(20):28765–28782.
59. Wang H, Lu B, Castillo J, et al. Tankyrase inhibitor sensitizes lung cancer cells to endothelial growth factor receptor (EGFR) inhibition via stabilizing angiomotins and inhibiting yap signaling. *J Biol Chem.* 2016;291(29):15256–15266.
60. Wang W, Li N, Li X, Tran MK, Han X, Chen J. Tankyrase inhibitors target YAP by stabilizing angiomotin family proteins. *Cell Rep.* 2015;13(3):524–532.
61. Gregorieff A, Liu Y, Inanlou MR, Khomchuk Y, Wrana JL. Yap-dependent reprogramming of Lgr5(+) stem cells drives intestinal regeneration and cancer. *Nature.* 2015;526(7575):715–718.
62. Smith RG, Burtner AP. Oral side-effects of the most frequently prescribed drugs. *Spec Care Dentist.* 1994;14(3):96–102.
63. Wang T, Glendinning J, Grushka M, Hummel T, Mansfield K. From the cover: drug-induced taste disorders in clinical practice and preclinical safety evaluation. *Toxicol Sci.* 2017;156(2):315–324.
64. Iwatsuki K, Liu HX, Gronder A, et al. Wnt signaling interacts with Shh to regulate taste papilla development. *Proc Natl Acad Sci USA.* 2007;104(7):2253–2258.
65. Liu F, Thirumangalathu S, Gallant NM, et al. Wnt-beta-catenin signaling initiates taste papilla development. *Nat Genet.* 2007;39(1):106–112.
66. Zhu X, Liu Y, Zhao P, et al. Gpr177-mediated Wnt signaling is required for fungiform placode initiation. *J Dent Res.* 2014;93(6):582–588.
67. Gaillard D, Xu M, Liu F, Millar SE, Barlow LA. Beta-catenin signaling biases multipotent lingual epithelial progenitors to differentiate and acquire specific taste cell fates. *PLoS Genet.* 2015;11(5):e1005208.
68. Perea-Martinez I, Nagai T, Chaudhari N. Functional cell types in taste buds have distinct longevities. *PLoS One.* 2013;8(1):e53399.
69. Ermilov AN, Kumari A, Li L, et al. Maintenance of taste organs is strictly dependent on epithelial Hedgehog/GLI signaling. *PLoS Genet.* 2016;12(11):e1006442.
70. Le Moigne M, Saint-Jean M, Jirka A, et al. Dysgeusia and weight loss under treatment with vismodegib: benefit of nutritional management. *Support Care Cancer.* 2016;24(4):1689–1695.
71. van Amerongen R, Berns A. Knockout mouse models to study Wnt signal transduction. *Trends Genetics.* 2006;22(12):678–689.
72. Staal FJ, Chhatta A, Mikkers H. Caught in a Wnt storm: complexities of Wnt signaling in hematopoiesis. *Exp Hematol.* 2016;44(6):451–457.
73. Luis TC, Naber BA, Roozen PP, et al. Canonical wnt signaling regulates hematopoiesis in a dosage-dependent fashion. *Cell Stem Cell.* 2011;9(4):345–356.
74. Kabiri Z, Numata A, Kawasaki A, Blank E, Tenen DG, Virshup DM. Wnts are dispensable for differentiation and self-renewal of adult murine hematopoietic stem cells. *Blood.* 2015;126(9):1086–1094.
75. Cobas M, Wilson A, Ernst B, et al. Beta-catenin is dispensable for hematopoiesis and lymphopoiesis. *J Exp Med.* 2004;199(2):221–229.
76. Luis TC, Weerkamp F, Naber BA, et al. Wnt3a deficiency irreversibly impairs hematopoietic stem cell self-renewal and leads to defects in progenitor cell differentiation. *Blood.* 2009;113(3):546–554.
77. Bowman TV, Trompouki E, Zon LI. Linking hematopoietic regeneration to developmental signaling pathways: a story of BMP and Wnt. *Cell Cycle.* 2012;11(3):424–425.

78. Congdon KL, Voermans C, Ferguson EC, et al. Activation of Wnt signaling in hematopoietic regeneration. *Stem Cells.* 2008;26(5):1202–1210.
79. Lento W, Ito T, Zhao C, et al. Loss of beta-catenin triggers oxidative stress and impairs hematopoietic regeneration. *Genes Dev.* 2014;28(9):995–1004.
80. Trompouki E, Bowman TV, Lawton LN, et al. Lineage regulators direct BMP and Wnt pathways to cell-specific programs during differentiation and regeneration. *Cell.* 2011;147(3):577–589.
81. Herault A, Binnewies M, Leong S, et al. Myeloid progenitor cluster formation drives emergency and leukaemic myelopoiesis. *Nature.* 2017;544(7648):53–58.
82. Abidin BM, Owusu Kwarteng E, Heinonen KM. Frizzled-6 regulates hematopoietic stem/progenitor cell survival and self-renewal. *J Immunol.* 2015;195(5):2168–2176.
83. Bond CE, McKeone DM, Kalimutho M, et al. RNF43 and ZNRF3 are commonly altered in serrated pathway colorectal tumorigenesis. *Oncotarget.* 2016;7(43):70589–70600.
84. Giannakis M, Hodis E, Jasmine Mu X, et al. RNF43 is frequently mutated in colorectal and endometrial cancers. *Nat Genet.* 2014;46(12):1264–1266.
85. Yan HH, Lai JC, Ho SL, et al. RNF43 germline and somatic mutation in serrated neoplasia pathway and its association with BRAF mutation. *Gut.* 2017;66(9):1645–1656.
86. Wu J, Jiao Y, Dal Molin M, et al. Whole-exome sequencing of neoplastic cysts of the pancreas reveals recurrent mutations in components of ubiquitin-dependent pathways. *Proc Natl Acad Sci USA.* 2011;108(52):21188–21193.
87. Ryland GL, Hunter SM, Doyle MA, et al. Mutational landscape of mucinous ovarian carcinoma and its neoplastic precursors. *Genome Med.* 2015;7(1):87.
88. Flanagan DJ, Phesse TJ, Barker N, et al. Frizzled7 functions as a Wnt receptor in intestinal epithelial Lgr5(+) stem cells. *Stem Cell Rep.* 2015;4(5):759–767.
89. Voloshanenko, O., Gmach, P., Winter, J., Kranz, D., Boutros, M. Mapping of Wnt-Frizzled interactions by multiplex CRISPR targeting of receptor gene families. *FASEB J.* 2017;31(11):4832–4844.
90. Dow LE, O'Rourke KP, Simon J, et al. Apc restoration promotes cellular differentiation and reestablishes crypt homeostasis in colorectal cancer. *Cell.* 2015;161(7):1539–1552.
91. Tanaka N, Mashima T, Mizutani A, et al. APC mutations as a potential biomarker for sensitivity to Tankyrase inhibitors in colorectal cancer. *Mol Cancer Ther.* 2017;16(4):752–762.
92. Crespo M, Vilar E, Tsai SY, et al. Colonic organoids derived from human induced pluripotent stem cells for modeling colorectal cancer and drug testing. *Nat Med.* 2017;23(7):878–884.
93. Zhong Y, Katavolos P, Nguyen T, et al. Tankyrase inhibition causes reversible intestinal toxicity in mice with a therapeutic index <1. *Toxicol Pathol.* 2016;44(2):267–278.
94. Voloshanenko O, Erdmann G, Dubash TD, et al. Wnt secretion is required to maintain high levels of Wnt activity in colon cancer cells. *Nat Commun.* 2013;4:2610.
95. Zhang L, Theodoropoulos PC, Eskiocak U, et al. Selective targeting of mutant adenomatous polyposis coli (APC) in colorectal cancer. *Sci Transl Med.* 2016;8(361). 361ra140.
96. Loregger A, Grandl M, Mejias-Luque R, et al. The E3 ligase RNF43 inhibits Wnt signaling downstream of mutated beta-catenin by sequestering TCF4 to the nuclear membrane. *Sci Signal.* 2015;8(393). ra90.
97. Mah AT, Yan KS, Kuo CJ. Wnt pathway regulation of intestinal stem cells. *J Physiol.* 2016;594(17):4837–4847.
98. Wang B, Zhao L, Fish M, Logan CY, Nusse R. Self-renewing diploid Axin2(+) cells fuel homeostatic renewal of the liver. *Nature.* 2015;524(7564):180–185.
99. Liu C, Li Y, Semenov M, et al. Control of beta-catenin phosphorylation/degradation by a dual-kinase mechanism. *Cell.* 2002;108(6):837–847.

100. Njei B. Changing pattern of epidemiology in intrahepatic cholangiocarcinoma. *Hepatology.* 2014;60(3):1107–1108.
101. Northcott PA, Buchhalter I, Morrissy AS, et al. The whole-genome landscape of medulloblastoma subtypes. *Nature.* 2017;547(7663):311–317.
102. Cadigan KM, Waterman ML. TCF/LEFs and Wnt signaling in the nucleus. *Cold Spring Harb Perspect Biol.* 2012;4:a007906.
103. Tang W, Dodge M, Gundapaneni D, Michnoff C, Roth M, Lum L. A genome-wide RNAi screen for Wnt/beta-catenin pathway components identifies unexpected roles for TCF transcription factors in cancer. *Proc Natl Acad Sci USA.* 2008;105 (28):9697–9702.
104. Angus-Hill ML, Elbert KM, Hidalgo J. Capecchi MR. T-cell factor 4 functions as a tumor suppressor whose disruption modulates colon cell proliferation and tumorigenesis. *Proc Natl Acad Sci USA.* 2011;108(12):4914–4919.
105. Bass AJ, Lawrence MS, Brace LE, et al. Genomic sequencing of colorectal adenocarcinomas identifies a recurrent VTI1A-TCF7L2 fusion. *Nat Genet.* 2011;43(10):964–968.
106. Yan D, Wiesmann M, Rohan M, et al. Elevated expression of axin2 and hnkd mRNA provides evidence that Wnt/beta-catenin signaling is activated in human colon tumors. *Proc Natl Acad Sci USA.* 2001;98(26):14973–14978.
107. Yan KS, Janda CY, Chang J, et al. Non-equivalence of Wnt and R-spondin ligands during Lgr5+ intestinal stem-cell self-renewal. *Nature.* 2017;545(7653):238–242.
108. Rheinbay E, Parasuraman P, Grimsby J, et al. Recurrent and functional regulatory mutations in breast cancer. *Nature.* 2017;547(7661):55–60.
109. Hayward NK, Wilmott JS, Waddell N, et al. Whole-genome landscapes of major melanoma subtypes. *Nature.* 2017;545(7653):175–180.
110. Wang Y, Krivtsov AV, Sinha AU, et al. The Wnt/beta-catenin pathway is required for the development of leukemia stem cells in AML. *Science.* 2010;327(5973):1650–1653.
111. Ricciardiello L, Ahnen DJ, Lynch PM. Chemoprevention of hereditary colon cancers: time for new strategies. *Nat Rev Gastroenterol Hepatol.* 2016;13(6):352–361.
112. Nguyen DX, Chiang AC, Zhang XH, et al. WNT/TCF signaling through LEF1 and HOXB9 mediates lung adenocarcinoma metastasis. *Cell.* 2009;138(1):51–62.
113. Scheel C, Eaton EN, Li SH, et al. Paracrine and autocrine signals induce and maintain mesenchymal and stem cell states in the breast. *Cell.* 2011;145(6):926–940.
114. Derksen PW, Tjin E, Meijer HP, et al. Illegitimate WNT signaling promotes proliferation of multiple myeloma cells. *Proc Natl Acad Sci USA.* 2004;101(16):6122–6127.
115. Lane SW, Wang YJ, Lo Celso C, et al. Differential niche and Wnt requirements during acute myeloid leukemia progression. *Blood.* 2011;118(10):2849–2856.
116. Chen L, Han X. Anti-PD-1/PD-L1 therapy of human cancer: past, present, and future. *J Clin Invest.* 2015;125(9):3384–3391.
117. Hanke N, Alizadeh D, Katsanis E, Larmonier N. Dendritic cell tumor killing activity and its potential applications in cancer immunotherapy. *Crit Rev Immunol.* 2013;33 (1):1–21.
118. Ma Y, Aymeric L, Locher C, Kroemer G, Zitvogel L. The dendritic cell-tumor cross-talk in cancer. *Curr Opin Immunol.* 2011;23(1):146–152.
119. Swafford D, Manicassamy S. Wnt signaling in dendritic cells: its role in regulation of immunity and tolerance. *Discov Med.* 2015;19(105):303–310.
120. Hong Y, Manoharan I, Suryawanshi A, et al. Deletion of LRP5 and LRP6 in dendritic cells enhances antitumor immunity. *Oncoimmunology.* 2016;5(4):e1115941.
121. Spranger S, Bao R, Gajewski TF. Melanoma-intrinsic beta-catenin signalling prevents anti-tumour immunity. *Nature.* 2015;523(7559):231–235.
122. Yaguchi T, Goto Y, Kido K, et al. Immune suppression and resistance mediated by constitutive activation of Wnt/beta-catenin signaling in human melanoma cells. *J Immunol.* 2012;189(5):2110–2117.

123. Manicassamy S, Reizis B, Ravindran R, et al. Activation of beta-catenin in dendritic cells regulates immunity versus tolerance in the intestine. *Science.* 2010;329 (5993):849–853.
124. Oderup C, LaJevic M, Butcher EC. Canonical and noncanonical Wnt proteins program dendritic cell responses for tolerance. *J Immunol.* 2013;190(12):6126–6134.
125. Shen CC, Kang YH, Zhao M, et al. WNT16B from ovarian fibroblasts induces differentiation of regulatory T cells through beta-catenin signal in dendritic cells. *Int J Mol Sci.* 2014;15(7):12928–12939.
126. Suryawanshi A, Manicassamy S. Tumors induce immune tolerance through activation of beta-catenin/TCF4 signaling in dendritic cells: a novel therapeutic target for cancer immunotherapy. *Oncoimmunology.* 2015;4(12):e1052932.
127. Suryawanshi A, Manoharan I, Hong Y, et al. Canonical Wnt signaling in dendritic cells regulates Th1/Th17 responses and suppresses autoimmune neuroinflammation. *J Immunol.* 2015;194(7):3295–3304.
128. Valencia J, Hernandez-Lopez C, Martinez VG, et al. Wnt5a skews dendritic cell differentiation to an unconventional phenotype with tolerogenic features. *J Immunol.* 2011;187(8):4129–4139.
129. Valencia J, Martinez VG, Hidalgo L, et al. Wnt5a signaling increases IL-12 secretion by human dendritic cells and enhances IFN-gamma production by CD4+ T cells. *Immunol Lett.* 2014;162(1 Pt A):188–199.
130. US Food and Drug Administration. Modification of the dosage regimen for nivolumab. 2016. Available from: https://www.fda.gov/Drugs/InformationOnDrugs/ApprovedDrugs/ucm520871.htm
131. Jeannet G, Boudousquie C, Gardiol N, Kang J, Huelsken J, Held W. Essential role of the Wnt pathway effector Tcf-1 for the establishment of functional CD8 T cell memory. *Proc Natl Acad Sci USA.* 2010;107(21):9777–9782.
132. Zhou X, Yu S, Zhao DM, Harty JT, Badovinac VP, Xue HH. Differentiation and persistence of memory CD8(+) T cells depend on T cell factor 1. *Immunity.* 2010;33 (2):229–240.
133. Wang C, Wu H, Katritch V, et al. Structure of the human smoothened receptor bound to an antitumour agent. *Nature.* 2013;497(7449):338–343.
134. Mosterd K. Intermittent vismodegib dosing to treat multiple basal-cell carcinomas. *Lancet Oncol.* 2017;18(3):284–286.

CHAPTER NINE

The Wnt Signaling Landscape of Mammary Stem Cells and Breast Tumors

Caroline M. Alexander[1]
University of Wisconsin-Madison, Madison, WI, United States
[1] Corresponding author. E-mail address: Alexander@oncology.wisc.edu

Contents

Abstract

Attention has been focused on Wnt signaling in the mouse mammary gland for several decades, firstly by the discovery of several Wnt loci among the oncogenes revealed by MMTV-based insertional mutagenesis screening of mouse mammary gland, and then by the remarkable visualization of Wnt-dependent specification of mammary placodes in embryonic skin. This review aims to summarize the impact of

Progress in Molecular Biology and Translational Science, Volume 153
ISSN 1877-1173
http://dx.doi.org/10.1016/bs.pmbts.2017.11.020

recent data for our understanding of the roles of Wnt signaling in these roles. The amount and identity of both familiar and novel Wnt signaling components is examined for mouse mammary epithelial cells. The hierarchical arrangement of mammary epithelial cell progenitors and stem cells inferred from the study of isolated cells is reinterpreted in an era that has demonstrated almost limitless cellular plasticity. Functional definitions of stem and progenitor activities are reevaluated with the discovery of novel stem cell activities and regulators, and we draw parallels with the arrangement of replication-competent cells in other tissues. Although Wnt signaling is highly oncogenic for mouse mammary epithelia, the data supporting Wnt signaling as a tumor driver for human breast cancer are still flimsy, and there is little support for the recruitment of normal Wnt-dependent breast stem cells as tumor precursor cells for either mouse or human. We discuss possible explanations for this paradox and questions still unanswered, including the potential impact of recent discoveries of Wnt-secreting microenvironments, oncogenic changes in the Rspo/Lgr/Ubiquitin ligase amplifier complex, as they could apply to breast tissues, and the feedback suppression of Wnt signaling that characterizes its developmental activity and may hide Wnt signatures in tumors.

1. INTRODUCTION

There are several excellent and recent reviews on how Wnt signaling regulates mammary gland development, stem cell activity and tumorigenesis.[1–6] Therefore, the aim of this chapter is to outline the conclusions emerging from studies on this topic, and to frame current questions.

In particular, the conceptual framework for mammary regeneration is now dominated by more general discoveries regarding induced pluripotency. The discoveries of the last decade have shown that stemness can be induced in any cell type, differentiated or not,[7] using simple protocols that rely on biological effectors. Perhaps surprisingly, these effectors are not directly connected to Wnt signaling, although one commonly used factor, Klf4, regulates β-catenin-dependent differentiation programs.[8] Therefore, models of normal or tumor development in mammals that rely on predetermined and irreversible stem cell hierarchies in solid tissues seem out of date.[9] Indeed the mammary gland provides a striking example of the environmental specification of fate; thus, testicular cells transferred to mammary gland were respecified to a fully functional mammary fate.[10]

Furthermore, activation of oncogenic signaling pathways can promote flexibility, and cells can acquire stem-like properties. An example of this is provided by the expression of constitutively activated phosphatidylinositol-

4,5-bisphosphate 3-kinase (PI3KCA) in lineage-specified luminal or basal cells: no matter which cell expressed this gene, it showed an expanded and flexible fate.[11] Thus, signaling pulses or interactions with at least 6 neighbor cells may "short-circuit" predetermined hierarchical relationships.

One of the frequently stated tenets of cancer stem cell theory is that the immortality and regenerative potential of normal stem cells can be exploited during their recruitment as tumor cells-of-origin. However, the acquisition of pluripotency with common oncogenic mutations makes this less theory less compelling. There may be orders of magnitude more cells that can be recruited from the nonstem cell pool, and therefore the study of the behavior of preexisting "stem cells" during transformation becomes less compelling as a rationale for quantifying risk of tumor development.

Although frequently assumed, the growth of adult breast tissues does not require a mammary stem cell (MaSC); the extensive cell growth and regression induced by cycling hormones and by multiple pregnancies are easily accommodated by the cell majority. This "cell majority" might be called "transit amplifying" cells in the parlance of developmental biologists.[12] In other words, these cells can divide many times and may only exit the cell cycle after interaction with a nonpermissive environment. Labeling studies show that the cell majority of both basal and luminal cell populations are replication-competent. Mathematical modeling by application of just a few biological rules shows that if each replication-competent cell can divide even 5 times, mammary organ turnover during many cycles of pregnancy can be accounted for.[13] This situation is therefore very different from that of the continuously regenerating lineages that include colorectal, skin, and hematopoietic tissue; for these lineages, stem cell deficiencies are represented as organ failure after a few weeks.[9] Perhaps a better parallel could be drawn between mammary gland and liver; thus, growth is focused around a pericentral Wnt-secreting endothelial niche in liver,[14] and diploid hepatocytes within the Wnt-exposed field are mostly quiescent but replication-competent. The cell majority shows high plasticity, and the organ is peppered with "stem" cells of different potency and origins, with variable impact on regeneration and repair.[15]

Several lineage tracing experiments in mammary gland in the last few years have illustrated the pattern of ductal and alveolar colonization of progenitors and stem cells during the cycles of growth and regression that are associated with puberty, estrus, and pregnancy. These can now be compared with the functional assays of "singlized," dissociated mammary epithelial cells that preceded them.[16–19]

Conclusions can be summarized as follows; dissociated mammary epithelial cells from the basal lineage are bipotent, and can divide to create basal and luminal cells that self-organize into fully functional ductal trees to regenerate the whole mammary gland.[20,21] In line with the view that plasticity is inducible in almost any cell, perhaps any replication-competent basal cell provided with a Wnt/Rspo source, or even recently exposed to Wnt/Rspo, could serve as a stem cell, if it remains undamaged by dissociation protocols.[22] Upon implantation, it is likely that MaSCs differentiate into luminal cell types to regenerate essential paracrine interactions to promote ductal morphogenesis.

However, lineage tracing gives a more surprising result: only rarely do MaSCs express their bipotency postnatally, and this bipotency is actively suppressed by the luminal cell partner.[16] Whereas dissociated luminal epithelial cells cannot regenerate a mammary gland, lineage tracing shows that they can divide as extensively as basal cells, to facilitate the cycles of growth and regression that characterize this hormone-dependent organ. Thus, the vast majority of growth of mammary gland is sustained by unipotent basal and luminal stem cells.[23] Another hormone-dependent gland, the prostate, also comprising interactive basal and luminal cells, shows similar regenerative behavior.[24,25]

Much of the molecular physiology associated with oncogenic Wnt signaling derives from the study of colorectal cancer; this has provided a template for the study of breast cancer. However, despite the early identification of gain of function Wnt oncogenes in mouse mammary gland, it has been difficult to show that Wnt signaling does initiate or promote the growth of human breast tumors. Similar mutations to those that drive the majority of colorectal tumors are simply not found in breast tumors. Why are Wnt pathway-related mutations not found in breast tumors? Potential answers are discussed in more detail later.

2. THE WNT LANDSCAPE OF MAMMARY EPITHELIAL CELLS

The molecules that regulate Wnt signaling may be tissue-specific. For example, Apc mutations that induce colorectal tumors at 100% penetrance and high multiplicity are only sufficient to confer increased susceptibility to mammary tumors.[26] An example of a "Wnt landscape" of mammary epithelial cells is shown in Fig. 1A–B. This shows the quantitative expression of

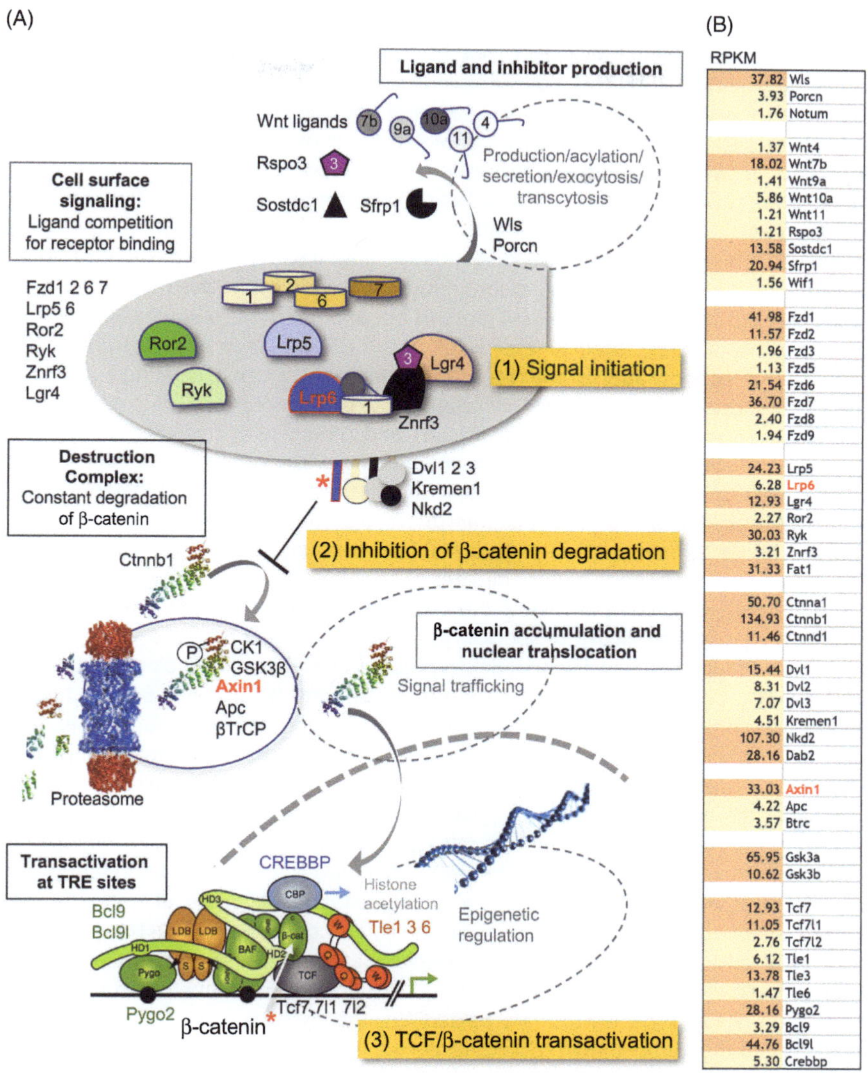

Fig. 1 The Wnt landscape in mouse mammary epithelial cells. (A) The components of the Wnt signaling pathway expressed by nontumorigenic mouse mammary epithelial cell line (EP cells) in basal growth conditions in culture are assembled in a summary schematic (for more detail on cell line, see Section 2). Assuming that mRNA expression results in protein expression, the landscape is described as follows: the extracellular signaling environment shows five Wnt ligands (colored in tones of *gray* with acyl tails). The translation and export of acylated Wnt proteins requires the chaperone Wls and the acyl-transferase, Porcn.[32] Acyl tails mediate Fzd binding and oligomerization.[33] Also shown is the Wnt signaling-related RSpondin ligand (Rspo3; *purple*) and two secreted inhibitors, Sfrp1 (that binds competitively to Fzd Wnt binding domain) and Sostdc1 (also known as WISE), known to interact with Lrp4 to inhibit Lrp-dependent Wnt

◀ signaling.[34–36] The array of receptors at the cell surface (represented by a *gray* disc) could include any of four Fzd species, together with either Lrp protein, Lrp5, and/or Lrp6. Other Wnt-binding receptors have different signaling outputs: not covered in this review are the effects of noncanonical Fzd- and Ror/Ryk-dependent signaling events, though these are clearly important.[37,38] Molecules known to be present in limiting amounts for Wnt signaling are indicated in *red* (Lrp6 and axin1). (1) Signaling is initiated in this example by a hypothetical complex comprising a ternary Lrp6/Wnt10a/Fzd1 complex in combination with Lgr4/Rspo3/Znrf3 ubiquitin ligase "amplifier." This is discussed more in Sections 2.1 and 4.6. (2) Levels of the Wnt signal transducer, β-catenin, are low due to continuous proteolysis by the destruction complex, shown here to include the scaffolding proteins Apc and Axin1, the ubiquitin ligase βTrCP (Btrc) and two kinases, casein kinase-1α (CK1), and GSK3β. These kinases promote the processive phosphorylation of the N-terminal domain of β-catenin, creating a substrate for βTrCP-induced ligation, and processing by the proteasome. Ligand/receptor complexation results in recruitment of axin1 and GSK3β away from the destruction complex, inducing the phosphorylation, and binding of the carboxy terminal PPPSPxP domains of Lrp6 (*red* starred). This serves to inhibit the degradation of β-catenin (gene name is Ctnnb1), allowing this signal transducer to accumulate, and translocate to the nucleus. (3) Recruitment of β-catenin to TCF factors bound in Wnt enhanceosomes at TCF-responsive elements (TRE) initiates the Wnt-induced changes of gene transcription.[39] Core components of the Wnt enhanceosome include three TCF DNA binding transcriptional activator species, Tcf1 (TCF7), Tcf3 (TCF7L1), and Tcf4 (TCF7L2) and three Graucho-related transcription repressors, Tle1, -3, and -6, along with other DNA binding proteins [including Pygopus2 (Pygo2)] and Bcl proteins Bcl9 and Bcl9l, serving to scaffold the complex. If there are specific properties for each family member, they are not understood. Several aspects of the signaling pathway are incompletely understood, especially transfer of Wnt and Rspo proteins from cell to cell and through cell layers, the role of trafficking in signal activation,[40] and the determination of Wnt transactivation targets by epigenetic cell type-specific regulation (these are shown circled in *gray dotted lines*). (B) The expression levels of each component were derived by RNASeq analysis, and show RPKM (reads per kilobase of transcript per million mapped reads); low expression is color coded *yellow* (1-10), mid-high expression levels are indicated in *beige*. Genes that were not significantly expressed in EP cells (defined here as expression level <1) include the noncanonical Wnt agonists, Wnt5a and Wnt5b, other lesser characterized Wnt ligands, such as Wnt6, Wnt7a, Wnt9b, Wnt10b, and the better known RSpondin species, Rspo1 and -2. None of the three Dkk inhibitor species are significantly expressed (Dkk1, -2, and -3). Other components notably absent include Norrin, Fzd4, Lrp4, Lgr5, Lgr6, Ror1, Apc2, Lef1, Tle2, and Tle4. Several species are commonly known by other names: Sostdc1 (WISE), Fat1 (atypical cadherin), Ctnna1 (α-catenin), Ctnnb1 (β-catenin), Ctnnd1 (p120), Btrc (βTRCP Ubiquitin ligase), Tcf7 (TCF1), Tcf7l1 (TCF3), Tcf7l2 (TCF4), Tle1/3/6 (Groucho repressors). As with any pathway, this one is intertwined with others, but these components are not included in this diagram.

mRNAs for components of the Wnt signaling pathway of a nontumorigenic, differentiation competent mammary epithelial cell line (EP cells).[27] This substrain of cells was rederived from HC11 cells,[28] in turn derived from the venerable COMMA-1D cell line,[29] and characterized over several decades. Cultures show mixed and modifiable basal and luminal fates,[27] and differentiate into alveoli that express milk proteins in response to prolactin and other mammotrophic hormones.[30] EP cells are nontumorigenic (though p53 mutant)[31]; indeed, despite expression of molecular elements of basal and luminal fate in vitro, they do not regenerate ductal trees after transfer to fat pad and therefore show no stem cell activity.

Wnt signaling is initiated by the binding of Wnt ligands to both Frizzled (Fzd) and *L*owDensity Lipoprotein-Receptor *R*elated *P*rotein-5 or -6 (Lrp5/6). Although the Lrp receptors are commonly denoted as interchangeable by the "Lrp5/6" designation, they are not functionally redundant. Indeed, Lrp5 and Lrp6 are commonly coexpressed, however, phenotypes that affect developmental morphogenesis or bone density map to only one of these sister genes (in this case, Lrp6 and Lrp5, respectively). For mammary epithelial cells, Lrp6 directly transmits a canonical Wnt signal, although Lrp5 is required to respond to most endogenously expressed Wnt ligands.[27,41] For mammary gland, Lrp6 is the principal trafficking coreceptor for the Fzd/Wnt ternary complex, and indeed cell surface Lrp6 may be a limiting component for canonical Wnt signaling.[1] Lrp5 is not phosphorylated efficiently at the PPSPxP motif in the carboxy-terminal domain, and is therefore a poor Wnt signaling receptor.[42] However, it is required in a complex with Lrp6 to react to endogenous mammary Wnt ligands (Wnt 9b and Wnt10b), and may be part of a broader trafficking function.[27]

There are typically several Fzd proteins expressed by any specific cell type; mammary epithelial cells express four at mid to high levels (Fzd1, -2, -6, and -7). Similarly, there is often redundant expression of Wnt ligands. In order of abundance, the Wnt ligands produced by EP cells are Wnt7b, Wnt10a, Wnt9a, Wnt4, and Wnt11. Although there is no specific pairing of ligands with Fzd species, within the competitive environment of multiple ligand/Fzd binding reactions at the cell surface, it is likely that single ligands bind single Fzd receptors.[5] This topic is discussed further in Section 2.2.

Binding of a Wnt ligand to the two cell surface receptors initiates the signal transduction cascade. The allosteric interaction of the ectodomain of Lrp with Fzd/Wnt promotes binding of key elements of the β-catenin destruction complex to the PPPSPxP domains of the carboxy-tail of Lrp6.[42] This inactivates the proteasomal targeting and destruction of β-catenin, allowing the accumulation and nuclear translocation of this primary signal transducer. By binding to the TCF group of transcription factors in the Wnt enhanceosome, this pathway initiates a highly cell type-specific transcriptional response. Almost no gene targets can serve as a common signature between different cell types; the most reliable specific marker of activation is Axin2 mRNA.

This gene induction reveals a key element of Wnt signaling, and indeed of most developmentally important pathways; their activation induces feedback inhibition at multiple levels.[43] This serves to limit the signal in time and space to allow a pulse of developmentally important morphogenesis. Thus, axin2 induction augments the activity of the destruction complex, where axin1 is present in limiting amounts. Similarly, other inhibitors are induced specifically by activation of the cascade, including Naked (here Nkd2)[44] and cell surface ubiquitin ligases, Znrf3, and Rnf43 (ZR). These ubiquitin ligases target Fzd receptors for destruction, disabling further signaling. If an "amplifier" complex is present, comprising an RSpondin (RSpo) ligand together with a cell surface associated *L*eucine *R*ich Repeat Containing *G* Protein-Coupled *R*eceptor (Lgr) family member, this will bind and inhibit the ZR ubiquitin ligases to amplify the Wnt signal by an order of magnitude.[45–48] Notice that when Wnt signaling is activated, the reaction at both proteomic (rapid) and genomic (slower) levels serves to suppress further Wnt responses, leading to difficulties of detection under chronic stimulation conditions.

Wnt signaling components identified recently and expressed in these cells include Ckap4, described as a receptor for the somewhat paradoxical Dkk1 (RPKM = 38.4)[49] and the protein C receptor (Procr), a Wnt target gene and mammary stem cell marker (RPKM = 1.4).[50] The full function of these proteins is not yet understood. Additional Wnt signaling components observed in vivo in mammary epithelia are: Wnt5a, Wnt5b, Ror1, Dkk1, Dkk3, Sfrp2, Sfrp4, and Nkd1, described in more detail in a previous review.[2] Additional components expressed by stroma include Wnt2 and Rspo1.

Overall, the expression of Wnt signaling components of EP cells accurately mirrors that of mouse mammary epithelial cells in vivo, and indeed these cells are responsive to the addition of exogenous ligands. This profile of Wnt signaling components in mammary epithelial cells raises the following points.

2.1 Autocrine Signaling is Insignificant, Despite Ligand Production

Wnt ligand expression is significant in EP cells. However, there is little evidence of active autocrine Wnt signaling. Thus, Axin2 mRNA is low in unstimulated cells, and unchanged by the inhibition of the essential Porcn-dependent Wnt acylation reaction.[27] Perhaps of significance for the interpretation of other studies, Wnt ligand expression is therefore not sufficient to imply active signaling, probably because of myriad feedback inhibitors of this key developmental pathway.

2.2 Ligand Competition for Receptor Binding

The diversity of Wnt pathway receptors expressed by mammary epithelial cells is dazzling. van Amerongen and Nusse proposed that this array of receptors is best considered as a competitive set of binding sites for Wnt ligands, where the ultimate signal reflects a composite of canonical signals from Fzd-Lrp pairs and noncanonical signals from unpaired Fzd species.[5] Little is known about how Lrp/Fzd pairing affects membrane localization and trafficking of Wnt signals.

None of the Wnt ligands known to be oncogenic for mouse mammary epithelia (Wnt1, Wnt3, and Wnt10b) are expressed in adult mammary epithelial cells,[51] although Wnt10b is developmentally important and shows canonical β-catenin/TCF transactivation activity.[52] Wnt ligands are not functionally interchangeable, and it is likely that these three loci have special tumorigenic attributes.

Rspo1 has been reported to be expressed by ER-negative luminal cells and to regulate mammary stem cell activity and alveolar development during pregnancy.[48,53] Rspo2 and -3 were also identified by MMTV-based insertional mutagenesis as mammary oncogenes; their expression during normal development is not known yet. This functional screen is biased toward oncogenes, because the integration of a viral long terminal repeat sequence upregulates expression of proximal loci.

Although this MMTV-based screen can identify dominant negative loss of function mutations, such as Notch/int3,[51] this is uncommon, perhaps explaining why Wnt negative regulators (such as the colorectal tumor suppressor Apc) were not identified.

Functional studies of Wnt ligands are often hindered by the lack of specific antibodies and difficulties of analyzing lipid-modified, insoluble recombinant proteins; the vast majority of such data therefore comes from study of Wnt3a, due to its unique solubility properties. This ligand has distinct receptor interactions and signaling properties (discussed by Goel *et al.*).[41] Few studies include a functional demonstration of activity for given expression vectors using model organisms (such as zebrafish). We therefore know little about Wnt7b- or Wnt9a-induced signaling or receptor partners to be able to predict their function in mammary epithelial cells. Wnt10a shows canonical Wnt signaling activity in ectodermal tissues, such as skin, teeth, and nails.[8] Note that although Wnt5a and Wnt5b are typically expressed by some luminal cells in mammary gland, they are not expressed significantly by EP cells. Wnt4 is discussed more in a later topic section on mammary stem cells.

2.3 Expression of Wnt Signaling Suppressors

There are high levels of the mRNA for a soluble inhibitor of Wnt signaling, Sfrp1, which competes for Wnt ligand binding to Fzd receptors. For example, loss of Sfrp1 has been observed for more than 80% of invasive breast carcinomas,[54] and has been proposed to regulate motility and noncanonical Wnt signaling (rather than canonical pathway activation). Indeed, there are a number of Wnt signaling components that are relatively enriched or depleted in subtypes of breast tumors.[55] However, there are several types of mammary epithelial cells, (discussed in more detail later), each with a different transcriptome, which typically differ with respect to their expression of Wnt signaling components. Often basal cells show high levels of Wnt signaling components. In tumors, the usual population diversity of mammary epithelial cell populations tends to be reduced or eliminated, so this change of pattern is predictable, but not necessarily functionally important.

Overall, there has been an increasing appreciation of the role of feedback control and the ubiquitinylation of elements of the Wnt signaling pathway, summarized by Lee and coworkers.[1] In EP cells, the mRNA for the transmembrane E3 ubiquitin ligase, Znrf3, is expressed (albeit at low levels). This ligase has been shown in other cell types to be induced in response to Wnt activation and to inhibit further signaling by targeting Fzd cell surface receptors for degradation. When Rspo ligands are present, they bind Lgr species (often shown for Lgr5 and Lgr6), which bind into a ternary complex with Znrf3/Rnf43 ligases to prevent Fzd downregulation and feedback inhibition.[56] The significance of these ubiquitin ligases for mammary gland development and tumorigenesis is not yet known.

Genes for other Wnt signaling inhibitors include Sostdc1 (Wise[34]) and high levels of Naked2, known to interact with Dvl (1, 2, 3) to inhibit productive signal transduction. The transactivation response is regulated by several cross-talking signaling partners of the β-catenin/TCF transcription complex,[57] both inhibitory and stimulatory. The Wnt "enhanceosome" for

EP cells is shown in Fig. 1, to illustrate the variants of this complex that exists in mammary epithelium. Thus, EP cells express three TCF factors, TCF1>TCF3>TCF4, and three Groucho repressors, Tle1, -3, and -6, together with the chromatin modifier, Pygopus2.[58]

2.4 The Wnt Signaling Response is Regulated by 3D Cues and Mechanotransduction

In common with other signaling pathways, it has been shown that the Wnt pathway regulates, and is regulated by, mechanotransduction cues via integrated pathways, such as YAP/TAZ.[59–61] The EP profiles presented here are defined for 2D cultures on tissue culture plastic in media with fetal bovine sera. With the introduction of cell polarization, components of the Wnt signaling pathway become organized across the apical to basolateral axis, separating components that are mixed in cell membranes in 2D cultures.[62] One of the principal developmental roles of Wnt signaling is to organize the axes of division and tissue expansion. Therefore, accurate Wnt responses will be better reconstructed in vitro by maintaining cell–cell interactions, for example, by studies of explanted mammary organoids.[63–65]

3. WNT SIGNALING AND MAMMARY STEM CELLS

3.1 Wnt Signaling is Highly Activated in Embryonic Mammary Glands

The activation of Wnt signaling in vivo during mammary embryogenesis through morphogenesis was originally visualized using the TOP-GAL Wnt transactivation reporter[66] (Fig. 2). At this stage of development, all the cells allocated to the placode could be defined as stem cells; they are bipotent, long-lived, and actively dividing. Without Wnt signaling, mammary placodes are not specified from ectoderm, and there is a complete failure of mammogenesis. After birth, Wnt/Rspo signaling declines (assayed by the same reporter), reducing both in magnitude and number of cells significantly labeled in the adult gland. Correspondingly, there is a decline in the flexibility of fate in adult mammary populations, and more differentiated and specified populations emerge.[3,17] These are designated as basal and luminal cells (described in more detail in Fig. 2 legend). Fetal cells often coexpress luminal and basal cytokeratins (so called "double positive" cells); whereas adult cells do not. It is not yet known whether Wnt signaling is important to

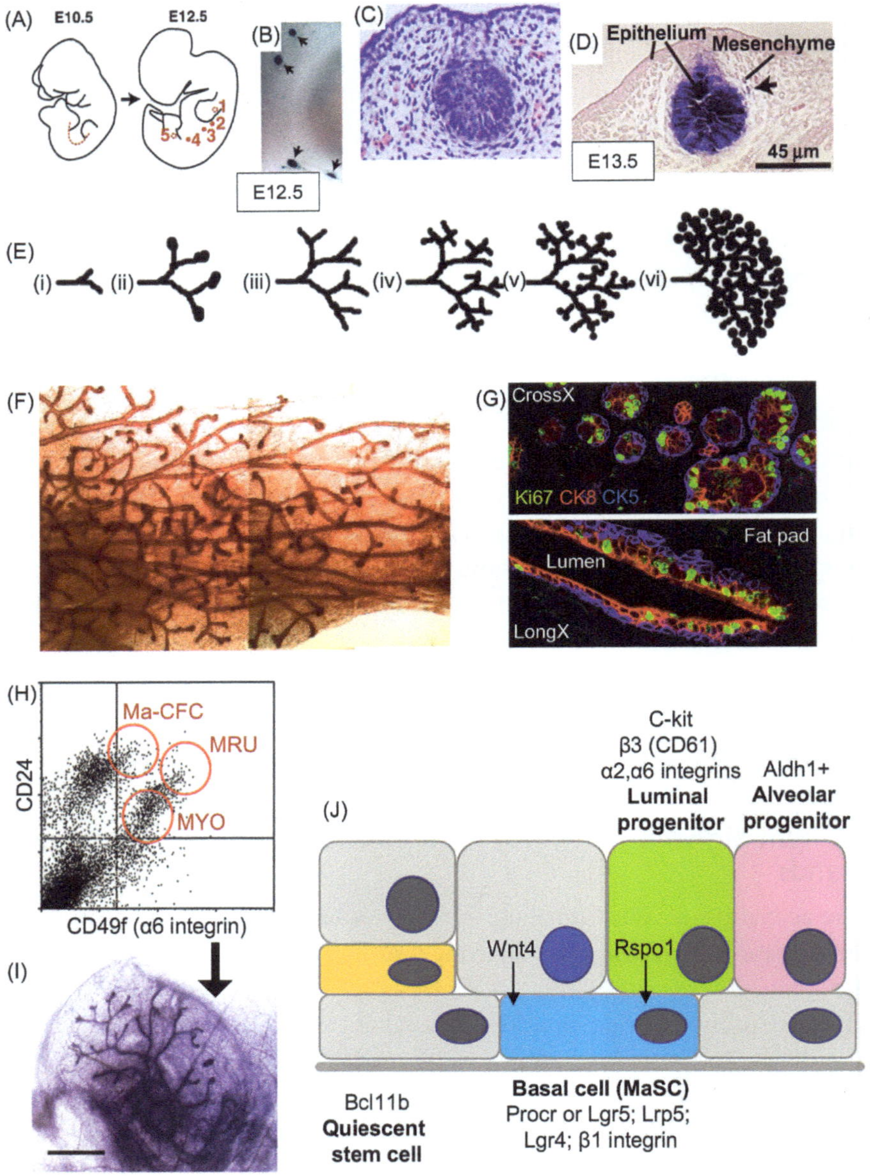

Fig. 2 Development of mouse mammary epithelial cell populations. During embryogenesis, Wnt signaling is required for the specification, expansion, and invagination of the mammary placodes. By embryonic day 12.5, 5 placodes per side are specified from the milk line (A) and are positive for TOP-gal (shown in *blue*), a reporter of catenin/TCF dependent Wnt transactivation (B). By E13.5d, a blastema of epithelial cells has pushed into an embryonic mesenchyme, the precursor of the mammary fat pad [(C), hematoxylin stained section]. Cells maintain high levels of Wnt

◀ signaling until the differentiated lineages are specified (D). Diagrams (A–D) are adapted from Chu *et al.*[66] During postnatal development, the embryonic rudiment expands to colonize the growing mammary fat pad (shown diagrammatically, adapted from Brisken and O'Malley).[68] (E) Ductal outgrowth is led by the proliferative engine provided by the dilated terminal end buds [E(ii)]. Full ductal branching is complete after puberty [E(iv)]. Pregnancy initiates alveolar development, starting from side-branches of the ductal tree [E(v)] and these alveoli stop dividing to become remarkable secretory powerhouses during lactation [E(vi)]. The ductal mammary epithelial cell population is typically visualized using carmine stain (pale background is the fat pad) (F). (G) Stains of sections of this structure show two layers; luminal cells, which differentiate to be the major milk producers during lactation, facing the lumen (stained *red* with anticytokeratin 8), and myoepithelial basal cells (*blue* stained with anticytokeratin 5). Myoepithelial cells are contractile, and lay down basement membrane, to provide a mechanical basis for milk ejection. Almost all cells of both lineages are replication competent; here cells in cycle are painted *green* (Ki67). The two cell types can be dissociated, sorted into two populations by flow cytometry [here using Cd24 and CD49f; (H)],[20] and tested prospectively for their growth potential. Some of the most marker-positive luminal cells can proliferate in culture [mammary colony forming cells (Ma-CFC)] and some of the most-marker positive basal cells can regenerate a new ductal tree and develop normally during pregnancy and lactation when transferred to a new fat pad in vivo (I). These are variously called mammary regenerating units (MRU) or mammary stem cells (MaSCs). (The mass population of myoepithelial/basal cells is marked MYO). Note that Ma-CFC does not score on the stem cell assay, and MaSCs contribute little to the culture assay (typically mammosphere cultures). Our current understanding of the maintenance of the mouse mammary stem cell is diagrammed (J). *Gray* cells indicate potentially replication-incompetent, terminally differentiated cells that do not contribute to regeneration; colors indicate replication-competent cells, summarized as stem and progenitor fractions. The relative proportions of each are not well understood, but the terminally differentiated group may be a small minority. Nuclei are painted *blue* if they express estrogen and progesterone receptors; these are typically classified as "mature" or terminally differentiated, but they process the hormone signal that governs hormone-dependent growth of the rest of the population.[68] Most of the markers noted here for stem/progenitor cell types have been shown to be functionally important, as summarized by Visvader and Stingl.[19] The Bcl11b-positive quiescent stem cell fraction was described by Clarke and coworkers.[69] A triad of cells that maintain the stem cell unit was deduced from Zeng and coworkers,[48,50] including an ER-positive mature luminal cell expressing Wnt4, and ER-negative luminal progenitor expressing RSpo1, which together signal to the Lrp5-positive MaSC basal cell. Note that the cell surface phenotypes that discriminate basal and luminal cells in vivo do not apply to cells transferred to culture, despite persistent basal/luminal marker expression; typically these two (or more) populations merge into one. Note also that the function and phenotype of human mammary stem and progenitor cell populations is not necessarily directly comparable to the same populations in mouse.[70]

the maintenance of adult gland; this requires a conditional approach to loss of function for a nonredundant Wnt component (such as Lrp6). However, adult mammary glands with no detectable stem cell activity have no problem undergoing multiple rounds of pregnancy and lactation.[67]

3.2 Cellular Heterogeneity of Mammary Epithelial Cells Populations

This heterogeneity was described decades ago from electron micrographs of mammary glands[71]; cells were shown to differ in size, cytoplasmic density, and nuclear morphology, and to occur in similar and dissimilar pairs, suggesting elastic fates after division. With the advent of flow cytometric separation of cells from mouse mammary gland, this heterogeneity has been confirmed, and several cell subpopulations from basal and luminal populations have been isolated and tested by functional assays of stemness, either in vitro or in vivo, though rarely using both. The current inventory of stem/progenitor fractions that score on different assays is shown in Fig. 2. There are basally located mammary stem cells, luminal and alveolar progenitors, and a quiescent Bcl11b+ stem cell that lies between luminal and basal cells. The alveolar lineage that develops during pregnancy in response to steroid hormone cues also comprises both basal and luminal cells. The functionality of alveolar subpopulations are rarely studied; they depend on pregnancy-level hormones for their survival.[68] Thus, although the ductal and alveolar mammary epithelial cells from pregnant mice may look alike, ductal cells (approximately 10% of total) do not undergo lysosomal-mediated cell death upon weaning.[72]

3.3 Which of These Stem Cells is Wnt-Dependent?

Basal stem cell activity is augmented by expression of β-catenin or Wnt1, and basal stem cell activity is absent in Lrp5 knockout mammary glands.[67,73–75] Note that there are two similarly named and therefore confusing cell surface receptors, Lgr5 and Lrp5, which may both be involved to promote Wnt signaling in mammary gland. Approximately half the basal cells express Lgr5,[76,77] and all express Lgr4 and Lrp5.[67,78] The phenotype of Lgr4 mutant mammary glands resembles that of Lrp5 mutants: mice show hypomorphic ductal outgrowth, and little adult regenerative activity. Indeed, both knockout mice show only minor (though significant) effects on other aspects of physiology (such as eye development and osteoblast differentiation). Furthermore, cultured $Lgr4^{-/-}$ organoids proliferate in response to

Wnt3A but not Rspo1.[78] Thus Lrp5 and Lgr4 cell surface receptors are required for the expression of stem cell activity.

Zeng and coworkers reported that this basal stem cell activity could be maintained by a combination of Rspo1 and Wnt4, produced by PR-negative and PR-positive luminal cells, respectively, in the presence of an estrogenic inducer[48,50] Rspo ligands interact with Lgr receptors to promote canonical Wnt signals by stabilizing their receptors (via inhibition of cell surface ubiquitin ligases).[46,47] This is an intriguing outcome that builds on prior observations and could align mammary stem cell function with the Rspo/Wnt interactions that maintain intestinal stem cells.[45] The Wnt4 ligand itself does not robustly generate a canonical signal in vitro and does not induce hyperplasia when overexpressed.[2] Although Wnt4 may not be sufficient to induce ductal hyperplasia, it is necessary for the survival of basal stem cells: thus $Wnt4^{-/-}$ mammary rudiments showed decreased regenerative capacity and decreased Wnt reporter activity in basal cells.[79] Similarly, although Rspo ligands promote Wnt signaling, they are not sufficient to induce Wnt signaling alone. The requirement for two groups of ligands and receptors (Rspo/Wnt and Lrp/Lgr), linked by an inhibitory ubiquitin ligase, may explain why the regenerative capacity of Lgr5-positive mammary epithelial cells has been so controversial, as recently summarized by Kolesnick and coworkers.[76] Further controversy arises from the description of a basal stem cell activity, that is, Lgr5-negative, and positive instead for the Wnt target gene, Procr (protein C receptor), also from Zeng and coworkers.[50] Perhaps Procr-positive cells represent a population newly exposed to Wnt signaling, with downregulated cell surface Lgr5 (Lgr5-negative during flow sorting).

The working hypothesis arising from these studies is that the regenerative capacity of this Lrp5- and Lgr5-positive basal cell community is maintained by the production of Rspo1 and Wnt4 by their luminal partners (Fig. 2). Stem cell functionality accumulates rapidly during estrus with the induction of Wnt4 cells.[80] It is unlikely that the accumulation of "stemness" is linked to cell division, and more likely that Wnt exposure is enough to induce stem cell activity in preexisting basal cells.

4. WNT SIGNALING IN BREAST CANCER

When Varmus and coworkers found that gain of function for Wnt or Rspo signaling was highly oncogenic for mouse mammary epithelia,[51,81] the hope was that this would translate to the discovery of a Wnt driver for human breast tumors. Thirty years later, this hope is fading; none of the canonical mutations that initiate and drive colorectal tumor growth are prevalent in breast cancer (Apc, β-catenin, or axin), and there is little genetic support of a gain of function by amplification or mutation of other pathway components.[82,83]

One of the main arguments for a gain of function for Wnt signaling in breast cancer is inferred by the disruption of normal β-catenin distribution from plasma membrane localization to a cytoplasmic/nuclear-proximal distribution in a substantial fraction of breast tumors.[55] However, this may relate more to E-cadherin than to Wnt signaling activity. Thus, E-cadherin is an important binding partner for β-catenin that controls cell-cell adhesion and motility, and is known to be lost (genetically) in the majority of lobular breast cancers, and reduced or redistributed in at least some of the cells in some 23%–60% of other tumors.[84–88] Furthermore, without additional signaling dysregulation, β-catenin released from its usual binding partner, E-cadherin, does not recruit this signal transducer to the Wnt signaling pool.[89–92] For example, despite biallelic loss of E-cadherin in lobular breast carcinomas, there is no gain of function for Wnt signaling.

The pool of β-catenin allocated to Wnt signaling is often too low to visualize by immunostaining. Nuclear b-catenin is not always visualized by immunochemical stains in Wnt-dependent mouse breast tissues and tumors, so false negatives are likely to be common in both mice and women. Despite these caveats, Geyer *et al.* have clearly demonstrated the localization and amount of β-catenin (using two different monoclonal antibodies) in a cohort of 245 breast tumors,[90] and showed that nuclear β-catenin is prevalent in a subpopulation of tumor epithelial cells in approximately one-third basaloid breast tumors. Basaloid breast tumors comprise about one-sixth total breast tumors, and show high levels of expression of basal cell markers (both protein and mRNA) alongside luminal cell markers.

Canonical Wnt inhibitors are in clinical trials in several settings,[6,93,94] however, none of these trials are for breast cancer. Experimentally, accumulating results suggest that these drugs are not generally effective at inhibiting growth of human PDX breast cancer strains, but more targeted drug trials may succeed.[95] Drugs include inhibitors of Wnt ligand production, such as the Protein-serine O-palmitoleoyltransferase (Porcn) inhibitor LGK974, cell surface signaling inhibitors (including Fzd8-Ig fusion/Ipafricept), inhibitors of the β-catenin/TCF nuclear complex (such as PRI-724), β-catenin nuclear import/export or less direct inhibitors of scaffolding factors, such as axin (tankyrase).[93,96,97]

There appears to be little that is fundamentally different about mouse mammary epithelial cells compared to human that could explain a discrepancy in the apparent oncogenicity of Wnt signaling for the two species. Indeed, human cells are susceptible to Wnt-mediated transformation in vitro.[98] GO ontology searches that identify Wnt signaling as a component of the tumor phenotype tend to be difficult to interpret, relying as they do on any gene affected by or interacting with the Wnt signaling cascade (e.g., the KEGG gene list relies on Axin1, Bcl9L, KDM6a, Lrrk2, Ryr2 and Wnt5a, Apc, Fzd2, Fzd7, Lrp5, Porcn, and Smo).[99] Only Axin2 appears to be a reliable direct Wnt transactivation reporter; other myc or cyclinD-based outputs rely on the indirect, growth promoting effects of Wnt signaling in mammary epithelial cell populations. Therefore, despite sweeping statements to the contrary in the literature, it seems unlikely that Wnt signaling will be an important target for therapeutic development for this disease.

However, there are still some stones unturned. If Wnt signaling is not sufficient as a tumor driver for the majority of breast tumors, perhaps it is a driver of one of the triple negative tumor types, or is necessary for promoting tumor initiation or for maintaining a minority cancer stem cell type. Wnt signals may be important for creating a tumorigenic microenvironment and creating drug-resistant niches. Perhaps, there is a correlation of Wnt-dependent tumors with activation of known collaborative pathways, or with activation of the Lgr/Rspo/RZ Wnt modifier complex. These options are considered later.

4.1 Wnt Signaling as a Driver of a Triple Negative Subtype

Analysis of breast tumor samples has focused on searching for evidence of gain of function of Wnt signaling. In particular, this search has been invigorated by the realization that this is not one disease but many, with different drivers. These different classes are described by the cancer genome project[86,100] and by the cancer cell line genome project,[101] with more focused characterization of breast cancer cell lines at the genetic[102–105] and metabolic levels.[106]

The principal drivers of breast tumors are the estrogen-ER-PR axis and HER2, both of which induce tumors comprising luminal-like tumor cells. The last group of diseases are called "triple negative" (ER−, PR−, HER2−), and comprise a grab-bag of tumors that show quite consistent mutational origins: thus 80% have mutant p53 and 22% show mutations in the KEGG PI3K/Akt/mTOR metabolic pathway.[86,100] Reis-Filho and coworkers showed that a subpopulation of tumor epithelial cells have nuclear β-catenin in approximately one-third basaloid breast tumors.[90] Consistent with their data, Wnt ligands have been shown to promote the growth of some well-known triple negative breast cancer cell lines (particularly MB231 and HCC1187 cells).[107,108] However, neither of the genetic profiles of these cell lines is wholly representative of human triple negative breast tumors (Ras mutant gene and HER2 overexpression, respectively). Furthermore, MB231 cells have recently been shown to depend upon an XBP1-HIF1α transcriptional response, derived from high levels of ER stress caused by dysregulated protein synthesis, combined with the hypoxic response.[109] Whether the Wnt and ER stress pathways are interdependent or epistatic with respect to one another is not yet understood.

With the widespread recognition of the importance of Rspo proteins for the induction of sustainable, high-level Wnt signaling responses, attention has shifted to their analysis in breast tumors. Thus Rspo2 and -4 ligand mRNAs are frequently overexpressed in triple negative tumors,[95] and their expression is associated with poor survival rates of breast cancer patients (perhaps because of the association with the triple negative subtype).

One of the characteristics expressed in the triple negative tumor cohort is a heterogeneity that recapitulates the normal breast epithelial cell population; these tumors include basal- and luminal-like tumor cells. This is also a feature of Wnt-induced mouse mammary tumors; in this case, the tumor initiating cells reside in the basal cell population.[110] Indeed, most triple negative tumors show enrichment of Wnt signaling components, probably because these molecules are more highly expressed by basal-like cells. The significance of this alone is not clear, but a functional study of the rate-limiting cell surface receptor for Wnt signaling, Lrp6, showed that Lrp6 mRNA is increased in triple negative tumors,[108] and that overexpression of Lrp6 is enough to drive growth (but not tumorigenesis) in vivo.

4.2 Could Oncogenic Wnt Expression be Transient?

There is a transient engagement of Wnt signaling developmentally as a means of pattern formation,[111] raising the possibility that a similar transient engagement may be important to tumor initiation. Relevant to this idea, Chodosh and coworkers tested whether Wnt signaling was required continuously in a mouse breast tumor model. Thus, expression of Wnt1 was switched off after tumor development, using a conditional doxycycline approach[112,113]; unlike a similar Myc-driven model, all Wnt-driven tumors regressed. This result makes it unlikely that transient expression is an important tumor promoter.

4.3 Does Wnt Sustain a Cancer Stem Cell Type?

Wnt signaling typically induces retrodifferentiation to increase the phenotypic plasticity of cells.[114] This allows pathway components that were previously separated into two cell types to be combined into one,[115] which in theory could promote autocrine cell growth, an independence from regulatory cues, and gain of stemness. Plasticity also allows colonization of atypical niches, and behaviors that might enable metastasis, for example, epithelial-mesenchymal transitions.[116] However, there is little data to support the recruitment of breast cancer stem cells from the normal, Wnt-maintained basal stem cell population. For example, for the consistently triple negative/basaloid tumor type of mutant Brca1 genetic origins, the tumor precursor cells were concluded to most likely derive from luminal cell precursors,[70] a conclusion substantiated by mouse studies using lineage-specific expression of Brca.[117] Indeed, basal cell fates can be induced by the expression of PIK3CA in luminal cells.[11] On the whole, cancer stem cell markers have proven difficult to replicate in different breast tumors; indeed, the phenotype of tumor-initiating cells may be quite specific to any individual tumor.[22,118]

However, other tumor types have been shown to be Wnt-dependent, despite the absence of canonical Wnt signaling pathway mutations, direct Wnt activation, or distinct Wnt genomic signatures. Thus, an elegant study from Jacks and coworkers showed that lung adenocarcinomas arising in mice were highly Wnt ligand-dependent, although tumors were induced in this model by different drivers, namely Ras activation and p53 mutation.[119] Analysis of the functional properties of the heterogeneous cell populations that comprise these tumors showed that Lgr5-positive tumor initiating cells (or "cancer stem cells") depend upon a majority of cells that synthesize Wnt ligands (and are Porcn-positive). Indeed, this signaling loop was found to be an essential paracrine tumor driver, and tumor growth was highly inhibited by the administration of Porcn inhibitors. This discovery in lung cancer may serve as a model for Wnt-dependency in breast tumors.

4.4 Is Wnt Signaling Important to the Tumorigenic Microenvironment?

Wnt production in stem cell niches maintains the architecture and differentiation pattern of normal somatic populations.[14,120] To test whether tumor microenvironments show Wnt activity, Wnt reporters have been visualized in PDX transplants,[121] and Wnt expressing stromal-tumor cell recombinants have been established in vivo.[122] So far, these studies have indicated that Wnt signaling is not activated in the PDX tumor cells themselves; furthermore, stromal cells overexpressing Wnt3a were found to either stimulate or inhibit tumor growth. The answer to this question is therefore not yet clear.

4.5 Is There Evidence of Appropriate Mutations in Collaborating Pathways?

Wnt signaling alone can induce growth and ductal hyperplasia in mouse mammary glands, but other mutations are required to induce frank tumors. The MMTV-based insertional mutagenesis screens first revealed that gain of function for Fgf signaling could serve this collaborative function, and a subsequent study suggested that this cooperative interaction worked at a translational level.[123] Gain of function for Fgf signaling has been observed in breast cancers through various mechanisms, including amplification of the receptor genes, aberrant ligand expression, receptor mutations, and translocations.[124] This is the basis for the development of clinical trials of FGFR inhibitors, however, there is no clear overlap between this group of tumors and those basal tumors with clear evidence of nuclear β-catenin.

4.6 Is the Lgr/Rspo/RZ Wnt Modifier Affected in Breast Tumors?

The Wnt landscape of mammary epithelial cells includes Lgr4, Znrf3, and RSpo3, and other family members may also be expressed in vivo (notably Lgr5, Rspo1). Rspo3 was one of the loci identified as an oncogene by MMTV-based insertional mutagenesis, so clearly this Wnt modifier pathway is active in mouse mammary gland. Znfr3 and its homolog, Rnf43 (collectively abbreviated here RZ proteins) are emerging as important tumor suppressor proteins for other tumor types; their mode of action is illustrated in Fig. 3. Thus, in addition to the discovery of RNF43 mutations in over 18% of colorectal cancers,[125,126] and mutations of the homolog ZNFR3 in adrenocortical carcinoma,[127] this amplifier pathway for Wnt activation has now been implicated in several tumor types not previously associated with gain of function for Wnt. For example, the majority of neoplastic cysts of the pancreas carry mutations in the suppressor ubiquitin ligase, RNF43,[128] and tumor cell lines with this mutation are addicted to Fzd5-dependent Wnt signaling and highly sensitized to Porcn inhibitors.[129,130] Other tumor types carrying mutations of either of the RZ genes include mucinous ovarian carcinoma, endometrial carcinoma, liver, and gastric neoplasias. Given the extensive feedback of the amplified Wnt pathways and cellular heterogeneity,

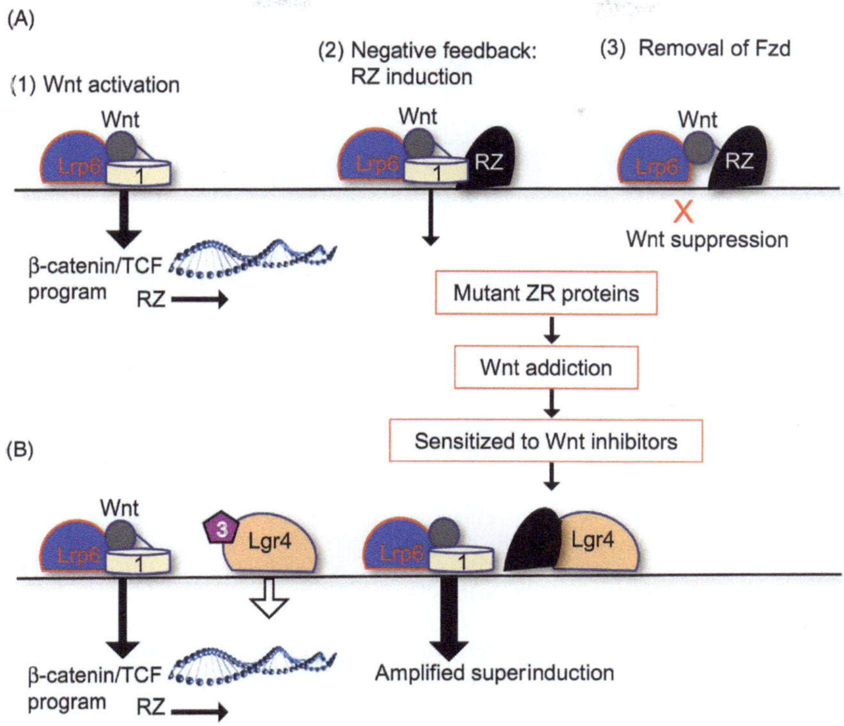

Fig. 3 Illustration of the mode of action of the Lgr/Rspo "amplifier" pathway. Details are described in Section 4.6. (A) Without Lgr/Rspo. (B) With Lgr/Rspo.

changes of activity of these Wnt modifiers may be hard to detect, and worthy of closer examination in breast tumors.

5. CONCLUSIONS

The mammary gland epithelial population comprises a diverse group of highly interactive cells that provides a remarkable opportunity to understand how Wnt signals provide cues to regulate regenerative cells, and organize neighborhoods of specialized cells to promote function. With better understanding of the sophistication of the molecular Wnt circuitry and the molecular features of the different cell types, there are considerable opportunities for future studies. However, it is an enduring paradox that this pathway, so oncogenic for other epithelia and for mouse mammary gland, has not yet been demonstrated to be a significant driver of breast tumor development in women.

ACKNOWLEDGMENTS

Many thanks for critical evaluation and advice from Dr. Ian Ellis (University of Nottingham), Dr. Jean-Paul Vincent (Crick Institute) and Dr. A.M.C. Brown (Weill Cornell Medical College).

REFERENCES

1. Saito-Diaz K, Chen TW, Wang X, et al. The way Wnt works: components and mechanism. *Growth Factors.* 2013;31(1):1–31.
2. Alexander CM, Goel S, Fakhraldeen SA, Kim S. Wnt signaling in mammary glands: plastic cell fates and combinatorial signaling. *Cold Spring Harb Perspect Biol.* 2012;4(10): a008037.
3. Makarem M, Spike BT, Dravis C, Kannan N, Wahl GM, Eaves CJ. Stem cells and the developing mammary gland. *J Mammary Gland Biol Neoplasia.* 2013;18(2):209–219.
4. Wend P, Holland JD, Ziebold U, Birchmeier W. Wnt signaling in stem and cancer stem cells. *Semin Cell Dev Biol.* 2010;21(8):855–863.
5. van Amerongen R, Nusse R. Towards an integrated view of Wnt signaling in development. *Development.* 2009;136(19):3205–3214.
6. Nusse R, Clevers H. Wnt/beta-catenin signaling, disease, and emerging therapeutic modalities. *Cell.* 2017;169(6):985–999.
7. Daley GQ. Stem cells and the evolving notion of cellular identity. *Philos Trans R Soc Lond B Biol Sci.* 2015;370(1680):20140376.
8. Xu M, Horrell J, Snitow M, et al. WNT10A mutation causes ectodermal dysplasia by impairing progenitor cell proliferation and KLF4-mediated differentiation. *Nat Commun.* 2017;8:15397.
9. Visvader JE, Clevers H. Tissue-specific designs of stem cell hierarchies. *Nat Cell Biol.* 2016;18(4):349–355.
10. Bruno RD, Fleming JM, George AL, Boulanger CA, Schedin P, Smith GH. Mammary extracellular matrix directs differentiation of testicular and embryonic stem cells to form functional mammary glands in vivo. *Sci Rep.* 2017;7:40196.
11. Van Keymeulen A, Lee MY, Ousset M, et al. Reactivation of multipotency by oncogenic PIK3CA induces breast tumour heterogeneity. *Nature.* 2015;525(7567): 119–123.
12. Jones PH, Simons BD, Watt FM. Sic transit gloria: farewell to the epidermal transit amplifying cell? *Cell Stem Cell.* 2007;1(4):371–381.
13. Paguirigan A, Beebe DJ, Alexander CM. Simulating mouse mammary gland development: cell ageing and its relation to stem and progenitor activity. *Cell Prolif.* 2007;40(1):106–124.
14. Wang B, Zhao L, Fish M, Logan CY, Nusse R. Self-renewing diploid Axin2(+) cells fuel homeostatic renewal of the liver. *Nature.* 2015;524(7564):180–185.
15. Kopp JL, Grompe M, Sander M. Stem cells versus plasticity in liver and pancreas regeneration. *Nat Cell Biol.* 2016;18(3):238–245.
16. Van Keymeulen A, Rocha AS, Ousset M, et al. Distinct stem cells contribute to mammary gland development and maintenance. *Nature.* 2011;479(7372):189–193.
17. van Amerongen R, Bowman AN, Nusse R. Developmental stage and time dictate the fate of Wnt/beta-catenin-responsive stem cells in the mammary gland. *Cell Stem Cell.* 2012;11(3):387–400.
18. Eirew P, Stingl J, Eaves CJ. Quantitation of human mammary epithelial stem cells with in vivo regenerative properties using a subrenal capsule xenotransplantation assay. *Nat Protoc.* 2010;5(12):1945–1956.
19. Visvader JE, Stingl J. Mammary stem cells and the differentiation hierarchy: current status and perspectives. *Genes Dev.* 2014;28(11):1143–1158.

20. Stingl J, Eirew P, Ricketson I, et al. Purification and unique properties of mammary epithelial stem cells. *Nature.* 2006;439(7079):993–997.
21. Shackleton M, Vaillant F, Simpson KJ, et al. Generation of a functional mammary gland from a single stem cell. *Nature.* 2006;439(7072):84–88.
22. Magee JA, Piskounova E, Morrison SJ. Cancer stem cells: impact, heterogeneity, and uncertainty. *Cancer Cell.* 2012;21(3):283–296.
23. Wuidart A, Ousset M, Rulands S, Simons BD, Van Keymeulen A, Blanpain C. Quantitative lineage tracing strategies to resolve multipotency in tissue-specific stem cells. *Genes Dev.* 2016;30(11):1261–1277.
24. Choi N, Zhang B, Zhang L, Ittmann M, Xin L. Adult murine prostate basal and luminal cells are self-sustained lineages that can both serve as targets for prostate cancer initiation. *Cancer Cell.* 2012;21(2):253–265.
25. Wang ZA, Mitrofanova A, Bergren SK, et al. Lineage analysis of basal epithelial cells reveals their unexpected plasticity and supports a cell-of-origin model for prostate cancer heterogeneity. *Nat Cell Biol.* 2013;15(3):274–283.
26. Moser AR, Luongo C, Gould KA, McNeley MK, Shoemaker AR, Dove WF. Apc^{Min}: a mouse model for intestinal and mammary tumorigenesis. *Eur J Cancer.* 1995;31(7–8):1061–1064.
27. Chin EN, Martin JA, Kim S, Fakhraldeen SA, Alexander CM. Lrp5 has a Wnt-independent role in glucose uptake and growth for mammary epithelial cells. *Mol Cell Biol.* 2015;36(6):871–885.
28. Ball RK, Friis RR, Schoenenberger CA, Doppler W, Groner B. Prolactin regulation of beta-casein gene expression and of a cytosolic 120-kd protein in a cloned mouse mammary epithelial cell line. *EMBO J.* 1988;7(7):2089–2095.
29. Danielson KG, Oborn CJ, Durban EM, Butel JS, Medina D. Epithelial mouse mammary cell line exhibiting normal morphogenesis in vivo and functional differentiation in vitro. *Proc Natl Acad Sci USA.* 1984;81(12):3756–3760.
30. Campbell SM, Taha MM, Medina D, Rosen JM. A clonal derivative of mammary epithelial cell line COMMA-D retains stem cell characteristics of unique morphological and functional heterogeneity. *Exp Cell Res.* 1988;177(1):109–121.
31. Merlo GR, Basolo F, Fiore L, Duboc L, Hynes NE. p53-Dependent and p53-independent activation of apoptosis in mammary epithelial cells reveals a survival function of EGF and insulin. *J Cell Biol.* 1995;128(6):1185–1196.
32. Yu J, Virshup DM. Updating the Wnt pathways. *Biosci Rep.* 2014;34(5):e00142.
33. Nile AH, Mukund S, Stanger K, Wang W, Hannoush RN. Unsaturated fatty acyl recognition by Frizzled receptors mediates dimerization upon Wnt ligand binding. *Proc Natl Acad Sci USA.* 2017;114(16):4147–4152.
34. Ahn Y, Sims C, Logue JM, Weatherbee SD, Krumlauf R. Lrp4 and wise interplay controls the formation and patterning of mammary and other skin appendage placodes by modulating Wnt signaling. *Development.* 2013;140(3):583–593.
35. Ohazama A, Porntaveetus T, Ota MS, Herz J, Sharpe PT. Lrp4: A novel modulator of extracellular signaling in craniofacial organogenesis. *Am J Med Genet A.* 2010;152A (12):2974–2983.
36. Ahn Y, Sanderson BW, Klein OD, Krumlauf R. Inhibition of Wnt signaling by Wise (Sostdc1) and negative feedback from Shh controls tooth number and patterning. *Development.* 2010;137(19):3221–3231.
37. Kessenbrock K, Smith P, Steenbeek SC, et al. Diverse regulation of mammary epithelial growth and branching morphogenesis through noncanonical Wnt signaling. *Proc Natl Acad Sci USA.* 2017;114(12):3121–3126.
38. Roarty K, Shore AN, Creighton CJ, Rosen JM. Ror2 regulates branching, differentiation, and actin-cytoskeletal dynamics within the mammary epithelium. *J Cell Biol.* 2015;208(3):351–366.

39. van Tienen LM, Mieszczanek J, Fiedler M, Rutherford TJ, Bienz M. Constitutive scaffolding of multiple Wnt enhanceosome components by Legless/BCL9. *Elife.* 2017;6:e20882.
40. Langton PF, Kakugawa S, Vincent JP. Making, exporting, and modulating Wnts. *Trends Cell Biol.* 2016;26(10):756–765.
41. Goel S, Chin EN, Fakhraldeen SA, Berry SM, Beebe DJ, Alexander CM. Both LRP5 and LRP6 receptors are required to respond to physiological Wnt ligands in mammary epithelial cells and fibroblasts. *J Biol Chem.* 2012;287(20):16454–16466.
42. Macdonald BT, Semenov MV, Huang H, He X. Dissecting molecular differences between Wnt coreceptors LRP5 and LRP6. *PLoS One.* 2011;6(8):e23537.
43. Cruciat CM, Niehrs C. Secreted and transmembrane wnt inhibitors and activators. *Cold Spring Harb Perspect Biol.* 2013;5(3):a015081.
44. Larraguibel J, Weiss AR, Pasula DJ, Dhaliwal RS, Kondra R, Van Raay TJ. Wnt ligand-dependent activation of the negative feedback regulator Nkd1. *Mol Biol Cell.* 2015;26(12):2375–2384.
45. Yan KS, Janda CY, Chang J, et al. Non-equivalence of Wnt and R-spondin ligands during Lgr5+ intestinal stem-cell self-renewal. *Nature.* 2017;545(7653):238–242.
46. Kumar KK, Burgess AW, Gulbis JM. Structure and function of LGR5: an enigmatic G-protein coupled receptor marking stem cells. *Protein Sci.* 2014;23(5):551–565.
47. Carmon KS, Gong X, Yi J, Thomas A, Liu Q. RSPO-LGR4 functions via IQGAP1 to potentiate Wnt signaling. *Proc Natl Acad Sci USA.* 2014;111(13):E1221–E1229.
48. Cai C, Yu QC, Jiang W, et al. R-spondin1 is a novel hormone mediator for mammary stem cell self-renewal. *Genes Dev.* 2014;28(20):2205–2218.
49. Kimura H, Fumoto K, Shojima K, et al. CKAP4 is a Dickkopf1 receptor and is involved in tumor progression. *J Clin Invest.* 2016;126(7):2689–2705.
50. Wang D, Cai C, Dong X, et al. Identification of multipotent mammary stem cells by protein C receptor expression. *Nature.* 2015;517(7532):81–84.
51. Callahan R, Smith GH. Common integration sites for MMTV in viral induced mouse mammary tumors. *J Mammary Gland Biol Neoplasia.* 2008;13(3):309–321.
52. Veltmaat JM, Van Veelen W, Thiery JP, Bellusci S. Identification of the mammary line in mouse by Wnt10b expression. *Dev Dyn.* 2004;229(2):349–356.
53. Kendrick H, Regan JL, Magnay FA, et al. Transcriptome analysis of mammary epithelial subpopulations identifies novel determinants of lineage commitment and cell fate. *BMC Genomics.* 2008;9:591.
54. Ugolini F, Charafe-Jauffret E, Bardou VJ, et al. WNT pathway and mammary carcinogenesis: loss of expression of candidate tumor suppressor gene SFRP1 in most invasive carcinomas except of the medullary type. *Oncogene.* 2001;20(41):5810–5817.
55. Howe LR, Brown AM. Wnt signaling and breast cancer. *Cancer Biol Ther.* 2004;3(1): 36–41.
56. Xie Y, Zamponi R, Charlat O, et al. Interaction with both ZNRF3 and LGR4 is required for the signalling activity of R-spondin. *EMBO Rep.* 2013;14(12):1120–1126.
57. Valenta T, Hausmann G, Basler K. The many faces and functions of beta-catenin. *EMBO J.* 2012;31(12):2714–2736.
58. Roarty K, Rosen JM. Pygopus 2: tilting the Wnt-Notch balance in mammary epithelial lineage determination. *Breast Cancer Res.* 2013;15(6):322.
59. Zanconato F, Cordenonsi M, Piccolo S. YAP/TAZ at the roots of cancer. *Cancer Cell.* 2016;29(6):783–803.
60. Dupont S, Morsut L, Aragona M, et al. Role of YAP/TAZ in mechanotransduction. *Nature.* 2011;474(7350):179–183.
61. Aragona M, Panciera T, Manfrin A, et al. A mechanical checkpoint controls multicellular growth through YAP/TAZ regulation by actin-processing factors. *Cell.* 2013;154(5):1047–1059.

62. Butler MT, Wallingford JB. Planar cell polarity in development and disease. *Nat Rev Mol Cell Biol.* 2017;18(6):375–388.
63. Nguyen-Ngoc KV, Shamir ER, Huebner RJ, Beck JN, Cheung KJ, Ewald AJ. 3D culture assays of murine mammary branching morphogenesis and epithelial invasion. *Methods Mol Biol.* 2015;1189:135–162.
64. Jamieson PR, Dekkers JF, Rios AC, Fu NY, Lindeman GJ, Visvader JE. Derivation of a robust mouse mammary organoid system for studying tissue dynamics. *Development.* 2017;144(6):1065–1071.
65. Linnemann JR, Meixner LK, Miura H, Scheel CH. An organotypic 3D assay for primary human mammary epithelial cells that recapitulates branching morphogenesis. *Methods Mol Biol.* 2017;1612:125–137.
66. Chu EY, Hens J, Andl T, et al. Canonical WNT signaling promotes mammary placode development and is essential for initiation of mammary gland morphogenesis. *Development.* 2004;131(19):4819–4829.
67. Badders NM, Goel S, Clark RJ, et al. The Wnt receptor, Lrp5, is expressed by mouse mammary stem cells and is required to maintain the basal lineage. *PLoS One.* 2009;4(8):e6594.
68. Brisken C, O'Malley B. Hormone action in the mammary gland. *Cold Spring Harb Perspect Biol.* 2010;2(12):a003178.
69. Cai S, Kalisky T, Sahoo D, et al. A quiescent Bcl11b high stem cell population is required for maintenance of the mammary gland. *Cell Stem Cell.* 2017;20(2):247–260. e245.
70. Lim E, Vaillant F, Wu D, et al. Aberrant luminal progenitors as the candidate target population for basal tumor development in BRCA1 mutation carriers. *Nat Med.* 2009;15(8):907–913.
71. Chepko G, Smith GH. Three division-competent, structurally-distinct cell populations contribute to murine mammary epithelial renewal. *Tissue Cell.* 1997;29(2):239–253.
72. Kreuzaler PA, Staniszewska AD, Li W, et al. Stat3 controls lysosomal-mediated cell death in vivo. *Nat Cell Biol.* 2011;13(3):303–309.
73. Liu Y, McDermott SP, Khwaja SS, Alexander CM. The transforming ability of Wnt effectors correlates with their ability to induce the accumulation of mammary progenitor cells. *Proc Natl Acad Sci USA.* 2004;101:4158–4163.
74. Li Y, Welm B, Podsypanina K, et al. Evidence that transgenes encoding components of the Wnt signaling pathway preferentially induce mammary cancers from progenitor cells. *Proc Natl Acad Sci USA.* 2003;100(26):15853–15858.
75. Teuliere J, Faraldo MM, Deugnier MA, et al. Targeted activation of beta-catenin signaling in basal mammary epithelial cells affects mammary development and leads to hyperplasia. *Development.* 2005;132(2):267–277.
76. Zhang L, Adileh M, Martin ML, et al. Establishing estrogen-responsive mouse mammary organoids from single Lgr5+ cells. *Cell Signal.* 2017;29:41–51.
77. Plaks V, Brenot A, Lawson DA, et al. Lgr5-expressing cells are sufficient and necessary for postnatal mammary gland organogenesis. *Cell Rep.* 2013;3(1):70–78.
78. Wang Y, Dong J, Li D, et al. Lgr4 regulates mammary gland development and stem cell activity through the pluripotency transcription factor Sox2. *Stem Cells.* 2013;31(9):1921–1931.
79. Rajaram RD, Buric D, Caikovski M, et al. Progesterone and Wnt4 control mammary stem cells via myoepithelial crosstalk. *EMBO J.* 2015;34(5):641–652.
80. Joshi PA, Jackson HW, Beristain AG, et al. Progesterone induces adult mammary stem cell expansion. *Nature.* 2010;465(7299):803–807.
81. Nusse R, Varmus HE. Many tumors induced by the mouse mammary tumor virus contain a provirus integrated in the same region of the host genome. *Cell.* 1982;31(1):99–109.

82. Yu QC, Verheyen EM, Zeng YA. Mammary development and breast cancer: a Wnt perspective. *Cancers.* 2016;8(7):65.
83. MacDonald BT, Tamai K, He X. Wnt/beta-catenin signaling: components, mechanisms, and diseases. *Dev Cell.* 2009;17(1):9–26.
84. Mahler-Araujo B, Savage K, Parry S, Reis-Filho JS. Reduction of E-cadherin expression is associated with non-lobular breast carcinomas of basal-like and triple negative phenotype. *J Clin Pathol.* 2008;61(5):615–620.
85. Parker C, Rampaul RS, Pinder SE, et al. E-cadherin as a prognostic indicator in primary breast cancer. *Br J Cancer.* 2001;85(12):1958–1963.
86. TCGA.Comprehensive molecular portraits of human breast tumours. *Nature.* 2012;490 (7418):61–70.
87. Canas-Marques R, Schnitt SJ. E-cadherin immunohistochemistry in breast pathology: uses and pitfalls. *Histopathology.* 2016;68(1):57–69.
88. Rakha EA, Patel A, Powe DG, et al. Clinical and biological significance of E-cadherin protein expression in invasive lobular carcinoma of the breast. *Am J Surg Pathol.* 2010;34(10):1472–1479.
89. Gumbiner BM. Propagation and localization of Wnt signaling. *Curr Op Gen Dev.* 1998;8:430–435.
90. Geyer FC, Lacroix-Triki M, Savage K, et al. β-Catenin pathway activation in breast cancer is associated with triple-negative phenotype but not with CTNNB1 mutation. *Mod Pathol.* 2011;24(2):209–231.
91. van de Wetering M, Barker N, Harkes IC, et al. Mutant E-cadherin breast cancer cells do not display constitutive Wnt signaling. *Cancer Res.* 2001;61(1):278–284.
92. Incassati A, Chandramouli A, Eelkema R, Cowin P. Key signaling nodes in mammary gland development and cancer: beta-catenin. *Breast Cancer Res.* 2010;12(6):213.
93. Lu B, Green BA, Farr JM, Lopes FC, Van Raay TJ. Wnt drug discovery: weaving through the screens, patents and clinical trials. *Cancers.* 2016;8(9):E82.
94. Anastas JN, Moon RT. WNT signalling pathways as therapeutic targets in cancer. *Nat Rev Cancer.* 2013;13(1):11–26.
95. Coussy F, Lallemand F, Vacher S, et al. Clinical value of R-spondins in triple-negative and metaplastic breast cancers. *Br J Cancer.* 2017;116(12):1595–1603.
96. Jamieson C, Sharma M, Henderson BR. Targeting the beta-catenin nuclear transport pathway in cancer. *Semin Cancer Biol.* 2014;27:20–29.
97. Polakis P. Wnt signaling and cancer. *Genes Dev.* 2000;14:1837–1851.
98. Ayyanan A, Civenni G, Ciarloni L, et al. Increased Wnt signaling triggers oncogenic conversion of human breast epithelial cells by a Notch-dependent mechanism. *Proc Natl Acad Sci USA.* 2006;103(10):3799–3804.
99. Weigelt B, Eberle C, Cowell CF, Ng CK, Reis-Filho JS. Metaplastic breast carcinoma: more than a special type. *Nat Rev Cancer.* 2014;14(3):147–148.
100. Lawrence MS, Stojanov P, Mermel CH, et al. Discovery and saturation analysis of cancer genes across 21 tumour types. *Nature.* 2014;505(7484):495–501.
101. Jiang G, Zhang S, Yazdanparast A, et al. Comprehensive comparison of molecular portraits between cell lines and tumors in breast cancer. *BMC Genomics.* 2016;17 (Suppl 7):525.
102. Mackay A, Tamber N, Fenwick K, et al. A high-resolution integrated analysis of genetic and expression profiles of breast cancer cell lines. *Breast Cancer Res Treat.* 2009;118 (3):481–498.
103. Hollestelle A, Elstrodt F, Nagel JH, Kallemeijn WW, Schutte M. Phosphatidylinositol-3-OH kinase or RAS pathway mutations in human breast cancer cell lines. *Mol Cancer Res.* 2007;5(2):195–201.
104. Hollestelle A, Nagel JH, Smid M, et al. Distinct gene mutation profiles among luminal-type and basal-type breast cancer cell lines. *Breast Cancer Res Treat.* 2010;121(1):53–64.

105. Riaz M, Elstrodt F, Hollestelle A, Dehghan A, Klijn JG, Schutte M. Low-risk susceptibility alleles in 40 human breast cancer cell lines. *BMC Cancer.* 2009;9:236.
106. Timmerman LA, Holton T, Yuneva M, et al. Glutamine sensitivity analysis identifies the xCT antiporter as a common triple-negative breast tumor therapeutic target. *Cancer Cell.* 2013;24(4):450–465.
107. Matsuda Y, Schlange T, Oakeley EJ, Boulay A, Hynes NE. WNT signaling enhances breast cancer cell motility and blockade of the WNT pathway by sFRP1 suppresses MDA-MB-231 xenograft growth. *Breast Cancer Res.* 2009;11(3):R32.
108. Liu CC, Prior J, Piwnica-Worms D, Bu G. LRP6 overexpression defines a class of breast cancer subtype and is a target for therapy. *Proc Natl Acad Sci USA.* 2010;107(11):5136–5141.
109. Chen X, Iliopoulos D, Zhang Q, et al. XBP1 promotes triple-negative breast cancer by controlling the HIF1alpha pathway. *Nature.* 2014;508(7494):103–107.
110. Kim S, Roopra A, Alexander CM. A phenotypic mouse model of basaloid breast tumors. *PLoS One.* 2012;7(2):e30979.
111. Clevers H, Nusse R. Wnt/beta-catenin signaling and disease. *Cell.* 2012;149(6):1192–1205.
112. Gunther EJ, Belka GK, Wertheim GB, et al. A novel doxycycline-inducible system for the transgenic analysis of mammary gland biology. *Faseb J.* 2002;16(3):283–292.
113. Gunther EJ, Moody SE, Belka GK, et al. Impact of p53 loss on reversal and recurrence of conditional Wnt-induced tumorigenesis. *Genes Dev.* 2003;17:488–501.
114. Scheel C, Weinberg RA. Cancer stem cells and epithelial-mesenchymal transition: concepts and molecular links. *Semin Cancer Biol.* 2012;22(5–6):396–403.
115. Kim S, Goel S, Alexander CM. Differentiation generates paracrine cell pairs that maintain basaloid mouse mammary tumors: proof of concept. *PLoS One.* 2011;6(4):e19310.
116. Wang X, Jung YS, Jun S, et al. PAF-Wnt signaling-induced cell plasticity is required for maintenance of breast cancer cell stemness. *Nat Commun.* 2016;7:10633.
117. Smalley MJ, Reis-Filho JS, Ashworth A. BRCA1 and stem cells: tumour typecasting. *Nat Cell Biol.* 2008;10(4):377–379.
118. Agliano A, Calvo A, Box C. The challenge of targeting cancer stem cells to halt metastasis. *Semin Cancer Biol.* 2017;44:25–42.
119. Tammela T, Sanchez-Rivera FJ, Cetinbas NM, et al. A Wnt-producing niche drives proliferative potential and progression in lung adenocarcinoma. *Nature.* 2017;545(7654):355–359.
120. Lien WH, Fuchs E. Wnt some lose some: transcriptional governance of stem cells by Wnt/beta-catenin signaling. *Genes Dev.* 2014;28(14):1517–1532.
121. Dobrolecki LE, Airhart SD, Alferez DG, et al. Patient-derived xenograft (PDX) models in basic and translational breast cancer research. *Cancer Metastasis Rev.* 2016;35(4):547–573.
122. Green JL, La J, Yum KW, et al. Paracrine Wnt signaling both promotes and inhibits human breast tumor growth. *Proc Natl Acad Sci USA.* 2013;110(17):6991–6996.
123. Pond AC, Herschkowitz JI, Schwertfeger KL, et al. Fibroblast growth factor receptor signaling dramatically accelerates tumorigenesis and enhances oncoprotein translation in the mouse mammary tumor virus-Wnt-1 mouse model of breast cancer. *Cancer Res.* 2010;70(12):4868–4879.
124. Brady N, Chuntova P, Bade LK, Schwertfeger KL. The FGF/FGFR axis as a therapeutic target in breast cancer. *Expert Rev Endocrinol Metab.* 2013;8(4):391–402.
125. Giannakis M, Hodis E, Jasmine Mu X, et al. RNF43 is frequently mutated in colorectal and endometrial cancers. *Nat Genet.* 2014;46(12):1264–1266.
126. Koo BK, Spit M, Jordens I, et al. Tumour suppressor RNF43 is a stem-cell E3 ligase that induces endocytosis of Wnt receptors. *Nature.* 2012;488(7413):665–669.

127. Zheng S, Cherniack AD, Dewal N, et al. Comprehensive pan-genomic characterization of adrenocortical carcinoma. *Cancer Cell.* 2016;30(2):363.
128. Wu J, Jiao Y, Dal Molin M, et al. Whole-exome sequencing of neoplastic cysts of the pancreas reveals recurrent mutations in components of ubiquitin-dependent pathways. *Proc Natl Acad Sci USA.* 2011;108(52):21188–21193.
129. Jiang X, Hao HX, Growney JD, et al. Inactivating mutations of RNF43 confer Wnt dependency in pancreatic ductal adenocarcinoma. *Proc Natl Acad Sci USA.* 2013;110(31):12649–12654.
130. Steinhart Z, Pavlovic Z, Chandrashekhar M, et al. Genome-wide CRISPR screens reveal a Wnt-FZD5 signaling circuit as a druggable vulnerability of RNF43-mutant pancreatic tumors. *Nat Med.* 2017;23(1):60–68.

CHAPTER TEN

The Central Role of Wnt Signaling and Organoid Technology in Personalizing Anticancer Therapy

Elizabeth Vincan*,†,1, Renate H.M. Schwab*, Dustin J. Flanagan*, Jean M. Moselen*, Bang M. Tran*, Nick Barker§, Toby J. Phesse*,‡

* Doherty Institute of Infection and Immunity, University of Melbourne, Melbourne, VIC, Australia
† Curtin University, Perth, WA, Australia
‡ European Cancer Stem Cell Research Institute, Cardiff University, Cardiff, United Kingdom
§ A*STAR Institute of Medical Biology, Singapore
1 Corresponding author. E-mail address: evincan@unimelb.edu.au

Contents

Abstract

The Wnt pathway is at the heart of organoid technology, which is set to revolutionize the cancer field. We can now predetermine a patient's response to any given anticancer therapy by exposing tumor organoids established from the patient's own tumor. This cutting-edge biomedical platform translates to patients being treated with the correct drug at the correct dose from the outset, a truly personalized and precise medical approach. A high throughput drug screen on organoids also allows drugs to be tested in limitless combinations. More recently, the tumor cells that are resistant to the therapy given to a patient were selected in culture using the patient's

Progress in Molecular Biology and Translational Science, Volume 153
ISSN 1877-1173
http://dx.doi.org/10.1016/bs.pmbts.2017.11.009

organoids. The resistant tumor organoids were then screened empirically to identify drugs that will kill the resistant cells. This information allows diagnosis in real-time to either prevent tumor recurrence or effectively treat the recurring tumor. Furthermore, the ability to culture stem cell–derived epithelium as organoids has enabled us to begin to understand how a stem cell becomes a cancer cell or to pin-point the genetic alteration that underlies a given genetic syndrome. Here we summarize these advances and the central role of Wnt signaling, and identify the next challenges for organoid technology.

1. INTRODUCTION

Stem cell–derived three-dimensional (3D) replicas of organs grown in tissue culture, termed organoids, have led to remarkable advances in stem cell and developmental biology, human disease, and regenerative medicine.[1] "Organoid" is a term originally used by developmental biologists working with tissue explants to unravel the mechanisms of organogenesis. The term literally means "organ-like." More recent use of the term organoid is defined as a 3D structure established from stem cells and consisting of organ-specific cell types that self-organize to mimic their tissue of origin.[2,3] Organoids can be initiated from two main types of stem cells (summarized in Table 1). The

Table 1 Two Ways of Generating Adult Tissue Organoids

	iPS	**Organ Restricted Stem Cell**
	One stem cell can generate cell types with characteristics of divers tissues	Tissue stem cell "remembers" tissue of origin
Step 1	Isolate adult mature cell	Isolate[a] adult stem cell
Step 2	Reprogram it to be a PS cell (i.e., *induced PS*)	Self-organize in culture to form "mini-tissues"
Step 3	Differentiate to different tissue cell types (embryonic cues)	Generate the tissue-specific cell types and architecture (adult cues)
	Model normal and disease tissue	Model normal and disease tissue
	Regenerative medicine/ transplantation	Regenerative medicine/ transplantation
		Applicable to diseased tissues, for example, solid tumors; personalized anticancer treatment
	Autologous diverse tissue cell types	

iPS, Pluripotent stem cell; PS, Pluripotent stem.
[a]It is not necessary to isolate the adult stem cells; tissue pieces also yield organoids.

first stem cell type is the pluripotent embryonic stem (ES) cell or the induced pluripotent stem cell (iPS). For iPS, adult cells are artificially reprogrammed to pluripotency,[4] and then differentiated toward different organ cell types using cues that have been identified to orchestrate the development of those organs during embryogenesis and organogenesis.[3,5] Diverse tissue and organ cell types can be derived from a pluripotent stem cell.

By contrast, the second type of stem cells, the tissue restricted adult stem cells, have a "memory" of their tissue of origin and self-organize and differentiate into structures that contain the different tissue-specific cell types; they recapitulate the characteristics of tissue function and architecture.[1] The culture conditions that were developed to establish adult tissue stem cell-derived organoids were then adapted to growing organoids from diseased tissues, such as cancers (Table 1). This innovation has led to one of the most important advances in cancer research—high throughput drug prescreening, in a clinically relevant time frame, on patient-derived tumor organoids to personalize treatment.[6,7] In this chapter we briefly summarize the discoveries that led to adult stem-cell-derived organoid technology, the central role of Wnt signaling in this advance and how this advance is poised to revolutionize anticancer treatment. We also highlight the next challenges for patient-derived tumor organoid technology in the quest for curative anticancer treatment. Improved survival is clearly a great outcome for anticancer treatment, but the ultimate goal is curative treatment where the tumor cells are eliminated.

2. WNT SIGNALING PATHWAY

The Wnt signal transduction pathway has several branches that are β-catenin-dependent and β-catenin-independent (i.e., the calcium and planar cell polarity pathways) and the core components of these are highly conserved through evolution.[8,9] Here we give a brief overview of the Wnt/β-catenin branch[10] as it is critical to stem cell function and organoid formation. In the absence of a Wnt ligand, β-catenin is primarily engaged in cell-cell adherens junctions. Any free newly synthesized β-catenin is rapidly recruited to a cytoplasmic destruction complex that contains several proteins, including adenomatous polyposis coli (APC), Axin and two kinases, casein kinase-1 (CK1), and glycogen synthase kinase-3 (GSK-3) that phosphorylate the recruited β-catenin. Subsequently, phosphorylated β-catenin is ubiquitylated by β-transduction repeat-containing protein (β-TrCP) targeting it for

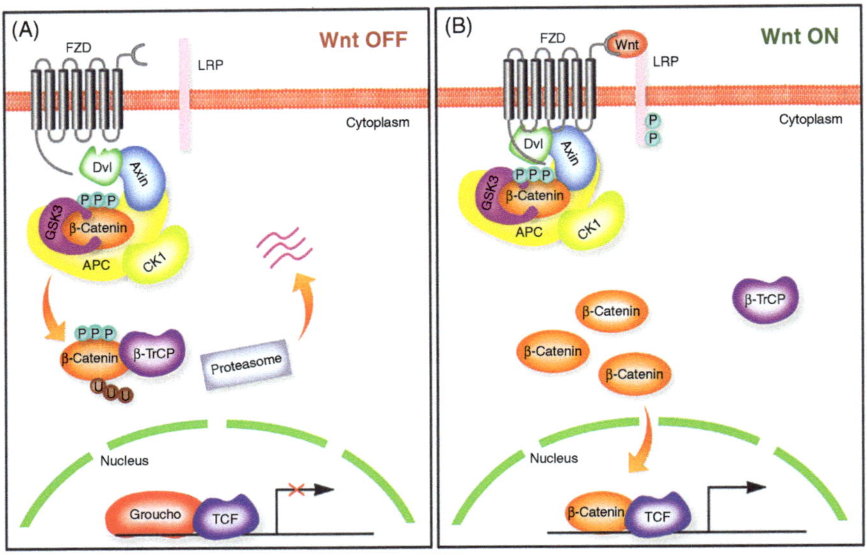

Fig. 1 Schematic of the Wnt/β-catenin signal transduction pathway. The pathway is inactive in the absence of Wnt [(A) "Wnt OFF"] and active upon Wnt binding [(B) "Wnt ON"]. *APC*, Adenomatous polyposis coli; *CK1*, casein kinase 1; *β-cat*, β-catenin; *FZD*, Frizzled; *GSK*, glycogen synthase kinase; *LRP*, low-density lipoprotein-related protein; *TCF*, T-cell factor/lymphocyte enhancer factor; *β-TrCP*, β-transduction repeat-containing protein.

proteasomal degradation (Fig. 1A). Upon Wnt binding to Frizzled (Fz) and its coreceptor LRP (low-density lipoprotein-related protein), the intracellular domain of LRP is phosphorylated and the destruction complex relocates to the receptor complex to transduce the signal. The mechanisms are still unclear, but ubiquitylation of β-catenin is inhibited and β-catenin escapes degradation. It accumulates in the cytoplasm and eventually translocates to the nucleus where it forms a transcriptionally active complex with T-cell factor (TCF)/lymphoid enhancer factor (LEF) transcription factors. β-catenin-mediated activation of transcription replaces Groucho (or TLE)-mediated repression of TCF/LEF to initiate the expression of Wnt/β-catenin target genes (Fig. 1B).[11]

The Wnt/β-catenin target genes that are induced are context-dependent and can mediate diverse outcomes in different cell types, even if the cell types are within the same tissue. For example, in the intestinal epithelium, active TCF/β-catenin transcription leads to the differentiation of Paneth cells[12] but it is also necessary for the proliferation of stem cells and early progenitors,[13] thus highlighting distinct target gene repertoires within the same tissue.[14]

An emerging theme in the field is that it is the fold change in the level of nuclear β-catenin rather than the absolute amount of β-catenin in the nucleus dictating activation of Wnt signaling.[15] Also, the pathway is not as simple as an "ON–OFF" switch; the level of Wnt activation dictates the ultimate cellular outcome. The latter led to the proposal of the "just right" or Goldilocks model of Wnt signaling[16] with extensive regulation of the signaling cascade at each step of the pathway, especially in cancer cells.[8,17]

Indeed, tight regulation of events at the plasma membrane certainly comes to the fore in cancer. This is perhaps not surprising given that the tumor cell microenvironment influences cell behavior and the Wnt pathway plays a key role in this regulation. This has been particularly well documented for colon cancer[18,19] (addressed further in Section 5). Several types of inhibitors of Wnt signaling are *bone fide* tumor suppressors in diverse cancers, including colon cancer.[20] Wnt pathway inhibitors that directly bind Wnt ligands, such as the Frizzled-related proteins (sFRPs) and Wnt inhibitory factor (WIF), can potentially affect any branch of the Wnt pathway. While other inhibitors like the DKK family act specifically on the Wnt/β-catenin branch as they block Wnt binding to LRP. sFRPs,[21–23] WIF,[24] and DKK (Dickkopf)[25] are epigentically silenced in colon cancer, for example, implicating a role for active Wnt signaling from the receptor complex.[20]

3. ORGANOIDS DERIVED FROM ADULT EPITHELIUM

Adult stem cell–derived organoids were first established from the epithelium lining the mouse intestine.[26] Several discoveries led to this game-changing achievement. The first was the demonstration in 1998 by Korinek and coworkers that Wnt signaling is critical for intestinal stem cells. Deletion of the gene that codes for Tcf4, the downstream effector of Wnt signaling (Fig. 1), from the developing mouse intestine led to a depletion of the putative epithelial stem cell compartment and postnatal death.[13] A similar depletion of the stem cell compartment was observed when other components of the Wnt pathway, for example, β-catenin,[27] were depleted from the epithelium or the Wnt-inhibitor Dkk-1 was overexpressed.[28,29]

The next important discovery almost a decade later was identifying leucine-rich repeat–containing G-protein–coupled receptor 5 (Lgr5), a highly expressed Wnt target gene in the intestinal crypt progenitor compartment,[30] as an exclusive marker of adult intestinal stem cells.[14] The epithelial lining of the intestine has the fastest turnover of any tissue in the adult, with the entire lining being replaced every several days. Using mice engineered to track Lgr5 positive (Lgr5$^+$) stem cells and their progeny led to the demonstration that this turnover of the epithelium is indeed maintained by Lgr5$^+$ stem cells that reside at the base of invaginations within the epithelium called the crypts of Lieberkuhn[14] (Fig. 2A).

The third important advance was the demonstration that, given the correct growth factors and environment, a single Lgr5$^+$ stem cell can self-renew and give rise to daughter cells that self-assemble to form a complex 3D structure containing all known intestinal epithelial cell types, thus forming ever-expanding 3D intestinal organoids in tissue culture[26] (represented in Fig. 2D). To demonstrate this, Lgr5$^+$ intestinal stem cells were purified using Lgr5-promoter driven EGFP expression and FACS sorting. The ability to grow "mini-guts" in tissue culture has enabled the identification and characterization of the stem cell niche factors,[31,32] and an understanding of the molecular mechanisms of Wnt signaling in stem cell maintenance, epithelial cell differentiation,[33] and aberrant Wnt signaling at the initiation of cancer.[34]

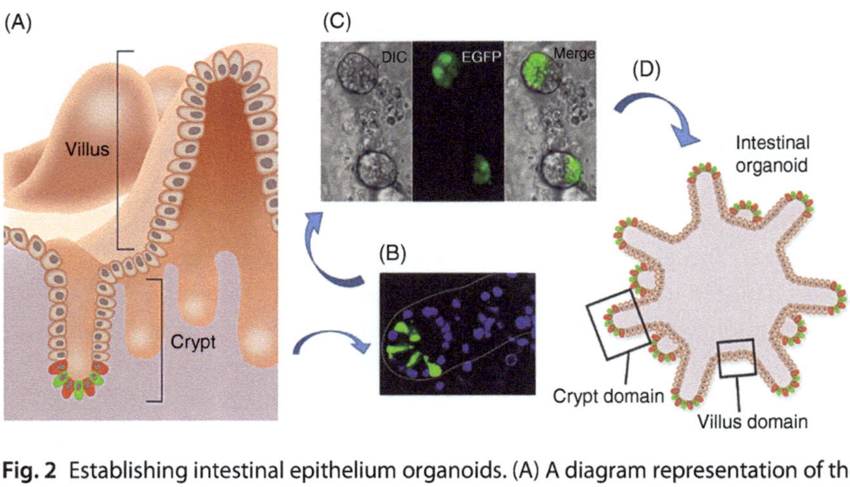

Fig. 2 Establishing intestinal epithelium organoids. (A) A diagram representation of the intestinal epithelium with crypts and villi. The intestinal stem cells *(green)* are flanked by Paneth cells *(red)* at the base of the crypts. Several crypts feed cells onto one villus, which is a finger-like projection into the lumen. (B) An isolated crypt from the Lgr5-EGFP-IRES-Cre mouse shows the *green* (EGFP$^+$) stem cells (this is a fresh crypt preparation). (C) After 1 day in culture, the isolated crypts form cysts with stem cells *(green cells)* polarized to one end of the cyst. (D) Over several days, the cells generated by the stem cells then self-organize into an organoid with domains that represent the crypt and villus of the epithelium. *DIC*, Differential interference contrast.

3.1 Lgr5 Stem Cells Generate 3D Organoid

To establish intestinal organoids, the crypts of the epithelium (Fig. 2A–B) are isolated and resuspended in a solubilized basement membrane preparation rich in extracellular matrix proteins (called Matrigel), which is liquid at 4°C but sets as a gel at 37°C. This provides the 3D matrix for the crypts to form organoids. Once the gel has set, it is overlaid with medium containing growth factors that recapitulate the in vivo crypt niche. Just three factors are necessary to maintain ever-expanding intestinal organoids: R-spondin to potentiate endogenous Wnt signals, the BMP inhibitor Noggin, and epidermal growth factor (EGF).[26] Within 24 h of plating, the crypts form cysts that are already polarized with stem cells to one pole (Fig. 2C); after several days in culture, organoids are formed with defined crypt and villus domains (Fig. 2D); the different epithelial cell types are represented in roughly equivalent relative proportions to the normal epithelium.[26] The organoids shed dead cells into the lumen and will eventually burst open, releasing dead cells, if not passaged. The intestinal organoids continue expanding and require fresh growth factors every other day, and passaging each week.[26] The organoids are genetically stable with continuous passage,[35] can be manipulated genetically using current genetic tools, such as CRISPR/Cas9 genetic editing,[36,37] Cre-LoxP mediated gene manipulation,[38] or transfection/transduction.[39] Intriguingly, when embedded within contracting collagen gels, intestinal organoids fuse to form macroscopic intestinal tubes that have a continuous lumen lined by villus cell types and crypt-like structures budding from the tube into the collagen.[40] Thus, lengths of intestine can be generated in tissue culture without the need for complex tissue-engineering scaffolds.

Organoids also hold great promise for regenerative medicine and transplantation, but one drawback is the cost of recombinant growth factors and animal cell line–derived matrix (Matrigel is derived from mouse sarcoma cell line). The cost of reagents was partly alleviated using conditioned media from cell lines that secrete the growth factors (e.g., Wnt3a, R-spondin, Noggin producing cell lines).[41] However, conditioned medium is not defined as it contains unknown components and thus is not applicable for generating tissue for transplantation and other clinical applications. To this end, several avenues are being investigated to overcome these barriers. For example, stabilizing Wnt in serum free medium. Unlike intestinal organoids, several other organoid types require Wnt in the growth factor cocktail. However,

Wnt ligands are hydrophobic and require serum for optimal activity; recombinant Wnts perform poorly compared to conditioned medium. A glycoprotein called afamin was recently identified as a component of serum that stabilizes Wnt and purified afamin improves the performance of recombinant Wnt.[42] Other groups have used phospholipids and cholesterol carriers to stabilize Wnt[43] or water soluble surrogate Wnt agonists that activate Wnt/β-catenin signaling.[44] Also, advances in bioengineering[45] has seen the development of modular synthetic hydrogel matrices that replace the need for Matrigel. Extracellular proteins were incorporated into the hydrogel networks and the components necessary for organoid formation, stem cell maintenance and cell differentiation have been defined.[46] Furthermore, these synthetic matrices allow the "stiffness" of the matrix to be varied, as the physical properties of the cellular microenvironment also affects cell behavior.[46] This is reminiscent of the well-documented effects of matrix "stiffness" on cancer cell behavior.[47]

4. ORGANOIDS DERIVED FROM PATIENT TUMORS

Once the culture techniques for growing intestinal organoids were established, variations on the same culture protocol led to the establishment of organoids from several other gastrointestinal tissues,[41,48] as well as many other stem-cell maintained adult tissues (several comprehensive recent reviews[49–51]). Clevers and coworkers also adapted the organoid protocols to grow patient-derived colon cancer tumor organoids. The mini-tumor organoids similarly recapitulate the features of actively growing colon cancers and can accurately predict the patient's response to treatment.[6] Consequently, mini-gut and mini-tumor organoid platforms provide powerful tools for drug discovery and predictive drug-response diagnostics for cancer treatment. Generally, about 60% of colon cancer patients respond to any particular therapeutic regimen. The nonresponders are then treated with alternative drug regimens. The ability to prescreen the drugs on an individual patient's tumor cells, both singly and in limitless combinations with other drugs, means that the patient is treated with the drug regimen that will work on their tumor from the outset.[6] This not only saves time between diagnosis and effective treatment, but also eliminates unnecessary treatment and the consequent side effects. Here we focus on colon cancer to highlight the advances and the next challenges of organoid technology in the cancer field.

4.1 Wnt Signaling in Colon Cancer

Colon cancer starts in the simple epithelium that lines the colon. Most colon cancers in humans, including somatic cancers, arise from adenomas (noncancerous tumors or polyps) that harbor truncating mutations in the *adenomatous polyposis coli* (*APC*) tumor suppressor gene.[52,53] These mutations in the *APC* gene lead to constitutive activation of the Wnt/β-catenin pathway[54,55] and the formation of adenomas in the epithelium of the colon. The adenomas can then progress to cancerous tumors through the accumulation of mutations to activate oncogenes and inactivate tumor suppressors.[56] Human colon cancers that do not harbor *APC* gene mutations often have oncogenic mutations in the β-catenin gene (*CTNNB1*).[53] Thus, the vast majority of colon cancers have mutations in the intracellular pathway components that activate Wnt/β-catenin signaling.

APC facilitates the phosphorylation and subsequent targeting of β-catenin for proteasomal degradation (Fig. 1). In colon cancer, mutations to the *APC* gene lead to a truncated APC protein and this facilitative function is lost, and consequently, the constitutive activation of the pathway.[54,55] Oncogenic mutations to *CTNNB1* alter the negative regulatory domain of β-catenin at the N-terminus, and again, lead to constitutive activation of the Wnt pathway.[57] Transgenic mice harboring these alterations to APC[58,59] or β-catenin[60] genes develop multiple intestinal adenomas with active Wnt signaling.

In addition to these mutations that activate the pathway, Wnt signaling is further regulated through multiple mechanisms in colon cancer.[8] The Frizzled (FZD) receptors and Wnt ligands are overexpressed in colon cancer and can modulate the pathway,[61,62] while naturally occurring inhibitors of Wnt-FZD interaction (e.g., sFRP) are epigenetically silenced and are *bona fide* tumor suppressors in human colon cancer.[20,21,23] Curiously, the net effect is to "constrain" the Wnt signaling pathway in the cancer cells to a submaximal level of activation as signaling can be decreased[23] and increased experimentally[63] and is hyper-activated in cancer cells engaged in tumor invasion.[64,65]

4.2 Genetic Dissection of Colon Cancer Development Using Organoids

In addition to constitutively active Wnt signaling at the initiation of colon cancer, progression from adenoma to carcinoma requires mutations in genes in other oncogenic pathways. The adenoma-carcinoma sequence was originally proposed by Vogelstein[56] and coworker based on the analysis of mutations present at each stage of tumor progression. Sequential acquisition of mutations that lead to functional loss of function of other tumor suppressors, such as *TP53* and *SMAD4*, as well as activating mutations in other oncogenes, such as *KRAS*. Intriguingly, the introduction of these mutations sequentially and in combination into normal epithelial cells using organoids and gene editing (CRISPR/Cas9) has not only identified the minimal mutations necessary for cancer development but has also provided an explanation of the growth factor requirements for organoid growth.[66] That is, each genetic mutation alleviates the need for a growth factor. Mutations to activate Wnt signaling alleviate the need for R-spondin and Wnt, mutation in *KRAS* alleviate the need for EGF and inactivating mutation in *SMAD4* alleviate the need for Noggin to inhibit BMP signaling. With the additional mutation of *TP53*, the quadruple mutant organoids grew without the need for growth factors and formed invasive subcutaneous tumors.[66] Using a similar strategy, Matano *et al.* showed that mutation in PI3K pathway (*PIK3CA*) can substitute for *KRAS* mutation.[67] These initial studies have been expanded upon to demonstrate that quadruple mutant organoids (i.e., *APC*, *KRAS*, *p53*, *SMAD4*) yield invasive tumors in an orthotopic mouse model.[68] These findings indicate that the loss of niche dependency leads to the ability to metastasize to secondary organs, at least in an experimental metastasis model.

4.3 Patient-Derived Tumor Organoids for Drug Prescreen and Biobanking

An important application of organoid technology has been the ability to establish tumor organoids from resected and biopsy samples and to adapt the mini-tumor organoids to high-throughput drug screens. This was first achieved by the Clevers lab in a retrospective study where patient drug response was compared to the response of the corresponding patient-derived organoids, and coupling this with genomics to identify gene-drug associations.[6] The patient-derived tumor organoids are a faithful replica of the patient's tumor and can be established from primary tumors and metastases.[69] Patient-derived organoids and the "omic" analyses of these (genomic, epigenomic, transcriptomic, proteomic), have revealed that normal organoids are more stable in culture than malignant tissues,[70,71] which might have been expected given that chromosomal instability is a common feature of cancer, but needed formal demonstration. Furthermore, by establishing a biobank of patient-derived tumor and normal tissue organoids, drug discovery and "clinical trials" are expedited. Novel drugs can be tested by simply thawing out a panel of tumor organoids and the high-throughput format allows for multiple combinations, titrations, etc. A not-for-profit foundation (HUB) has been established by the Clevers group to advance this technology (http://hub4organoids.eu/).

Patient-derived xenografts (PDX) have become the gold standard for "personalized anticancer treatment" (numerous recent reviews, e.g., Byrne *et al.*).[72] However, PDX models are limited in their application to personalized medicine for several seasons. The success rate of establishing xenograft tumors from patient material is low, the time to establish tumor xenografts is slow, and the cost of mouse models is just a few of the caveats. PDX also does not lend itself to high throughput. Delivering information in a clinically relevant timeframe is a real hindrance to using PDX models as a diagnostic tool. Patient-derived organoids fulfill this unmet need for personalizing anticancer diagnosis and treatment. Biobanks, such as the HUB are being established around the world, for example, the nonprofit organization Australian Living Organoid Alliance (ALOA).

DasGupta and coworkers recently took the high-throughput screen on patient-derived organoids to a new dimension.[73] They generated a library of patient derived organoids and PDX models from head and neck squamous cell carcinomas and used these to select for tumor cells that are resistant to the standard treatment given to the patient. The resistant organoids were then comprehensively interrogated to identify patient-specific gene signatures that could potentially underlie the resistance to therapy. Using this strategy they identified that selection for Yes-associated protein-1 (YAP-1) positive cells paralleled failed therapy; implicating YAP-1 is a putative biomarker for resistance.[73] Tumors are heterogeneous and the ability to select resistant cells using organoid culture coupled with the power of "omics" analyses allows diagnosis in real-time. Such a high throughput drug screen is not possible with any other patient-derived model. Numerous other cancers and tissues have been effectively modeled using organoids and these have been covered by a number of comprehensive recent reviews.[74] Next we will highlight the new challenges for patient-derived organoids.

5. MODELING DORMANT TUMOR CELLS: THE NEXT FRONTIER FOR TUMOR ORGANOIDS

One limitation for mini-tumor organoids as a drug screen is that the ever-expanding tumor organoids mimic the actively growing tumor cells. The mini-tumor organoids do not mimic the dormant tumor cell state. The key to curative cancer treatment is to therapeutically target and eliminate the disseminated dormant tumor cells that eventually reestablish tumors at secondary sites and are ultimately the cause of death. Some therapies will target actively dividing, as well as dormant tumor cells; but we need to be able to establish "dormant" organoid cells from the patient's mini-tumor organoids to screen for these. This is the next challenge for the organoid platform, but we have clues about how to do this from other model systems.

5.1 Reversible Phenotype Transitions Underlie Metastasis

Although most human colon cancers are relatively well differentiated with an epithelial phenotype, in localized areas, termed the "invasive front," the tumor cells take on a more mesenchymal phenotype that is associated with migratory and invasive properties, and the cells shut-down cell proliferation.[18] This phenotypic change, termed epithelial-to-mesenchymal transition (EMT), is thought to enable the tumor cells to dissociate from the tumor mass and disseminate to other organs in the body. The disseminated tumor cells are dormant and acquire resistance to therapies, particularly therapies that target actively dividing cells, such as chemotherapy and radiation therapy. The EMT program also induces stem cell-specific gene expression, thus promoting self-renewal properties.[19] Dissemination can occur early in the disease process, and tumor cells can sit dormant for many years. However, for the tumor cells to reinstate tumor growth at the secondary site, the cells must undergo the reverse transition, mesenchymal-to-epithelial transition (MET), because the secondary tumors recapitulate the differentiated epithelial phenotype of the primary tumor.[18,75]

5.2 An In Vitro Model of Tumor Morphogenesis

5.2.1 Modeling Dormant Tumor Cells

These reversible phenotype transitions have recently been modeled in a human colon cancer cell line that grows as an organoid sphere in tissue culture.[63,76] The parental cell line LIM1863[77] grows as spheres of epithelial cells that are highly polarized along the baso-lateral axis and are organized around a central lumen. These spheres can spontaneously anchor to the tissue culture plastic and form an adherent monolayer patch. After 3–4 days in culture, cells in the monolayer patches reorganize to reform the spheres that eventually float freely in the tissue culture medium and the whole process starts again. The parental cell line was adapted to efficiently undergo this spontaneous, reversible transition between monolayer and organoid sphere and the adapted cell line is called LIM1863-*Mph* (for *morph*ogenetic)[63] (Fig. 3A). Immunofluorescence confocal microscopy for the junction protein ZO-1 clearly shows the transition between monolayer and organoid sphere (Fig. 3B).[78] Molecular and phenotypic analysis of the cells during these transitions revealed that the features of EMT and MET that underscore colon cancer metastasis are faithfully recapitulated in this model system.[63,76,78,79] Importantly, the monolayer cells (EMT state) decrease cell proliferation and are resistant to agents that block cell proliferation (e.g., Mitomycin C)[63] and thus mimic the properties of chemo resistant mesenchymal invasive front cells. The epithelial spheres can be induced to undergo EMT with TGFβ and TNFα treatment; however, this transition is not reversible.[76] Nonetheless cytokine induced or spontaneous monolayer formation and the reverse transition in the LIM1983-*Mph* cells provide clues about the underlying mechanisms of dormancy which could be adopted for the tumor mini-organoid platforms. Notably, the monolayer cells are

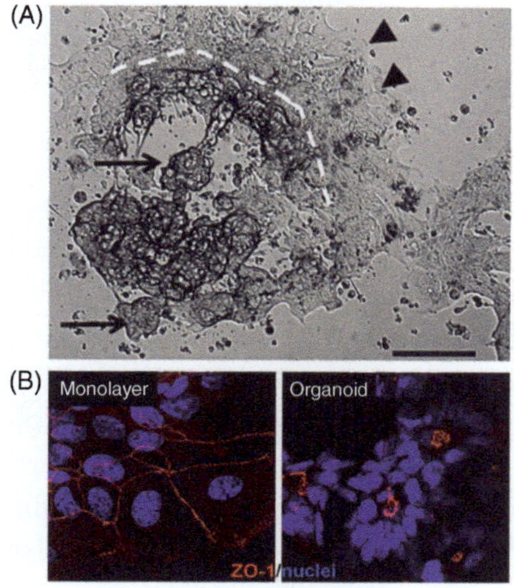

Fig. 3 LIM-1863-Mph: a colon cancer tumor morphogenesis model. (A) A differential interference contrast (DIC) image of a LIM1863-Mph monolayer patch undergoing MET and tumor organoid formation. Assembled organoids that are about to dislodge from the patch *(arrow)* are generated by monolayer cells *(arrow heads)* that spontaneously reorganize at the center of the monolayer patch (along the *white dashed line*, scale bar is 100 μ). (B) The transition between organoid and monolayer is readily detected using immunofluorescence confocal microscopy to detect ZO-1 *(red)*. The organoid lumen is lined by the apical surface of the cells and thus ZO-1 staining is seen as a tight rosette (shown are stacked Z-sections at the cross section of organoid). This is in stark contrast to the monolayer cells where it shows the cell outline. Nuclei are *blue* (DAPI stained).

resistant to the PI3K inhibitor LY49002. As noted earlier, PI3K is one of the genetic insults that converts normal cells to cancer cells in an organoid model,[67] yet the LIM1863-*Mph* cells are resistant to a PI3K inhibitor when in the mesenchymal state even though they are sensitive to it in the epithelial state.[63]

The LIM1863-*Mph* tumor morphogenesis model highlights one caveat for patient-derived tumor organoids—modeling reversible tumor dormancy. Indeed, studies to date reveal that metastasis relies on subtle changes rather than "driver" gene mutations. For example, in one study mutant organoids engineered from human normal epithelium to carry the driver mutations seen in colon cancer formed micrometastases when injected into the spleen of mice, but failed to colonize the liver, the usual metastatic site for colon cancer. In contrast, mutant organoids derived from human adenomas formed liver metastases when the same drivers were introduced.[67] Similarly, tumor organoids derived from colon metastases metastasize better than their matched primary tumor, despite having indistinguishable genetic mutations and niche requirements.[7] Consequently, human genomics needs to be coupled to epigenomics and phenomics if we are to unravel the mechanisms of tumor dormancy and mechanisms of metastasis.

5.2.2 Acute High Wnt/β-Catenin Signaling in MET

Another intriguing feature of the LIM1863-*Mph* is the dynamic regulation of the levels of nuclear β-catenin, the hallmark of active Wnt/β-catenin signaling (Fig. 1B). The organoid cells and the monolayer cells have very low, but detectable levels of nuclear β-catenin.[63] This is not surprising as the LIM1863 cells harbor truncating mutations in the *APC* gene.[80] However, as the monolayer cells start to transition back to epithelial morphology and start to lift off the tissue culture plastic and reorganize themselves into spheres, there is a dramatic transient increase in nuclear β-catenin, which is concomitant with a sharp increase in cell division (Ki-67 staining).[63] Cells in the organoid spheres continue to divide (maintain strong Ki-67 staining) but the level of nuclear β-catenin decreases to just detectable levels. As expected, the organoid sphere cells are susceptible to agents that target cell proliferation (Mitomycin C)[63] and the PI3K inhibitor LY49002 (data not shown). Collectively, this indicates that emergence from a mesenchymal monolayer ("dormant") state (i.e., MET) requires a sharp burst of Wnt/β-catenin signaling.[63] This was confirmed recently by an independent study that showed lithium chloride (LiCl), a known activator of Wnt/β-catenin signaling,[63,81] induced MET in primary colon cancer cell cultures.[82]

This requirement for Wnt/β-catenin signaling for MET provides several novel avenues to combat the formation of metastases. Experimentally, it was shown that Wnt is necessary for MET in the LIM1863-*Mph* cells[78] and the Wnt receptor Frizzled-7 was identified as the necessary Wnt receptor;[63] thus therapeutic targeting of Frizzled-7 would target both actively dividing and dormant tumor cells. Indeed, inhibition of Frizzled-7 mediated signaling in colon cancer cells potently blocked colon cancer xenograft growth.[83]

Another avenue might be to activate Wnt/β-catenin signaling in the dormant tumor cells to induce MET and render them susceptible to conventional chemotherapy and radiotherapy. Activating Wnt signaling may sound risky, but lithium could potentially serve this purpose. Lithium is an FDA-approved and preferred drug for the treatment of mood disorders, and evidence is emerging about its potential use as an anticancer drug in colon[84–86] and other cancers.[87] Importantly, there is no increased risk, in fact a slight decrease, of cancer in psychiatric patients treated with lithium.[88] Repurposing lithium to activate the Wnt/β-catenin pathway immediately before administering conventional chemotherapy and radiotherapy might have a beneficial effect in colon cancer.

6. CONCLUSIONS

In this chapter we have tried to highlight the current state of the tumor organoid field, which is advancing at an astronomical rate. Mini-gut and mini-tumor organoids have revolutionized our understanding of the molecular mechanisms that underlie transition from normal epithelial stem cell to cancer cell, identify the molecular drivers of cancer cells, and predict their susceptibility to anticancer drugs. The ability to conduct drug screens on dormant patient tumor cells is the next frontier and vital to our quest to improve cancer patient survival rates toward a cure.

Indeed, we propose that the most important challenge for organoid technology and high throughput drug screens is to model patient-derived dormant tumor cells, because the barrier to curative treatment is metastasis, the cause of death in most cases. Tumor cells can spread to other organs early in the disease, long before diagnosis, and can lay dormant in the secondary organs for years.[89] Organoids with dormant tumor cells have the potential to identify therapies to eliminate these cells. We have clues from model systems, such as LIM1863-*Mph* described earlier but also possibly from $Lgr5^+$ stem cells. It was recently shown that $Lgr5^+$ stem cells can be made quiescent in vitro, and that the quiescent state is reversible.[90] We already know that Lgr5$^+$ colon cancer cells maintain tumor growth and progression,[91,92] akin to the role of Lgr5$^+$ intestinal stem cells maintaining the intestinal epithelium. Thus, Lgr5$^+$ cancer cells behave-like cancer stem cells. Notably, cancer stem cell plasticity, where Lgr5 negative cells can revert to being Lgr5$^+$, has also been recently demonstrated in a mouse model,[92] again reminiscent of intestinal stem cell plasticity where daughter cells can revert to the Lgr5$^+$ state to repopulate the crypt.[93–95] Perhaps the Lgr5$^+$ cancer cells will also provide the clues to tumor dormancy.

In 1960s Gurdon demonstrated that an adult nucleus can be the blueprint for an organism,[96] and the eventual adoption of this knowledge in 2006 to induce pluripotency in adult cells by Yamanaka and coworker,[4] and the ability to derive organoids from these pluripotent stem cells, has led to a steady stream of organoid-based publications. However, the demonstration in 2009 that adult stem cells can be coerced to generate their tissue of origin in vitro has led to an exponential increase in organoid-based publications.[1] This explosion in knowledge is set to be fueled by combining the two varieties of stem cell-derived organoids (Table 1). Adult tissue stem cell-derived organoids do not contain other cell types in the organ, for example, no immune cells, neuronal cells or stromal cells, for example. However, by adding patient iPS-derived cell types to patient adult stem cell-derived organoids, we can start to build more complex tissues or even organs, because autologous iPS cells can provide cells types that are present in the organ but are not derived from the adult tissue-restricted stem cell. For example, recently iPS and adult stem cell–derived organoids have been combined to generate stem-cell derived intestinal tissue with a functional enteric nervous system.[97] This is just the beginning—very exciting times ahead.

ACKNOWLEDGMENTS

We thank Gavin Mitchell for generating the artwork, and Nancy Amin and Jordane Malaterre for the confocal images. Funding is gratefully acknowledged from the National Health and Medical Research Council of Australia (566679 & APP1099302); Melbourne Health project grants (605030 & PG-002) and early career researcher grant (GIA-033); Cancer Council Victoria project grant (APP1020716) and Fellowship; and Cardiff University.

REFERENCES

1. Clevers H. Modeling development and disease with organoids. *Cell.* 2016;165(7): 1586–1597.
2. Eiraku M, Sasai Y. Self-formation of layered neural structures in three-dimensional culture of ES cells. *Curr Opin Neurobiol.* 2012;22(5):768–777.
3. Lancaster MA, Knoblich JA. Organogenesis in a dish: modeling development and disease using organoid technologies. *Science.* 2014;345(6194):1247125.

4. Takahashi K, Yamanaka S. Induction of pluripotent stem cells from mouse embryonic and adult fibroblast cultures by defined factors. *Cell.* 2006;126(4):663–676.
5. Shi Y, Inoue H, Wu JC, Yamanaka S. Induced pluripotent stem cell technology: a decade of progress. *Nat Rev Drug Discov.* 2017;16(2):115–130.
6. van de Wetering M, Francies HE, Francis JM, et al. Prospective derivation of a living organoid biobank of colorectal cancer patients. *Cell.* 2015;161(4):933–945.
7. Fujii M, Shimokawa M, Date S, et al. A colorectal tumor organoid library demonstrates progressive loss of niche factor requirements during tumorigenesis. *Cell Stem Cell.* 2016;18(6):827–838.
8. Phesse T, Flanagan D, Vincan E. Frizzled7: a promising achilles' heel for targeting the Wnt receptor complex to treat cancer. *Cancers.* 2016;8(5).
9. Niehrs C. The complex world of WNT receptor signalling. *Nat Rev Mol Cell Biol.* 2012;13(12):767–779.
10. Nusse R, Clevers H. Wnt/beta-catenin signaling, disease, and emerging therapeutic modalities. *Cell.* 2017;169(6):985–999.
11. Daniels DL, Weis WI. Beta-catenin directly displaces Groucho/TLE repressors from Tcf/Lef in Wnt-mediated transcription activation. *Nat Struct Mol Biol.* 2005; 12(4):364–371.
12. van Es JH, Jay P, Gregorieff A, et al. Wnt signalling induces maturation of Paneth cells in intestinal crypts. *Nat Cell Biol.* 2005;7(4):381–386.
13. Korinek V, Barker N, Moerer P, et al. Depletion of epithelial stem-cell compartments in the small intestine of mice lacking Tcf-4. *Nat Genetics.* 1998;19(4):379–383.
14. Barker N, van Es JH, Kuipers J, et al. Identification of stem cells in small intestine and colon by marker gene Lgr5. *Nature.* 2007;449(7165):1003–1007.
15. Goentoro L, Kirschner MW. Evidence that fold-change, and not absolute level, of beta-catenin dictates Wnt signaling. *Mol Cell.* 2009;36(5):872–884.
16. Albuquerque C, Breukel C, van der Luijt R, et al. The 'just-right" signaling model: APC somatic mutations are selected based on a specific level of activation of the beta-catenin signaling cascade. *Hum Mol Genetics.* 2002;11(13):1549–1560.
17. Driehuis E, Clevers H. WNT signalling events near the cell membrane and their pharmacological targeting for the treatment of cancer. *Br J Pharmacol.* 2017 10.1111/ bph.13758.
18. Brabletz T, Jung A, Reu S, et al. Variable beta-catenin expression in colorectal cancers indicates tumor progression driven by the tumor environment. *Proc Natl Acad Sci USA.* 2001;98(18):10356–10361.
19. Brabletz T, Jung A, Spaderna S, Hlubek F, Kirchner T. Opinion: migrating cancer stem cells—an integrated concept of malignant tumour progression. *Nat Rev Cancer.* 2005; 5(9):744–749.
20. Vincan E, Barker N. The upstream components of the Wnt signalling pathway in the dynamic EMT and MET associated with colorectal cancer progression. *Clin Exp Metastasis.* 2008;25(6):657–663.
21. Caldwell GM, Jones C, Gensberg K, et al. The Wnt antagonist sFRP1 in colorectal tumorigenesis. *Cancer Res.* 2004;64(3):883–888.
22. Caldwell GM, Jones CE, Taniere P, et al. The Wnt antagonist sFRP1 is downregulated in premalignant large bowel adenomas. *Br J Cancer.* 2006;94(6):922–927.
23. Suzuki H, Watkins DN, Jair KW, et al. Epigenetic inactivation of SFRP genes allows constitutive WNT signaling in colorectal cancer. *Nat Genetics.* 2004;36(4):417–422.
24. Taniguchi H, Yamamoto H, Hirata T, et al. Frequent epigenetic inactivation of Wnt inhibitory factor-1 in human gastrointestinal cancers. *Oncogene.* 2005;24(53):7946–7952.
25. Aguilera O, Fraga MF, Ballestar E, et al. Epigenetic inactivation of the Wnt antagonist DICKKOPF-1 (DKK-1) gene in human colorectal cancer. *Oncogene.* 2006;25(29): 4116–4121.

26. Sato T, Vries RG, Snippert HJ, et al. Single Lgr5 stem cells build crypt-villus structures in vitro without a mesenchymal niche. *Nature.* 2009;459(7244):262–265.
27. Ireland H, Kemp R, Houghton C, et al. Inducible Cre-mediated control of gene expression in the murine gastrointestinal tract: effect of loss of beta-catenin. *Gastroenterology.* 2004;126(5):1236–1246.
28. Pinto D, Gregorieff A, Begthel H, Clevers H. Canonical Wnt signals are essential for homeostasis of the intestinal epithelium. *Genes Dev.* 2003;17(14):1709–1713.
29. Kuhnert F, Davis CR, Wang HT, et al. Essential requirement for Wnt signaling in proliferation of adult small intestine and colon revealed by adenoviral expression of Dickkopf-1. *Proc Natl Acad Sci USA.* 2004;101(1):266–271.
30. van de Wetering M, Sancho E, Verweij C, et al. The beta-catenin/TCF-4 complex imposes a crypt progenitor phenotype on colorectal cancer cells. *Cell.* 2002;111(2):241–250.
31. Sato T, van Es JH, Snippert HJ, et al. Paneth cells constitute the niche for Lgr5 stem cells in intestinal crypts. *Nature.* 2011;469(7330):415–418.
32. Farin HF, Van Es JH, Clevers H. Redundant sources of wnt regulate intestinal stem cells and promote formation of paneth cells. *Gastroenterology.* 2012;143(6):1518–1529. e1517.
33. Barker N, Clevers H. Tracking down the stem cells of the intestine: strategies to identify adult stem cells. *Gastroenterology.* 2007;133(6):1755–1760.
34. Barker N, Ridgway RA, van Es JH, et al. Crypt stem cells as the cells-of-origin of intestinal cancer. *Nature.* 2009;457(7229):608–611.
35. Sato T, Stange DE, Ferrante M, et al. Long-term expansion of epithelial organoids from human colon, adenoma, adenocarcinoma, and Barrett's epithelium. *Gastroenterology.* 2011;141(5):1762–1772.
36. Schwank G, Clevers H. CRISPR/Cas9-mediated genome editing of mouse small intestinal organoids. *Methods Mol Biol.* 2016;1422:3–11.
37. Schwank G, Koo BK, Sasselli V, et al. Functional repair of CFTR by CRISPR/Cas9 in intestinal stem cell organoids of cystic fibrosis patients. *Cell Stem Cell.* 2013;13(6): 653–658.
38. Flanagan DJ, Phesse TJ, Barker N, et al. Frizzled7 functions as a Wnt receptor in intestinal epithelium stem cells. *Stem Cell Reports.* 2015;4(5):759–767.
39. Koo, BK, Sasselli, V, Clevers, H,Koo BK, Sasselli V, Clevers H. Retroviral gene expression control in primary organoid cultures. *Curr Protoc Stem Cell Biol.* 2013;27:Unit 5A 6. doi: 10.1002/9780470151808.sc05a06s27
40. Sachs N, Tsukamoto Y, Kujala P, Peters PJ, Clevers H. Intestinal epithelial organoids fuse to form self-organizing tubes in floating collagen gels. *Development.* 2017;144(6): 1107–1112.
41. Broutier L, Andersson-Rolf A, Hindley CJ, et al. Culture and establishment of self-renewing human and mouse adult liver and pancreas 3D organoids and their genetic manipulation. *Nat Protoc.* 2016;11(9):1724–1743.
42. Mihara E, Hirai H, Yamamoto H, et al. Active and water-soluble form of lipidated Wnt protein is maintained by a serum glycoprotein afamin/alpha-albumin. *eLife.* 2016;5:e11621.
43. Tuysuz N, van Bloois L, van den Brink S, et al. Lipid-mediated Wnt protein stabilization enables serum-free culture of human organ stem cells. *Nat Commun.* 2017;8:14578.
44. Janda CY, Dang LT, You C, et al. Surrogate Wnt agonists that phenocopy canonical Wnt and beta-catenin signalling. *Nature.* 2017;545(7653):234–237.
45. Gjorevski N, Ranga A, Lutolf MP. Bioengineering approaches to guide stem cell-based organogenesis. *Development.* 2014;141(9):1794–1804.
46. Gjorevski N, Sachs N, Manfrin A, et al. Designer matrices for intestinal stem cell and organoid culture. *Nature.* 2016;539(7630):560–564.
47. Levental KR, Yu H, Kass L, et al. Matrix crosslinking forces tumor progression by enhancing integrin signaling. *Cell.* 2009;139(5):891–906.

48. Flanagan DJ, Schwab RH, Tran BM, Moselen JM, Phesse TJ, Vincan E. Isolation and culture of adult intestinal, gastric, and liver organoids for cre-recombinase-mediated gene deletion. *Methods Mol Biol.* 2016. (Epub ahead of print).
49. Bartfeld S, Clevers H. Stem cell-derived organoids and their application for medical research and patient treatment. *J Mol Med.* 2017;95(7):729–738.
50. Dutta D, Heo I, Clevers H. Disease modeling in stem cell-derived 3D organoid systems. *Trends Mol Med.* 2017;23(5):393–410.
51. Jamieson PR, Dekkers JF, Rios AC, Fu NY, Lindeman GJ, Visvader JE. Derivation of a robust mouse mammary organoid system for studying tissue dynamics. *Development.* 2017;144(6):1065–1071.
52. Clevers H, Nusse R. Wnt/beta-catenin signaling and disease. *Cell.* 2012;149(6): 1192–1205.
53. Fearon ER. Molecular genetics of colorectal cancer. *Annu Rev Pathol.* 2011;6:479–507.
54. Korinek V, Barker N, Morin PJ, et al. Constitutive transcriptional activation by a beta-catenin-Tcf complex in APC$^{-/-}$ colon carcinoma. *Science.* 1997;275(5307):1784–1787.
55. Morin PJ, Sparks AB, Korinek V, et al. Activation of beta-catenin-Tcf signaling in colon cancer by mutations in beta-catenin or APC. *Science.* 1997;275(5307):1787–1790.
56. Fearon ER, Vogelstein B. A genetic model for colorectal tumorigenesis. *Cell.* 1990; 61(5):759–767.
57. Ding Y, Dale T. Wnt signal transduction: kinase cogs in a nano-machine? *Trends Biochem Sci.* 2002;27(7):327–329.
58. Su LK, Kinzler KW, Vogelstein B, et al. Multiple intestinal neoplasia caused by a mutation in the murine homolog of the APC gene. *Science.* 1992;256(5057):668–670.
59. Shibata H, Toyama K, Shioya H, et al. Rapid colorectal adenoma formation initiated by conditional targeting of the APC gene. *Science.* 1997;278(5335):120–123.
60. Harada N, Tamai Y, Ishikawa T, et al. Intestinal polyposis in mice with a dominant stable mutation of the beta-catenin gene. *EMBO J.* 1999;18(21):5931–5942.
61. Dimitriadis A, Vincan E, Mohammed IM, Roczo N, Phillips WA, Baindur-Hudson S. Expression of Wnt genes in human colon cancers. *Cancer Lett.* 2001;166(2):185–191.
62. Vincan E. Frizzled/WNT signalling: the insidious promoter of tumour growth and progression. *Front Biosci.* 2004;9:1023–1034.
63. Vincan E, Darcy PK, Farrelly CA, Faux MC, Brabletz T, Ramsay RG. Frizzled-7 dictates three-dimensional organization of colorectal cancer cell carcinoids. *Oncogene.* 2007; 26(16):2340–2352.
64. Brabletz T, Jung A, Hermann K, Gunther K, Hohenberger W, Kirchner T. Nuclear overexpression of the oncoprotein beta-catenin in colorectal cancer is localized predominantly at the invasion front. *Pathol Res Pract.* 1998;194(10):701–704.
65. Vincan E, Swain RK, Brabletz T, Steinbeisser H. Frizzled7 dictates embryonic morphogenesis: implications for colorectal cancer progression. *Front Biosci.* 2007;12:4558–4567.
66. Drost J, van Jaarsveld RH, Ponsioen B, et al. Sequential cancer mutations in cultured human intestinal stem cells. *Nature.* 2015;521(7550):43–47.
67. Matano M, Date S, Shimokawa M, et al. Modeling colorectal cancer using CRISPR-Cas9-mediated engineering of human intestinal organoids. *Nat Med.* 2015;21(3):256–262.
68. Fumagalli A, Drost J, Suijkerbuijk SJ, et al. Genetic dissection of colorectal cancer progression by orthotopic transplantation of engineered cancer organoids. *Proc Natl Acad Sci USA.* 2017;114(12):E2357–E2364.
69. Weeber F, van de Wetering M, Hoogstraat M, et al. Preserved genetic diversity in organoids cultured from biopsies of human colorectal cancer metastases. *Proc Natl Acad Sci USA.* 2015;112(43):13308–13311.
70. Lugli N, Dionellis VS, Ordonez-Moran P, et al. Enhanced rate of acquisition of point mutations in mouse intestinal adenomas compared to normal tissue. *Cell Rep.* 2017; 19(11):2185–2192.

71. Cristobal A, van den Toorn HW, van de Wetering M, Clevers H, Heck AJ, Mohammed S. Personalized proteome profiles of healthy and tumor human colon organoids reveal both individual diversity and basic features of colorectal cancer. *Cell Rep.* 2017; 18(1):263–274.
72. Byrne AT, Alferez DG, Amant F, et al. Interrogating open issues in cancer precision medicine with patient-derived xenografts. *Nat Rev Cancer.* 2017;17(4):254–268.
73. Chia S, Low JL, Zhang X, et al. Phenotype-driven precision oncology as a guide for clinical decisions one patient at a time. *Nat Commun.* 2017;8(1):435.
74. Fatehullah A, Tan SH, Barker N. Organoids as an in vitro model of human development and disease. *Nat Cell Biol.* 2016;18(3):246–254.
75. Thiery JP. Epithelial-mesenchymal transitions in tumour progression. *Nat Rev Cancer.* 2002;2(6):442–454.
76. Vincan E, Brabletz T, Faux MC, Ramsay RG. A human three-dimensional cell line model allows the study of dynamic and reversible epithelial-mesenchymal and mesenchymal-epithelial transition that underpins colorectal carcinogenesis. *CellsTissues Organs.* 2007;185(1–3):20–28.
77. Whitehead RH, Jones JK, Gabriel A, Lukies RE. A new colon carcinoma cell line (LIM1863) that grows as organoids with spontaneous differentiation into crypt-like structures in vitro. *Cancer Res.* 1987;47(10):2683–2689.
78. Schwab RHM, Amin N, Flanagan DJ, Johanson TM, Phesse TJ, Vincan E. Wnt is necessary for mesenchymal to epithelial transition in colorectal cancer cells. *Dev Dyn.* 2017 10.1002/dvdy.24527.
79. Vincan E, Whitehead RH, Faux MC. Analysis of Wnt/FZD-mediated signalling in a cell line model of colorectal cancer morphogenesis. *Methods Mol Biol.* 2008;468:263–273.
80. Zhang HH, Walker F, Kiflemariam S, et al. Selective inhibition of proliferation in colorectal carcinoma cell lines expressing mutant APC or activated B-Raf. *Int J Cancer.* 2009;125(2):297–307.
81. Kao KR, Elinson RP. The legacy of lithium effects on development. *Biol Cell.* 1998; 90(8):585–589.
82. Costabile V, Duraturo F, Delrio P, et al. Lithium chloride induces mesenchymaltoepithelial reverting transition in primary colon cancer cell cultures. *Int J Oncol.* 2015; 46(5):1913–1923.
83. Vincan E, Darcy PK, Smyth MJ, et al. Frizzled-7 receptor ectodomain expression in a colon cancer cell line induces morphological change and attenuates tumor growth. *Differentiation.* 2005;73(4):142–153.
84. de Araujo WM, Robbs BK, Bastos LG, et al. PTEN overexpression cooperates with lithium to reduce the malignancy and to increase cell death by apoptosis via PI3K/Akt suppression in colorectal cancer cells. *J Cell Biochem.* 2016;117(2):458–469.
85. Vidal F, de Araujo WM, Cruz AL, Tanaka MN, Viola JP, Morgado-Diaz JA. Lithium reduces tumorigenic potential in response to EGF signaling in human colorectal cancer cells. *Int J Oncol.* 2011;38(5):1365–1373.
86. Li H, Huang K, Liu X, et al. Lithium chloride suppresses colorectal cancer cell survival and proliferation through ROS/GSK-3beta/NF-kappaB signaling pathway. *Oxid Med Cell Longev.* 2014;2014:241864.
87. Nowicki MO, Dmitrieva N, Stein AM, et al. Lithium inhibits invasion of glioma cells; possible involvement of glycogen synthase kinase-3. *Neuro-oncology.* 2008;10(5): 690–699.
88. Cohen Y, Chetrit A, Cohen Y, Sirota P, Modan B. Cancer morbidity in psychiatric patients: influence of lithium carbonate treatment. *Med Oncol.* 1998;15(1):32–36.
89. Naumov GN, MacDonald IC, Weinmeister PM, et al. Persistence of solitary mammary carcinoma cells in a secondary site: a possible contributor to dormancy. *Cancer Res.* 2002;62(7):2162–2168.

90. Basak O, Beumer J, Wiebrands K, Seno H, van Oudenaarden A, Clevers H. Induced quiescence of Lgr5+ stem cells in intestinal organoids enables differentiation of hormone-producing enteroendocrine cells. *Cell Stem Cell.* 2017;20(2):177–190. e174.
91. Shimokawa M, Ohta Y, Nishikori S, et al. Visualization and targeting of LGR5+ human colon cancer stem cells. *Nature.* 2017;545(7653):187–192.
92. de Sousa e Melo F, Kurtova AV, Harnoss JM, et al. A distinct role for Lgr5+ stem cells in primary and metastatic colon cancer. *Nature.* 2017;543(7647):676–680.
93. van Es JH, Sato T, van de Wetering M, et al. Dll1+ secretory progenitor cells revert to stem cells upon crypt damage. *Nat Cell Biol.* 2012;14(10):1099–1104.
94. Basak O, van de Born M, Korving J, et al. Mapping early fate determination in Lgr5+ crypt stem cells using a novel Ki67-RFP allele. *EMBO J.* 2014;33(18):2057–2068.
95. Buczacki SJ, Zecchini HI, Nicholson AM, et al. Intestinal label-retaining cells are secretory precursors expressing Lgr5. *Nature.* 2013;495(7439):65–69.
96. Gurdon JB. Adult frogs derived from the nuclei of single somatic cells. *Dev Biol.* 1962;4:256–273.
97. Workman MJ, Mahe MM, Trisno S, et al. Engineered human pluripotent-stem-cell-derived intestinal tissues with a functional enteric nervous system. *Nat Med.* 2017; 23(1):49–59.

CHAPTER ELEVEN

Wnt Signaling in Hematological Malignancies

Stephanie Grainger*,†, David Traver*, Karl Willert*,†,1
* University of California, La Jolla, San Diego, CA, United States
† Sanford Consortium for Regenerative Medicine, La Jolla, San Diego, CA, United States
[1] Corresponding author. E-mail address: kwillert@ucsd.edu

Contents

Abstract

Leukemia and lymphoma are a wide encompassing term for a diverse set of blood malignancies that affect people of all ages and result in approximately 23,000 deaths in the United States per year (Siegel RL, Miller KD, Jemal A. Cancer statistics, 2016. *CA Cancer J Clin.* 2016;66(1):7–30.). Hematopoietic stem cells (HSCs) are tissue-specific stem cells at the apex of the hierarchy that gives rise to all of the terminally differentiated blood cells, through progressively restricted progenitor populations, a process that is known to be Wnt-responsive. In particular, the progenitor populations are subject to uncontrolled expansion during oncogenic processes, namely the common myeloid progenitor and common lymphoid progenitor, as well as the myeloblast and lymphoblast. Unregulated growth of these cell-types leads to mainly three types of blood cancers (i.e., leukemia, lymphoma, and myeloma), which frequently exhibit deregulation of the Wnt signaling pathway. Generally, leukemia is caused by the expansion of myeloid progenitors, leading to an overproduction of white blood cells; as such, patients are unable to make sufficient numbers of red blood cells and platelets. Likewise, an overproduction of lymphocytes leads to clogging of the lymph

Progress in Molecular Biology and Translational Science, Volume 153
ISSN 1877-1173
http://dx.doi.org/10.1016/bs.pmbts.2017.11.002

system and impairment of the immune system in lymphomas. Finally, cancer of the plasma cells in the blood is called myeloma, which also leads to immune system failure. Within each of these three types of blood cancers, there are multiple subtypes, usually characterized by their timeline of onset and their cell type of origin. Of these, 85% of leukemias are encompassed by the four most common diseases, that is, acute myeloid leukemia (AML), chronic myeloid leukemia (CML), acute lymphoblastic leukemia (ALL), and chronic lymphocytic leukemia (CLL); AML accounts for the majority of leukemia-related deaths (Siegel RL, Miller KD, Jemal A. Cancer statistics, 2016. *CA Cancer J Clin.* 2016;66(1):7–30.). Through understanding how HSCs are normally developed and maintained, we can understand how the normal functions of these pathways are disrupted during blood cancer progression; the Wnt pathway is important in regulation of both normal and malignant hematopoiesis. In this chapter, we will discuss the role of Wnt signaling in normal and aberrant hematopoiesis. Our understanding the relationship between Wnt and HSCs will provide novel insights into therapeutic targets.

1. HEMATOPOIETIC STEM CELL (HSC) DEVELOPMENT

The development of hematopoietic stem cells (HSCs) is a dynamic process, involving migration through several niches in distinct anatomical locations. The first long-term bona fide HSCs arise directly from specialized hemogenic endothelium in all species examined to date, in a process known as the endothelial-to-hematopoietic transition.[1] As this transition occurs, HSCs begin to express markers of hematopoietic fate, such as the transcriptional regulator *Runx1*.[2,3] In mammals, the first HSCs arise from the aorta, umbilical artery, and vitelline artery in the midgestation embryo[1,4,5] (around embryonic day 10 in mouse). These later migrate to the fetal liver, and finally the bone marrow, where HSCs are maintained throughout the lifespan of the animal.[6] It is thought that at each of these progressive locations, the HSCs receive induction cues, such as FGF, Notch, and Wnt, which are important for their maturation and function.[7] During this dynamic developmental process, these HSCs are subject to inputs from various surrounding niches and a variety of developmental signaling pathways, reviewed elsewhere.[8–10] These early inputs give rise to later HSC function, and disruption of these cues fundamentally alter the identity and behavior of these cells.[11,12] The final result of these developmental cues are long-term HSCs (LT-HSCs), housed in the bone marrow; these lie at the apex of the hematopoietic hierarchy, seeding the entire blood system for the duration of the organism's life.

2. HSC HOMEOSTASIS

LT-HSCs are defined by their ability to repopulate the hematopoietic system of an organism for a long duration.[13] In the bone marrow, these are largely quiescent, but give rise to a subpopulation of short-term HSCs (ST-HSCs), which are limited in their capacity for self-renewal and are able to repopulate an organism only for a few weeks.[14] HSCs in turn give rise to the lineage restricted common progenitors of the myeloid and lymphoid fates, which together can differentiate into all mature blood cells.[15,16] (Fig. 1). Recent single-cell analyses have indicated that all of these progenitor types have low levels of coexpression of multiple lineages, suggesting overlapping cell fates at the transcriptional level, and that these fate decisions are rather plastic.[17] Additionally, HSCs with long-term single lineage reconstitution fate have been identified, suggesting that the binary lineage tree of hematopoietic fate is more flexible than previously established.[18] It is through careful regulation of inductive cues that HSCs are able to self-renew and differentiate into the full complement of blood. Disruption of these regulatory cues can accordingly result in cancers of the blood, and our understanding of these cues during normal hematopoiesis will be critical to determining how they are coopted during oncogenesis.

Our understanding of the native processes governing hematopoietic homeostasis has been hindered by technical difficulties, including the relative rarity of these cells in vivo, where it is estimated that HSCs form approximately 1/20,000 nucleated hematopoietic cells.[14] As a result, it has been difficult to elucidate the molecular mechanisms governing their homeostasis, as purification of a large enough number of cells for analysis is challenging. It is apparent that HSCs receive both cell-intrinsic and -extrinsic cues to govern their maintenance reviewed in Ref. 19. In the bone marrow, HSCs reside in close proximity to endothelial cells, osteoblasts, and mesenchymal stem cells, which are proposed to supply the niche cues governing self-renewal and differentiation[20] (Fig. 1), though this is incompletely understood. Among others, the Wnt signaling cascade is known to play important roles in these processes.

3. WNT SIGNALING

Wnt genes encode lipid-modified, secreted signaling molecules, which regulate a diverse set of developmental and homeostatic programs,

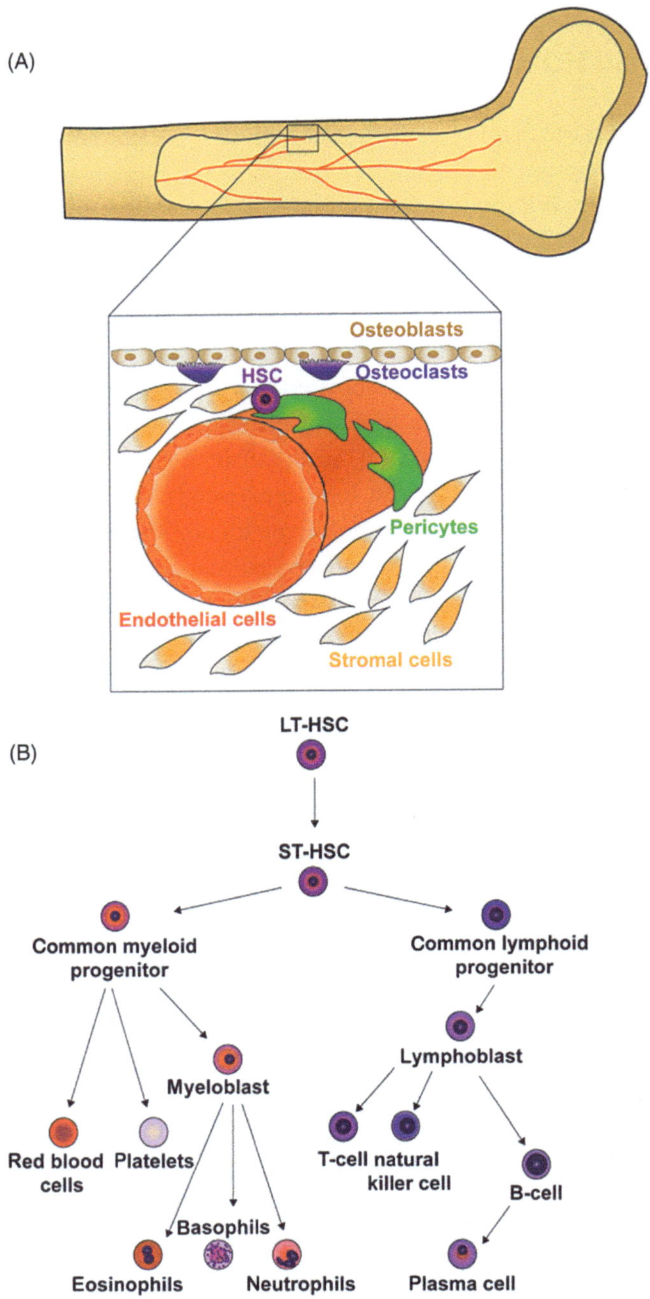

Fig. 1 Adults hematopoietic stem cells (HSCs) reside in the bone marrow. (A) The marrow of long bones is the niche for homeostasis of HSCs, where they reside in close proximity to many cell types. For example, the blood vessels, which carry

ranging from axial patterning to stem cell identity and self-renewal reviewed in Ref. 21.[22–24] Their broad range of effects are in part due to the complex genetic nature of these factors with 19 independent Wnt genes in the mammalian genome and multiple genes encoding structurally and functionally distinct receptors, including Frizzled (Fzd), lipoprotein related protein (Lrp), receptor tyrosine kinase-like orphan receptor (Ror), and receptor-like tyrosine kinase (Ryk). Activation of the Wnt signaling cascade has been loosely grouped into either the canonical (β-catenin dependent) or noncanonical (β-catenin independent) pathways, though it is important to note that these pathways do not necessarily work independently or in opposition to each other. Furthermore, a given Wnt may activate distinct signaling cascades depending on expression of downstream signaling components.[21,25–29]

In the absence of Wnt signals (Fig. 2), the DNA binding proteins of the lymphoid enhancing factor (LEF)/T-cell factor (TCF) family recruit corepressors, such as Groucho/transducing-like enhancer (GRG/TLE) and CtBP1, to silence expression of Wnt target genes. The transcriptional activator β-catenin is constitutively tagged for proteasomal degradation by the so-called destruction complex, consisting in part of GSK3β, APC, and Axin. In the presence of a Wnt ligand, Fzd receptors and Lrp coreceptors oligomerize at the membrane, leading to dissociation/relocation of the destruction complex, thus releasing β-catenin from proteasomal targeting and allowing it to enter the nucleus where it displaces GRG/TLE from LEF/TCF and activates expression of Wnt target genes.[30] Activation of this arm of the Wnt signal often leads to maintenance or specification of cell fate (e.g., stem cell self-renewal or differentiation).[21] Wnt signaling can be modulated with secreted negative regulators; such as secreted frizzed-related proteins (SFRP),[31] which are hypothesized to sequester the Wnt ligand; Notum,[32–34] which inactivates Wnt through deacylation; and Dickkopf (Dkk), which bind Lrp to prevent oligomerization with Fzd.[35]

The β-catenin independent pathways remain poorly understood. In its downstream signal, there are a variety of effectors, including RhoA, JNK, and calcium influx pathways[21,36–38] (Fig. 2). Different arms of this cascade

◀ circulating HSCs and their progeny; the bone-building osteoblasts; the bone-degrading osteoclasts; the supportive pericytes, and other bone marrow stromal cells. (B) In the niche, the quiescent long-term (LT) HSCs give rise to short-term (ST) HSCs, which in turn can gives rise to the common myeloid or lymphoid progenitors, which can differentiate into all cell types of the blood, and many cells from the immune system.

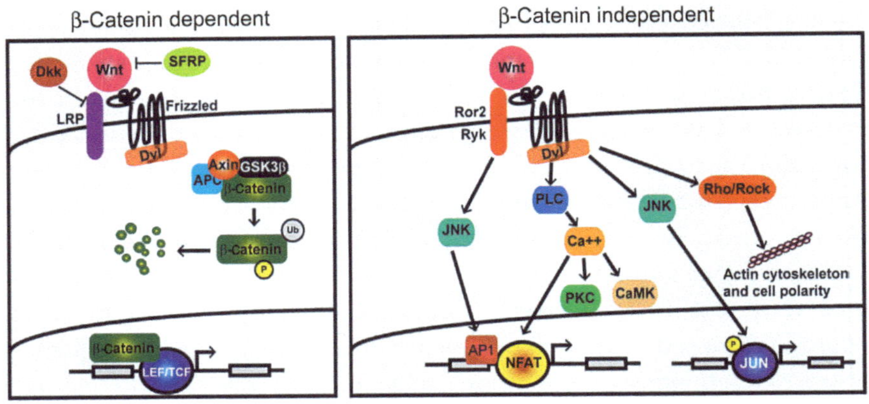

Fig. 2 Wnt signaling pathways. Schematic representations of the β-catenin dependent and independent Wnt signaling cascades. In the β-catenin dependent cascade, Wnt-Fzd-LRP interaction culminates in the nuclear translocation of β-catenin to bind lymphoid enhancing factor (LEF)/T-cell factor (TCF) transcription factors and drive gene expression. In the β-catenin independent pathway, there are several arms of signaling. For instance, Wnt–Fzd interaction leads to either Rho/Rock or Rac/JNK signaling, culminating in changes in the actin cytoskeleton or gene expression, respectively. Alternatively, WNT-Fzd-Ror2/Ryk interaction lead to changes in intracellular Ca++ levels, which affect protein kinase C (PKC), calmodullin-dependent protein kinase (CaMK) signaling cascades, or NFAT/AP1 driven gene expression changes. *Dkk*, Dickkopf; *Fzd*, Frizzled; *LRP*, lipoprotein related protein; *Ryk*; receptor-like tyrosine kinase; *SFRP*, secreted frizzed-related proteins.

can proceed with or without a Fzd receptor, and use different coreceptors, such as Ror and Ryk.[27,39] The result of activation of these pathways is often polarization of the cell for division, migration or extension, or even the polarization of whole tissues (i.e., planar cell polarity).[21] However, there is also evidence for β-catenin independent Wnt signaling in cell fate specification,[40] indicating that these seemingly distinct pathways have overlapping functions and are possibly tightly integrated.

4. WNT IN HSC DEVELOPMENT AND HOMEOSTASIS

The role of Wnt in HSC development and homeostasis has been intensely debated by the field, in part due to conflicting data in the literature.[41] Furthermore, the genetic complexity of the ligands and receptors, as well as the dynamic nature of HSC development, including the multiple cell types involved in self-renewal and differentiation lends itself to a complex series of inputs. In particular, harnessing the particular intricacies of timing and signal dosage may be the key to unlocking our understanding of Wnt in HSC ontogeny and maintenance.

Almost all *Wnts* and *Fzds* are expressed, at least to some degree, in the various hematopoietic sites during development,[42] indicating a potential role for these during HSC ontogeny; however, the specific requirements, if any, for these are only beginning to be elucidated. For example, a role for Wnt16 in the specification of HSCs in zebrafish has been established,[40] whereas Wnt5a seems to regulate lymphopoiesis.[43,44] In addition, a requirement for Wnt9a in the intraaortic amplification of HSCs was recently shown.[45] The specific requirements for Fzd receptors is less clear, potentially due to functional redundancy, or our incomplete understanding of Wnt-Fzd specificity requirements. It is known that Fzd9 is required for lymphoid development and maturation, for example, but its potential role in the HSC itself is unclear.[46]

The expression of Wnt components in the adult bone marrow are only slightly more refined, with a handful of these being expressed in putative HSC niche cells: osteoblasts, vascular cells, and mesenchymal stem cells.[47–51] Wnt function dependent on β-catenin in HSCs is driven through LEF/TCF DNA elements, as in other cells.[52] Accordingly, conditional deletion of β-catenin in the hematopoietic population impairs the long-term self-renewal[53] and regenerative capacities of HSCs.[54] Amplification of the Wnt signal, through viral expression of a constitutively active β-catenin results in increased numbers of HSCs in vitro and conversely, inhibition of Wnt signaling through viral expression of *Axin2* or addition of a soluble form of a Fzd ligand binding domain results in loss of HSC reconstitution in vivo,[52,55] potentially due to premature differentiation. Early work showed contradictory evidence for the role of Wnt/β-catenin signaling in the HSC system: loss of Wnt function through Wnt3a deletion,[56] β-catenin mutation,[53] or overexpression of the antagonist Dkk1[57] depleted the HSC pool in vivo, whereas activation of Wnt signaling through stabilized forms of β-catenin or the Wnt target prostaglandin E2, resulted in increased HSC number.[58,59] However, others observed a depletion of the progenitor pool upon β-catenin overexpression.[60,61] These contradictions are likely explained by dosage-dependent effects of Wnt signaling on different populations of blood cells[62]; using an allelic series of APC mutants, it was possible to show that a low level of Wnt signaling maintains a proliferative HSC phenotype, whereas a high level of Wnt activation resulted in total impairment of repopulation capacity and exhaustion of the stem cell pool.[62]

There is also evidence for specific function of Wnt ligands in adult HSCs. For example, *Wnt3a*, *Wnt5a*, and *Wnt10b* are all expressed in the bone marrow niche; interestingly, expression of *Wnt10b* is upregulated in response to injury to drive proliferation of bone marrow hematopoietic cells.[63] Fzd receptors likely play specific roles in the adult niche. For example, Fzd8 regulates the long-term quiescence of HSCs by regulating the noncanonical downstream calcium pathway.[64] HSCs from mice deficient in the β-catenin independent receptor *Ryk* have reduced quiescence, decreased self-renewal, and increased apoptosis.[65] During HSC ageing, there is a shift from β-catenin dependent to independent Wnt signaling, which seems to underlie the loss of self-renewal and lineage skewing seen in aged HSCs.[66] Altogether, these studies indicate that a careful balance of Wnt signals is required to regulate HSC specification, amplification, homeostasis and ageing. These effects of Wnt on normal HSCs are often hijacked in cancer processes, including the establishment and progression of blood cancers (Table 1).

5. WNT SIGNALING IN LEUKEMIA

The balance of self-renewal and differentiation in HSCs is critical to survival of an organism. For a cancer to progress, there must be a sustained increase in proliferation and also a block in differentiation. As such, mutations affecting cell identity or proliferation, such as *FLT3*, *TP53*, *RUNX1*,

Table 1 Known Wnt and Fzd Knockdown and Knockout (KO) Phenotypes in HSCs.

Genes	Knockdown or Knockout Phenotypes	References
Wnt1	Decreased number of thymocytes; some functional redundancy with Wnt4 (mouse KO)	67
Wnt3a	Homozygous mutants are lethal at E12.5; loss of HSCs and progenitors and impairment of long-term engraftment (mouse KO)	56
Wnt4	Decreased number of thymocytes; some functional redundancy with Wnt1 (mouse KO)	67
Wnt5a	Increased B-cells (mouse KO, cell culture model)	43,44
Wnt9a	Decreased intraaortic expansion of HSCs and progenitors (zebrafish knockdown and KO)	45
Wnt16	Loss of HSC specification (zebrafish knockdown)	40
Fzd5	Extraembryonic vascular development is abnormal (mouse KO)	68
Fzd9	Decreased B-cells (mouse KO)	46

HSC, Hematopoietic stem cell.

and *KMT2A* are commonly found in leukemias.[69] In addition to these mutations, chromosomal translocations, especially affecting chromatin modulators, such as *MLLT3-KMT2A*, *BCR-ABL*, and *RUNX1-MECOM* are associated with, and are sometimes sufficient to cause leukemia.[69,70] Differences in genetic causes for leukemias also give rise to cancers with different mechanisms of action, prognostic outcomes, and treatment regimes, making it crucial for us to understand the molecular cues driving these events to derive effective treatments.

In different subsets of leukemias, *WNT* expression and signaling in the bone marrow microenvironment are perturbed, often in the absence of a direct mutation in *WNT* component genes. For example, in chronic lymphocytic leukemia (CLL) B-cells, the WNT signal and expression of *WNT3* and the transcription factor *LEF1* are amplified compared to normal B-cells.[71] In addition, WNT3, WNT4, WNT5B, WNT6, WNT7B, WNT9A, WNT10A, WNT14, and WNT16 are all highly expressed in CLL B-cells.[72,73] Similarly, in E2A-Pbx1 acute lymphoblastic leukemia (ALL) cells, *WNT16* is robustly expressed, although it can be scarcely detected in normal pre-B cells; this expression is dependent on E2A-Pbx1, suggesting that Wnt activation occurs downstream of this translocation.[74] Expression of other leukemic translocation products also induces the expression of WNT components in hematopoietic cell lines.[75]

There is potential for therapeutic targeting of WNT in leukemias. The β-catenin independent coreceptor ROR1 is expressed in CLL leukemic cells, but not nonleukemic leukocytes, allowing for targeting of these cells with an anti-ROR1 monoclonal antibody (mAb, also known as cirmtuzumab).[76] High levels of ROR1 are associated with an accelerated form of CLL.[77] Addition of Wnt5a enhances the proliferation and migration of CLL cells through RhoA/Rac1; this effect can be blocked with cirmtuzumab,[78] and cannot be achieved in non-ROR1 expressing leukemic cells.[77] Clinically, combining cirmtuzumab with a B-cell receptor blocking agent is significantly more effective at clearing leukemic cells in vivo than either drug in isolation.[79] Taken together, these studies indicate how WNT targeting can be used to kill leukemic cells.

On the other hand, not all WNT activation is detrimental. For example, mice hemizygous for Wnt5a, which regulates the calcium signaling cascade, have enhanced B-cell proliferation and develop spontaneous myeloid leukemia and B-cell lymphomas, possibly because Wnt5a antagonizes the β-catenin signal.[44,80] In fact, WNT5A has tumor suppressing activity as loss of function mutations and/or loss of expression are found in a majority of primary human leukemia cells,[44] suggesting that this relief of β-catenin signaling repression may be necessary for some leukemias to progress. In another instance, exogenous Wnt3a reduced the proliferative capacity of B-cells in vitro[81]; however, this may be due to the dosage-dependent effects of Wnt on HSCs.

Many upstream regulators of Wnt signaling, such as members of the *Dickkopf* (*Dkk*) or *SFRP* families have been reported to be transcriptionally silenced due to promoter hypermethylation in human samples,[82–86] thereby presumably elevating endogenous Wnt signaling activity, which may promote proliferation and oncogenic transformation. In RUNX1-ETO mediated (acute myeloid leukemia) AML, decreased expression of *SFRP1* results from direct transcriptional repression from the *RUNX1-ETO* gene product, and leads to an increase in Wnt activation.[87] Although these are classically thought to be negative regulators of the Wnt signal, this is not always the case. This decrease in antagonist expression actually seems to have a positive effect on patient survival in *BCR-ABL* ALL,[88] suggesting that SFRPs may act to activate the WNT signal in leukemic cells. SFRPs have been suggested to act as carrier proteins for WNTs in other systems, acting to expand the range of the WNT signal, which may also play a role in leukemia.[89] Taken together, these studies exemplify that Wnt signaling is ectopically activated in many subsets of leukemia, leading to increased proliferative and self-renewal capacities of the leukemic stem cells.

Some leukemic stem cells are dependent on Wnt for self-renewal and/or survival. For example, β-catenin enhances the self-renewal and proliferation capacity of *BCR-ABL* chronic myeloid leukemia (CML) cells, which also show an increase in β-catenin signaling during their expansion; this effect is lost when the Wnt signal is dampened with AXIN overexpression.[90] This Wnt-mediated self-renewal is due to a splicing defect in *GSK3*β, which impairs its ability to bind AXIN, and target β-catenin for degradation.[90,91] Furthermore, mouse *BCR-ABL* CML cells refractory to standard treatment are dependent on β-catenin, indicating a potential role for Wnt in the evolution of these types of hematological malignancies.[92] This dependence on β-catenin for cancer progression is not limited to CML; it is also seen in subsets of AML,[93] CLL,[71] and ALL.[94,95] In addition, Wnt is also known to synergize with other leukemic mutations in mouse models, immortalized cells, and patient samples,[96–100] likely due to the increased WNT signal resulting from aberrant transgenes. Finally, the dependence of leukemic stem cells on Wnt is not limited to the β-catenin pathway, as *BCR-ABL* CML cells also rely on the Wnt-mediated calcium pathway for survival[101]; inhibition of numerous components of Wnt/Ca^{2+}/NFAT sensitizes cells to BCR-ABL inhibition. Altogether, these studies point to WNT as a potential therapeutic target; however, consideration of the effective dosage of the WNT signal and the particular cascade affected must be carefully considered.

6. WNT SIGNALING IN LYMPHOMA

Lymphomas are blood cancers that arise in the lymph nodes, as opposed to leukemias that mainly arise in the bone marrow. There are many diseases that fall under the umbrella term of lymphoma, including Hodgkin's Lymphoma, Mantle Cell Lymphoma (MCL), and Burkitt's lymphoma, to name a few. As these cancers arise in a plethora of anatomical locations, it has been difficult to discern the molecular cues regulating their initiation and progression. It is known that lymphomas differ in the genetic causes and also their activation of, and reliance on, WNT signaling. For example, patient samples of many different types of lymphoma have differentially increased levels of nuclear (active) β-catenin, but the majority of these do not contain mutations in APC or β-catenin,[102–105] suggesting that activation of the pathway is induced by autocrine or paracrine cues. β-Catenin independent signaling has also been proposed to play a role in anaplastic large cell lymphoma, although the extent to which this applies to other lymphomas is unclear.[106]

Many WNT components are found to be dysregulated in various types of lymphoma. For example, MCL is associated with a chromosomal translocation, which leads to an increase in expression of the context-dependent Wnt target gene *cyclinD1*, which is not sufficient to drive lymphoma on its own.[107,108] A microarray study compared gene expression in MCL cells and naïve B-cells (their wild-type counterparts), and found that several WNT pathway components, including *TCF7*, *FZD7*, *LRP5*, *AXIN1*, *APC*, *DVL3* were upregulated in MCL cells, indicating a potential requirement for WNT signaling in the disease progression.[109] Furthermore, MCL cells frequently show inactivation of GSK3β, which leads to an increase in nuclear β-catenin.[110] These studies fall in line with the general paradigm that an increase in the canonical arm of the Wnt signal leads to an increase in tumorigenicity.

Other lymphomas also have dysregulated Wnt; for example, there is evidence for overexpression of the transcription factors TCF1 and LEF1 in subset of T-cell lymphomas and small B-cell lymphomas.[111,112] This upregulation also occurs in lymphomas that are chemoresistant,[113] suggesting that WNT could be involved in the evolution of these cancer stem cells (the slow-dividing cells that initiate tumors) as well. It is worthy of note that TCF1 has also been implicated as a tumor suppressor since 50% of *Tcf1*$^{-/-}$ mice spontaneously develop thymic tumors.[114] The authors suggest that this may be due to TCF/LEF1 specific functions in this process, mainly TCF1 directly inhibiting *LEF1* transcription and leading to an overall increase in the Wnt signal. Like in leukemia, the absence of Wnt5a, which is thought to antagonize the β-catenin signal, can lead to cancer. Speaking to this, loss of heterozygosity in Wnt5a$^{+/-}$ mice develop B-cell lymphomas that are clonal in origin.[44] These studies also highlight the importance of Wnt signaling as an initiating event in lymphoma, which has also been proposed based on expression studies in humans.[115] However, the reliance of cancer stem cells on Wnt in lymphomas is less clear than that of leukemic stem cells. In a mouse model of lymphoma, which develops to 100% penetrance in all lymphoid organs, the disease is reliant on upregulation of the Wnt signal,[116] demonstrating that at least some lymphoma cancer cells are reliant on Wnt for survival. Finally, different WNTs also differentially affect these cancer cells. For example, a recent study highlights that WNT5A is a motility factor in Hodgkin's lymphoma; addition of WNT5A increases the motility of Hodgkin's lymphoma cells in culture, however, neither WNT10A and WNT10B nor WNT16 are able to elicit the same response.[117] Developing a clear understanding of how different Wnts affect downstream pathways in these diseases will be crucial toward developing therapeutic interventions for patients.

7. WNT SIGNALING IN MYELOMA

Myeloma encompasses cancers of the plasma cells, of which multiple myeloma (MM) is the most common and is characterized by an accumulation of plasma cells in the bone marrow. WNT activation in the absence of β-catenin or APC mutation has been observed in multiple myelomas cells, which were dependent on this signal for continued growth.[118] In fact, treatment of these with the compound PKF115-584 (which targets the β-catenin/TCF complex) results in cytotoxicity of primary patient MM cells and established cell lines, without effect on normal plasma cells.[119] *Wnt2b*, *Wnt5a*, *Wnt7a*, *Wnt10b*, *Wnt11*, and *Wnt16* are all expressed in MM cell lines[120]; however, it is worthy of noting that cell-autonomous Wnts are not necessarily involved in MM and the effects may not be limited to these particular Wnts. There is some evidence for sensitizing MM cells to autocrine and paracrine Wnt signals from the bone marrow niche.[121] In addition to the β-catenin dependent signal, Wnt3a is able to induce the RhoA signal in MM cells, including changes in cytoskeleton and morphology, indicative of an effect on planar cell polarity.[122] These alterations lead to an increase in the migration and invasion ability of these cells, and similar effects can be achieved using Wnt1 or Wnt4; these effects are dependent on the RhoA arm of Wnt signaling.[120] These data support roles for both β-catenin dependent and independent WNT signaling in MM cells.

MM leads to the osteolytic bone disease, resulting from the destruction of bone (osteoclast induction) and repression of its repair (osteoblast suppression). These in turn lead to further bone marrow niche dysfunction and these effects are also in part due to loss of Wnt regulation. For example, the WNT inhibitors SFRP2 and DKK1 are secreted by several MM cell lines, as well as primary patient cells.[123] SFRP2 inhibits bone mineralization in vitro,[124] and both *Dkk1* and *SFRP2* expression are associated with advanced destructive bone lesions in patient samples.[123,124] In addition, soluble Dkk1 can inhibit the differentiation of osteoblasts in vitro, suggesting that inhibition of Wnt through these soluble factors can inhibit bone formation and repair.[123,125] Severe combined immunodeficient (SCID) mice implanted with MM tumor cells spontaneously develop a loss in bone mineral density, which can be overcome with anti-Dkk1 antibody treatment; this is due to increased levels of osteoblasts and decreased osteoclasts, leading to an overall induction in bone density.[126–128] In addition to its role in MM initiation, MM tumor progression during chemoresistance involves the adhesion of MM cells to the bone marrow stroma, where expression of Wnt3 correlates with increased adhesion, likely progressing through the RhoA signaling cascade.[129] Taken altogether, these data point to the importance in Wnt regulating the interplay between MM and bone homeostasis, which when disrupted leads to osteolytic bone disease.

8. CONCLUSIONS AND PERSPECTIVES

The link between aberrant WNT signaling and many cancers is well established. However, in the case of blood malignancies, this requirement has been less clear, possibly due to the dosage-dependent effects of WNT on the stem cell niche and the dynamic nature of blood development and homeostasis. These studies together show that WNT plays a role in leukemias and lymphomas. Of note, these observations indicate a potential role for planar cell polarity in the pathogenesis of MM. Planar cell polarity has largely been studied in the context of whole tissue polarity, in contrast to the single cell nature of the blood; however, the involvement of the RhoA arm of Wnt signaling in MM suggests that this may be a mechanism exploited by tumorigenic single cells as well. One study has also pointed to a role for planar cell polarity in CLL.[130] Further studies will reveal the involvement of Wnt/planar cell polarity in tumorigenesis of hematopoietic cells. In addition to the roles described earlier, aberrant WNT seems to be a common mechanism for drug resistance in leukemia and lymphoma,[131–133] suggesting that these tumors may be "WNT addicted." In one example, pretreatment of purified patient ALL blasts to suppress the WNT signal was sufficient to restore chemosensitivity in relapsed patient cell lines,[134] an indication of one potential therapeutic avenue targeting WNT. A thorough understanding of how WNT regulates normal and malignant hematopoiesis will be key to future therapeutic interventions.

REFERENCES

1. Kissa K, Herbomel P. Blood stem cells emerge from aortic endothelium by a novel type of cell transition. *Nature.* 2010;464:112–115.
2. Burns CE, Traver D, Mayhall E, Shepard JL, Zon LI. Hematopoietic stem cell fate is established by the Notch-Runx pathway. *Genes Dev.* 2005;19(19):2331–2342.
3. North TE, Stacy T, Matheny CJ, Speck NA, de Bruijn MF. Runx1 is expressed in adult mouse hematopoietic stem cells and differentiating myeloid and lymphoid cells, but not in maturing erythroid cells. *Stem Cells.* 2004;22:158.
4. Boisset JC, van Cappellen W, Andrieu-Soler C, Galjart N, Dzierzak E, Robin C. In vivo imaging of haematopoietic cells emerging from the mouse aortic endothelium. *Nature.* 2010;464(7285):116–120.
5. Bertrand JY, Chi NC, Santoso B, Teng S, Stainier DY, Traver D. Haematopoietic stem cells derive directly from aortic endothelium during development. *Nature.* 2010;464(7285):108–111.
6. Murayama E, Kissa K, Zapata A, et al. Tracing hematopoietic precursor migration to successive hematopoietic organs during zebrafish development. *Immunity.* 2006;25:963–975.
7. Ciriza J, Thompson H, Petrosian R, Manilay JO, Garcia-Ojeda ME. The migration of hematopoietic progenitors from the fetal liver to the fetal bone marrow: lessons learned and possible clinical applications. *Exp Hematol.* 2013;41(5):411–423.
8. Kim AD, Stachura DL, Traver D. Cell signaling pathways involved in hematopoietic stem cell specification. *Exp Cell Res.* 2014;329(2):227–233.
9. Robertson AL, Avagyan S, Gansner JM, Zon LI. Understanding the regulation of vertebrate hematopoiesis and blood disorders—big lessons from a small fish. *FEBS Lett.* 2016;590(22):4016–4033.
10. Kauts ML, Vink CS, Dzierzak E. Hematopoietic (stem) cell development: how divergent are the roads taken? *FEBS Lett.* 2016;590(22):3975–3986.

11. Butko E, Pouget C, Traver D. Complex regulation of HSC emergence by the Notch signaling pathway. *Dev Biol.* 2016;409(1):129–138.
12. Lee Y, Decker M, Lee H, Ding L. Extrinsic regulation of hematopoietic stem cells in development, homeostasis and diseases. *Wiley Interdiscip Rev Dev Biol.* 2017.
13. Spangrude GJ, Heimfield DS, Weissman IL. Purification and characterization of mouse hematopoietic stem cells. *Science.* 1988;241:58.
14. Morrison SJ, Weissman IL. The long-term repopulating subset of hematopoietic stem cells is deterministic and isolatable by phenotype. *Immunity.* 1994;1(8):661–673.
15. Akashi K, Traver D, Miyamoto T, Weissman IL. A clonogenic common myeloid progenitor that gives rise to all myeloid lineages. *Nature.* 2000;404(6774):193–197.
16. Kondo M, Weissman IL, Akashi K. Identification of clonogenic common lymphoid progenitors in mouse bone marrow. *Cell.* 1997;91(5):661–672.
17. Olsson A, Venkatasubramanian M, Chaudhri VK, et al. Single-cell analysis of mixed-lineage states leading to a binary cell fate choice. *Nature.* 2016;537(7622):698–702.
18. Notta F, Zandi S, Takayama N, et al. Distinct routes of lineage development reshape the human blood hierarchy across ontogeny. *Science.* 2016;351(6269):aab2116.
19. Mendelson A, Frenette PS. Hematopoietic stem cell niche maintenance during homeostasis and regeneration. *Nat Med.* 2014;20(8):833–846.
20. Morrison SJ, Scadden DT. The bone marrow niche for haematopoietic stem cells. *Nature.* 2014;505(7483):327–334.
21. Loh KM, van Amerongen R, Nusse R. Generating cellular diversity and spatial form: Wnt signaling and the evolution of multicellular animals. *Dev Cell.* 2016;38(6): 643–655.
22. Niehrs C. The complex world of WNT receptor signalling. *Nat Rev Mol Cell Biol.* 2012;13(12):767–779.
23. Acebron SP, Niehrs C. β-Catenin-independent roles of Wnt/LRP6 signaling. *Trends Cell Biol.* 2016;26(12):956–967.
24. Yang Y, Mlodzik M. Wnt-Frizzled/planar cell polarity signaling: cellular orientation by facing the wind (Wnt). *Annu Rev Cell Dev Biol.* 2015;31:623–646.
25. Rocheleau CE, Downs WD, Lin R, et al. Wnt signaling and an APC-related gene specify endoderm in early *C. elegans* embryos. *Cell.* 1997;90(4):707–716.
26. Habib SJ, Chen BC, Tsai FC, et al. A localized Wnt signal orients asymmetric stem cell division in vitro. *Science.* 2013;339(6126):1445–1448.
27. Mikels AJ, Nusse R. Purified Wnt5a protein activates or inhibits beta-catenin-TCF signaling depending on receptor context. *PLoS Biol.* 2006;4(4):e115.
28. He X, Saint-Jeannet J-P, Wang Y, Nathans J, Dawid I, Varmus H. A member of the Frizzled protein family mediating axis induction by Wnt-5A. *Science.* 1997;275: 1652–1654.
29. Tao Q, Yokota C, Puck H, et al. Maternal wnt11 activates the canonical wnt signaling pathway required for axis formation in *Xenopus* embryos. *Cell.* 2005;120(6):857–871.
30. Li VS, Ng SS, Boersema PJ, et al. Wnt signaling through inhibition of beta-catenin degradation in an intact axin1 complex. *Cell.* 2012;149(6):1245–1256.
31. Rattner A, Hsieh JC, Smallwood PM, et al. A family of secreted proteins contains homology to the cysteine-rich ligand-binding domain of frizzled receptors. *Proc Natl Acad Sci USA.* 1997;94(7):2859–2863.
32. Kakugawa S, Langton PF, Zebisch M, et al. Notum deacylates Wnt proteins to suppress signalling activity. *Nature.* 2015;519(7542):187–192.
33. Zhang X, Cheong SM, Amado NG, et al. Notum is required for neural and head induction via Wnt deacylation, oxidation, and inactivation. *Dev Cell.* 2015;32 (6):719–730.
34. Giraldez AJ, Copley RR, Cohen SM. HSPG modification by the secreted enzyme Notum shapes the wingless morphogen gradient. *Dev Cell.* 2002;2(5):667–676.

35. Glinka A, Wu W, Delius H, Monaghan AP, Blumenstock C, Niehrs C. Dickkopf-1 is a member of a new family of secreted proteins and functions in head induction. *Nature.* 1998;391(6665):357–362.
36. Boutros M, Paricio N, Strutt DI, Mlodzik M. Dishevelled activates JNK and discriminates between JNK pathways in planar polarity and wingless signaling. *Cell.* 1998;94 (1):109–118.
37. Angers S, Moon RT. Proximal events in Wnt signal transduction. *Nat Rev Mol Cell Biol.* 2009;10(7):468–477.
38. Habas R, Kato Y, He X. Wnt/Frizzled activation of Rho regulates vertebrate gastrulation and requires a novel Formin homology protein Daam1. *Cell.* 2001;107(7): 843–854.
39. Yoshikawa S, McKinnon RD, Kokel M, Thomas JB. Wnt-mediated axon guidance via the Drosophila Derailed receptor. *Nature.* 2003;422(6932):583–588.
40. Clements WK, Kim AD, Ong KG, Moore JC, Lawson ND, Traver D. A somitic Wnt16/Notch pathway specifies haematopoietic stem cells. *Nature.* 2011;474:220–224.
41. Staal FJ, Chhatta A, Mikkers H. Caught in a Wnt storm: complexities of Wnt signaling in hematopoiesis. *Exp Hematol.* 2016;44(6):451–457.
42. Corrigan PM, Dobbin E, Freeburn RW, Wheadon H. Patterns of Wnt/Fzd/LRP gene expression during embryonic hematopoiesis. *Stem Cells Dev.* 2009;18:759–772.
43. Malhotra S, Baba Y, Garrett KP, Staal FJ, Gerstein R, Kincade PW. Contrasting responses of lymphoid progenitors to canonical and noncanonical Wnt signals. *J Immunol.* 2008;181(6):3955–3964.
44. Liang H, Chen Q, Coles AH, et al. Wnt5a inhibits B cell proliferation and functions as a tumor suppressor in hematopoietic tissue. *Cancer Cell.* 2003;4(5):349–360.
45. Grainger S, Richter J, Palazon RE, et al. Wnt9a is required for the aortic amplification of nascent hematopoietic stem cells. *Cell Rep.* 2016;17(6):1595–1606.
46. Ranheim EA, Kwan HC, Reya T, Wang YK, Weissman IL, Francke U. Frizzled 9 knock-out mice have abnormal B-cell development. *Blood.* 2005;105(6):2487–2494.
47. Spencer GJ, Utting JC, Etheridge SL, Arnett TR, Genever PG. Wnt signalling in osteoblasts regulates expression of the receptor activator of NF-kappaB ligand and inhibits osteoclastogenesis in vitro. *J Cell Sci.* 2006;119(Pt 7):1283–1296.
48. Dufourcq P, Descamps B, Tojais NF, et al. Secreted frizzled-related protein-1 enhances mesenchymal stem cell function in angiogenesis and contributes to neovessel maturation. *Stem Cells.* 2008;26(11):2991–3001.
49. Li H, Daculsi R, Grellier M, Bareille R, Bourget C, Amedee J. Role of neural-cadherin in early osteoblastic differentiation of human bone marrow stromal cells cocultured with human umbilical vein endothelial cells. *Am J Physiol.* 2010;299(2):C422–C430.
50. Qiu W, Chen L, Kassem M. Activation of non-canonical Wnt/JNK pathway by Wnt3a is associated with differentiation fate determination of human bone marrow stromal (mesenchymal) stem cells. *Biochem Biophys Res Commun.* 2011;413(1):98–104.
51. Planutiene M, Planutis K, Holcombe RF. Lymphoid enhancer-binding factor 1, a representative of vertebrate-specific Lef1/Tcf1 sub-family, is a Wnt-beta-catenin pathway target gene in human endothelial cells which regulates matrix metalloproteinase-2 expression and promotes endothelial cell invasion. *Vasc Cell.* 2011;3:28.
52. Reya T, Duncan AW, Ailles L, et al. A role for Wnt signalling in self-renewal of haematopoietic stem cells. *Nature.* 2003;423:409–414.
53. Zhao C, Blum J, Chen A, et al. Loss of β-catenin impairs the renewal of normal and CML stem cells in vivo. *Cancer Cell.* 2007;12:528–541.
54. Lento W, Ito T, Zhao C, et al. Loss of beta-catenin triggers oxidative stress and impairs hematopoietic regeneration. *Genes Dev.* 2014;28(9):995–1004.
55. Ruiz-Herguido C, Guiu J, D'Altri T, et al. Hematopoietic stem cell development requires transient Wnt/β-catenin activity. *J Exp Med.* 2012;209:1457–1468.

56. Luis TC, Weerkamp F, Naber BAE, et al. Wnt3a deficiency irreversibly impairs hematopoietic stem cell self-renewal and leads to defects in progenitor cell differentiation. *Blood.* 2009;113:546–554.
57. Fleming HE, Janzen V, Lo Celso C, et al. Wnt signaling in the niche enforces hematopoietic stem cell quiescence and is necessary to preserve self-renewal in vivo. *Cell Stem Cell.* 2008;2:274–283.
58. North TE, Goessling W, Walkley CR, et al. Prostaglandin E2 regulates vertebrate haematopoietic stem cell homeostasis. *Nature.* 2007;447:1007–1011.
59. Goessling W, North TE, Loewer S, et al. Genetic interaction of PGE2 and Wnt signaling regulates developmental specification of stem cells and regeneration. *Cell.* 2009;136:1136–1147.
60. Scheller M, Huelsken J, Rosenbauer F, et al. Hematopoietic stem cell and multilineage defects generated by constitutive β-catenin activation. *Nat Immunol.* 2006;7:1037–1047.
61. Kirstetter P, Anderson K, Porse BT, Jacobsen SE, Nerlov C. Activation of the canonical Wnt pathway leads to loss of hematopoietic stem cell repopulation and multilineage differentiation block. *Nat Immunol.* 2006;7(10):1048–1056.
62. Luis TC, Naber BAE, Roozen PPC, et al. Canonical wnt signaling regulates hematopoiesis in a dosage-dependent fashion. *Cell Stem Cell.* 2011;9:345–356.
63. Congdon KL, Voermans C, Ferguson EC, et al. Activation of Wnt signaling in hematopoietic regeneration. *Stem Cells.* 2008;26:1202–1210.
64. Sugimura R, He Xi C, Venkatraman A, et al. Noncanonical Wnt signaling maintains hematopoietic stem cells in the niche. *Cell.* 2012;150:351–365.
65. Famili F, Perez LG, Naber BA, Noordermeer JN, Fradkin LG, Staal FJ. The non-canonical Wnt receptor Ryk regulates hematopoietic stem cell repopulation in part by controlling proliferation and apoptosis. *Cell Death Dis.* 2016;7(11):e2479.
66. Florian MC, Nattamai KJ, Dorr K, et al. A canonical to non-canonical Wnt signalling switch in haematopoietic stem-cell ageing. *Nature.* 2013;503(7476):392–396.
67. Mulroy T, McMahon JA, Burakoff SJ, McMahon AP, Sen J. Wnt-1 and Wnt-4 regulate thymic cellularity. *Eur J Immunol.* 2002;32(4):967–971.
68. Ishikawa T, Tamai Y, Zorn AM, et al. Mouse Wnt receptor gene Fzd5 is essential for yolk sac and placental angiogenesis. *Development.* 2001;128(1):25–33.
69. Dohner H, Weisdorf DJ, Bloomfield CD. Acute myeloid leukemia. *N Engl J Med.* 2015;373(12):1136–1152.
70. Shtivelman E, Lifshitz B, Gale RP, Canaani E. Fused transcript of abl and bcr genes in chronic myelogenous leukaemia. *Nature.* 1985;315(6020):550–554.
71. Gutierrez Jr A, Tschumper RC, Wu X, et al. LEF-1 is a prosurvival factor in chronic lymphocytic leukemia and is expressed in the preleukemic state of monoclonal B-cell lymphocytosis. *Blood.* 2010;116(16):2975–2983.
72. Lu D, Zhao Y, Tawatao R, et al. Activation of the Wnt signaling pathway in chronic lymphocytic leukemia. *Proc Natl Acad Sci USA.* 2004;101(9):3118–3123.
73. Memarian A, Hojjat-Farsangi M, Asgarian-Omran H, et al. Variation in WNT genes expression in different subtypes of chronic lymphocytic leukemia. *Leuk Lymphoma.* 2009;50(12):2061–2070.
74. McWhirter JR, Neuteboom ST, Wancewicz EV, Monia BP, Downing JR, Murre C. Oncogenic homeodomain transcription factor E2A-Pbx1 activates a novel WNT gene in pre-B acute lymphoblastoid leukemia. *Proc Natl Acad Sci USA.* 1999;96(20):11464–11469.
75. Muller-Tidow C, Steffen B, Cauvet T, et al. Translocation products in acute myeloid leukemia activate the Wnt signaling pathway in hematopoietic cells. *Mol Cell Biol.* 2004;24(7):2890–2904.
76. Fukuda T, Chen L, Endo T, et al. Antisera induced by infusions of autologous Ad-CD154-leukemia B cells identify ROR1 as an oncofetal antigen and receptor for Wnt5a. *Proc Natl Acad Sci USA.* 2008;105(8):3047–3052.

77. Cui B, Ghia EM, Chen L, et al. High-level ROR1 associates with accelerated disease progression in chronic lymphocytic leukemia. *Blood.* 2016;128(25):2931–2940.
78. Yu J, Chen L, Cui B, et al. Wnt5a induces ROR1/ROR2 heterooligomerization to enhance leukemia chemotaxis and proliferation. *J Clin Invest.* 2016;126(2): 585–598.
79. Yu J, Chen L, Cui B, et al. Cirmtuzumab inhibits Wnt5a-induced Rac1 activation in chronic lymphocytic leukemia treated with ibrutinib. *Leukemia.* 2017;31(6): 1333–1339.
80. Ying J, Li H, Chen YW, Srivastava G, Gao Z, Tao Q. WNT5A is epigenetically silenced in hematologic malignancies and inhibits leukemia cell growth as a tumor suppressor. *Blood.* 2007;110(12):4130–4132.
81. Nygren MK, Dosen G, Hystad ME, Stubberud H, Funderud S, Rian E. Wnt3A activates canonical Wnt signalling in acute lymphoblastic leukaemia (ALL) cells and inhibits the proliferation of B-ALL cell lines. *Br J Haematol.* 2007;136(3):400–413.
82. Liu TH, Raval A, Chen SS, Matkovic JJ, Byrd JC, Plass C. CpG island methylation and expression of the secreted frizzled-related protein gene family in chronic lymphocytic leukemia. *Cancer Res.* 2006;66(2):653–658.
83. Roman-Gomez J, Jimenez-Velasco A, Agirre X, et al. Transcriptional silencing of the Dickkopfs-3 (Dkk-3) gene by CpG hypermethylation in acute lymphoblastic leukaemia. *Br J Cancer.* 2004;91(4):707–713.
84. Roman-Gomez J, Jimenez-Velasco A, Cordeu L, et al. WNT5A, a putative tumour suppressor of lymphoid malignancies, is inactivated by aberrant methylation in acute lymphoblastic leukaemia. *Eur J Cancer.* 2007;43(18):2736–2746.
85. Pehlivan M, Sercan Z, Sercan HO. sFRP1 promoter methylation is associated with persistent Philadelphia chromosome in chronic myeloid leukemia. *Leuk Res.* 2009;33 (8):1062–1067.
86. Reins J, Mossner M, Neumann M, et al. Transcriptional down-regulation of the Wnt antagonist SFRP1 in haematopoietic cells of patients with different risk types of MDS. *Leuk Res.* 2010;34(12):1610–1616.
87. Cheng CK, Li L, Cheng SH, et al. Secreted-frizzled related protein 1 is a transcriptional repression target of the t(8;21) fusion protein in acute myeloid leukemia. *Blood.* 2011;118(25):6638–6648.
88. Martin V, Agirre X, Jimenez-Velasco A, et al. Methylation status of Wnt signaling pathway genes affects the clinical outcome of Philadelphia-positive acute lymphoblastic leukemia. *Cancer Sci.* 2008;99(9):1865–1868.
89. Mii Y, Taira M. Secreted Frizzled-related proteins enhance the diffusion of Wnt ligands and expand their signalling range. *Development.* 2009;136(24):4083–4088.
90. Jamieson CH, Ailles LE, Dylla SJ, et al. Granulocyte-macrophage progenitors as candidate leukemic stem cells in blast-crisis CML. *N Engl J Med.* 2004;351(7): 657–667.
91. Abrahamsson AE, Geron I, Gotlib J, et al. Glycogen synthase kinase 3beta missplicing contributes to leukemia stem cell generation. *Proc Natl Acad Sci USA.* 2009;106(10): 3925–3929.
92. Hu Y, Chen Y, Douglas L, Li S. β-Catenin is essential for survival of leukemic stem cells insensitive to kinase inhibition in mice with BCR-ABL-induced chronic myeloid leukemia. *Leukemia.* 2009;23(1):109–116.
93. Siapati EK, Papadaki M, Kozaou Z, et al. Proliferation and bone marrow engraftment of AML blasts is dependent on beta-catenin signalling. *Br J Haematol.* 2011;152(2): 164–174.
94. Mazieres J, You L, He B, et al. Inhibition of Wnt16 in human acute lymphoblastoid leukemia cells containing the t(1;19) translocation induces apoptosis. *Oncogene.* 2005;24 (34):5396–5400.

95. Khan NI, Bradstock KF, Bendall LJ. Activation of Wnt/beta-catenin pathway mediates growth and survival in B-cell progenitor acute lymphoblastic leukaemia. *Br J Haematol.* 2007;138(3):338–348.
96. Tickenbrock L, Schwable J, Wiedehage M, et al. Flt3 tandem duplication mutations cooperate with Wnt signaling in leukemic signal transduction. *Blood.* 2005;105 (9):3699–3706.
97. Tickenbrock L, Hehn S, Sargin B, et al. Activation of Wnt signaling in cKit-ITD mediated transformation and imatinib sensitivity in acute myeloid leukemia. *Int J Hematol.* 2008;88(2):174–180.
98. Kaveri D, Kastner P, Dembele D, Nerlov C, Chan S, Kirstetter P. β-Catenin activation synergizes with Pten loss and Myc overexpression in Notch-independent T-ALL. *Blood.* 2013;122(5):694–704.
99. Sadras T, Perugini M, Kok CH, et al. Interleukin-3-mediated regulation of beta-catenin in myeloid transformation and acute myeloid leukemia. *J Leukoc Biol.* 2014;96(1):83–91.
100. Ugarte GD, Vargas MF, Medina MA, et al. Wnt signaling induces transcription, spatial proximity, and translocation of fusion gene partners in human hematopoietic cells. *Blood.* 2015;126(15):1785–1789.
101. Gregory MA, Phang TL, Neviani P, et al. Wnt/Ca2+/NFAT signaling maintains survival of Ph+ leukemia cells upon inhibition of Bcr-Abl. *Cancer Cell.* 2010;18 (1):74–87.
102. Zhang D, O'Neil MF, Cunningham MT, Fan F, Olyaee M, Li L. Abnormal Wnt signaling and stem cell activation in reactive lymphoid tissue and low-grade marginal zone lymphoma. *Leuk Lymphoma.* 2010;51(5):906–910.
103. Groen RW, Oud ME, Schilder-Tol EJ, et al. Illegitimate WNT pathway activation by beta-catenin mutation or autocrine stimulation in T-cell malignancies. *Cancer Res.* 2008;68(17):6969–6977.
104. Gelebart P, Anand M, Armanious H, et al. Constitutive activation of the Wnt canonical pathway in mantle cell lymphoma. *Blood.* 2008;112(13):5171–5179.
105. Ge X, Lv X, Feng L, Liu X, Wang X. High expression and nuclear localization of beta-catenin in diffuse large B-cell lymphoma. *Mol Med Rep.* 2012;5(6):1433–1437.
106. Hegazy SA, Alshareef A, Gelebart P, et al. Disheveled proteins promote cell growth and tumorigenicity in ALK-positive anaplastic large cell lymphoma. *Cell Signal.* 2013;25 (1):295–307.
107. Bosch F, Jares P, Campo E, et al. PRAD-1/cyclin D1 gene overexpression in chronic lymphoproliferative disorders: a highly specific marker of mantle cell lymphoma. *Blood.* 1994;84(8):2726–2732.
108. Bodrug SE, Warner BJ, Bath ML, Lindeman GJ, Harris AW, Adams JM. Cyclin D1 transgene impedes lymphocyte maturation and collaborates in lymphomagenesis with the myc gene. *EMBO J.* 1994;13(9):2124–2130.
109. Rizzatti EG, Falcao RP, Panepucci RA, et al. Gene expression profiling of mantle cell lymphoma cells reveals aberrant expression of genes from the PI3K-AKT, WNT and TGFbeta signalling pathways. *Br J Haematol.* 2005;130(4):516–526.
110. Chung R, Peters AC, Armanious H, Anand M, Gelebart P, Lai R. Biological and clinical significance of GSK-3beta in mantle cell lymphoma—an immunohistochemical study. *Int J Clin Exp Pathol.* 2010;3(3):244–253.
111. Dorfman DM, Greisman HA, Shahsafaei A. Loss of expression of the WNT/beta-catenin-signaling pathway transcription factors lymphoid enhancer factor-1 (LEF-1) and T cell factor-1 (TCF-1) in a subset of peripheral T cell lymphomas. *Am J Pathol.* 2003;162(5):1539–1544.
112. Tandon B, Peterson L, Gao J, et al. Nuclear overexpression of lymphoid-enhancer-binding factor 1 identifies chronic lymphocytic leukemia/small lymphocytic lymphoma in small B-cell lymphomas. *Mod Pathol.* 2011;24(11):1433–1443.

113. Koivula S, Valo E, Raunio A, Hautaniemi S, Leppa S. Rituximab regulates signaling pathways and alters gene expression associated with cell death and survival in diffuse large B-cell lymphoma. *Oncol Rep.* 2011;25(4):1183–1190.
114. Tiemessen MM, Baert MR, Schonewille T, et al. The nuclear effector of Wnt-signaling, Tcf1, functions as a T-cell-specific tumor suppressor for development of lymphomas. *PLoS Biol.* 2012;10(11):e1001430.
115. Kimura Y, Arakawa F, Kiyasu J, et al. The Wnt signaling pathway and mitotic regulators in the initiation and evolution of mantle cell lymphoma: Gene expression analysis. *Int J Oncol.* 2013;43(2):457–468.
116. Wang Y, Liu Y, Wu C, et al. Epm2a suppresses tumor growth in an immunocompromised host by inhibiting Wnt signaling. *Cancer Cell.* 2006;10(3):179–190.
117. Linke F, Zaunig S, Nietert MM, et al. WNT5A: a motility-promoting factor in Hodgkin lymphoma. *Oncogene.* 2017;36(1):13–23.
118. Derksen PW, Tjin E, Meijer HP, et al. Illegitimate WNT signaling promotes proliferation of multiple myeloma cells. *Proc Natl Acad Sci USA.* 2004;101(16): 6122–6127.
119. Sukhdeo K, Mani M, Zhang Y, et al. Targeting the beta-catenin/TCF transcriptional complex in the treatment of multiple myeloma. *Proc Natl Acad Sci USA.* 2007;104 (18):7516–7521.
120. Qiang YW, Walsh K, Yao L, et al. Wnts induce migration and invasion of myeloma plasma cells. *Blood.* 2005;106(5):1786–1793.
121. van Andel H, Ren Z, Koopmans I, et al. Aberrantly expressed LGR4 empowers Wnt signaling in multiple myeloma by hijacking osteoblast-derived R-spondins. *Proc Natl Acad Sci USA.* 2017;114(2):376–381.
122. Qiang YW, Endo Y, Rubin JS, Rudikoff S. Wnt signaling in B-cell neoplasia. *Oncogene.* 2003;22(10):1536–1545.
123. Tian E, Zhan F, Walker R, et al. The role of the Wnt-signaling antagonist DKK1 in the development of osteolytic lesions in multiple myeloma. *N Engl J Med.* 2003;349 (26):2483–2494.
124. Oshima T, Abe M, Asano J, et al. Myeloma cells suppress bone formation by secreting a soluble Wnt inhibitor, sFRP-2. *Blood.* 2005;106(9):3160–3165.
125. Gunn WG, Conley A, Deininger L, Olson SD, Prockop DJ, Gregory CA. A crosstalk between myeloma cells and marrow stromal cells stimulates production of DKK1 and interleukin-6: a potential role in the development of lytic bone disease and tumor progression in multiple myeloma. *Stem Cells.* 2006;24(4):986–991.
126. Yaccoby S, Ling W, Zhan F, Walker R, Barlogie B, Shaughnessy Jr JD. Antibody-based inhibition of DKK1 suppresses tumor-induced bone resorption and multiple myeloma growth in vivo. *Blood.* 2007;109(5):2106–2111.
127. Heath DJ, Chantry AD, Buckle CH, et al. Inhibiting Dickkopf-1 (Dkk1) removes suppression of bone formation and prevents the development of osteolytic bone disease in multiple myeloma. *J Bone Miner Res.* 2009;24(3):425–436.
128. Fulciniti M, Tassone P, Hideshima T, et al. Anti-DKK1 mAb (BHQ880) as a potential therapeutic agent for multiple myeloma. *Blood.* 2009;114(2):371–379.
129. Kobune M, Chiba H, Kato J, et al. Wnt3/RhoA/ROCK signaling pathway is involved in adhesion-mediated drug resistance of multiple myeloma in an autocrine mechanism. *Mol Cancer Ther.* 2007;6(6):1774–1784.
130. Kaucka M, Plevova K, Pavlova S, et al. The planar cell polarity pathway drives pathogenesis of chronic lymphocytic leukemia by the regulation of B-lymphocyte migration. *Cancer Res.* 2013;73(5):1491–1501.
131. Riether C, Schurch CM, Flury C, et al. Tyrosine kinase inhibitor-induced CD70 expression mediates drug resistance in leukemia stem cells by activating Wnt signaling. *Sci Transl Med.* 2015;7(298):298ra119.

132. Eiring AM, Khorashad JS, Anderson DJ, et al. β-Catenin is required for intrinsic but not extrinsic BCR-ABL1 kinase-independent resistance to tyrosine kinase inhibitors in chronic myeloid leukemia. *Leukemia.* 2015;29(12):2328–2337.
133. Dharmapuri G, Doneti R, Philip GH, Kalle AM. Celecoxib sensitizes imatinib-resistant K562 cells to imatinib by inhibiting MRP1-5, ABCA2 and ABCG2 transporters via Wnt and Ras signaling pathways. *Leukemia Res.* 2015;39(7):696–701.
134. Dandekar S, Romanos-Sirakis E, Pais F, et al. Wnt inhibition leads to improved chemosensitivity in paediatric acute lymphoblastic leukaemia. *Br J Haematol.* 2014;167 (1):87–99.

FURTHER READING

Siegel RL, Miller KD, Jemal A. Cancer statistics, 2016. *CA Cancer J Clin.* 2016;66(1):7–30.

INDEX

C

D

I

J

K

Q

R

S

T

X

Y

Z

FSC
www.fsc.org
MIX
Papier | Fördert
gute Waldnutzung
FSC® C083411

Kolibri 360 GmbH
Ferdinand-Jühlke-Straße 7
99095 Erfurt, Deutschland
produktsicherheit@kolibri360.de